WORLD FOOD

THE WILEY BICENTENNIAL—KNOWLEDGE FOR GENERATIONS

*E*ach generation has its unique needs and aspirations. When Charles Wiley first opened his small printing shop in lower Manhattan in 1807, it was a generation of boundless potential searching for an identity. And we were there, helping to define a new American literary tradition. Over half a century later, in the midst of the Second Industrial Revolution, it was a generation focused on building the future. Once again, we were there, supplying the critical scientific, technical, and engineering knowledge that helped frame the world. Throughout the 20th Century, and into the new millennium, nations began to reach out beyond their own borders and a new international community was born. Wiley was there, expanding its operations around the world to enable a global exchange of ideas, opinions, and know-how.

For 200 years, Wiley has been an integral part of each generation's journey, enabling the flow of information and understanding necessary to meet their needs and fulfill their aspirations. Today, bold new technologies are changing the way we live and learn. Wiley will be there, providing you the must-have knowledge you need to imagine new worlds, new possibilities, and new opportunities.

Generations come and go, but you can always count on Wiley to provide you the knowledge you need, when and where you need it!

WILLIAM J. PESCE
PRESIDENT AND CHIEF EXECUTIVE OFFICER

PETER BOOTH WILEY
CHAIRMAN OF THE BOARD

WORLD FOOD
Production and Use

Alfred R. Conklin, Jr.
Wilmington College of Ohio

Thomas Stilwell
Wilmington College of Ohio

BICENTENNIAL
1807
WILEY
2007
BICENTENNIAL

WILEY-INTERSCIENCE
A JOHN WILEY & SONS, INC., PUBLICATION

Published by John Wiley & Sons, Inc., Hoboken, New Jersey
Published simultaneously in Canada

For general information on our other products and services or for technical support, please
contact our Customer Care Department within the United States at (800) 762-2974,
outside the United States at (317) 572-3993 or fax (317) 572-4002.

Wiley also publishes its books in a variety of electronic formats. Some content that appears in print
may not be available in electronic formats. For more information about Wiley products, visit
our web site at www.wiley.com.

Library of Congress Cataloging-in-Publication Data:

Conklin, Alfred Russel, 1941-
    World food : production and use/Alfred R. Conklin, Thomas Stilwell.
        p. cm.
    Includes index.
    ISBN 978-0-470-04382-0 (cloth)
    1. Food supply—Textbooks. 2. Agriculture—Textbooks. 3. Nutrition—Textbooks.
    I. Stilwell, Thomas C. II. Title.
        TX354.C65 2007
        338.1′9- -dc22   2007002341

Printed in the United States of America

10 9 8 7 6 5 4 3 2 1

CONTENTS

11 INCREASING FOOD SUPPLIES   383

12 GENETICALLY MODIFIED CROPS AND ANIMALS   407

PREFACE

The authors intend this book to be a textbook for any world food course or for any person or group that wishes to learn more about world food. Included is a CD that contains all the pictures, tables, and drawings suitable for presentation. The CD also includes recipes for "finger foods" that can be used as part of a class. We encourage educators to use this book as a text, a resource for class discussion, and as a resource for student presentations.

This book is about world food and includes all those components related to food, its production, and use. Humans have been involved in agriculture for thousands of years as indicated in the table below.

In thinking and working on world food and its production, it is essential to first know what the major food sources are and their nutritional values. With this knowledge, along with basic concepts of soil, water, and ways in which agriculture production can be increased, one can begin to make informed decisions about world hunger and its alleviation and agriculture development.

Jared Diamond[*] has written a provocative opinion piece proposing that agriculture is the greatest mistake humans have made. Basically, he argues that when people turned from hunting and gathering and adopted agriculture, there came increased

Domestication of Plants and Animals		
Area of World	Domesticated Plants and Animals	Date BC
Southwest Asia	Wheat, barley, lentil, pea, melon, olive, sheep, goat, cattle	8,500
China	Rice, millet, soybeans, mung bean, pig	7,500
Mesoamerica	Corn, beans, squash, avocado, turkey	3,500
Andes and Amazon	Potato, peanut, sweet potato, squash, llama, guinea pig	3,500
Eastern North America	Sunflower, artichoke, squash	2,500
Sahel	Sorgum, millet, rice, pea, guinea fowl	5,000
West Africa	Yams, watermelon, groundnut	3,000
New Guinea	Sugarcane, banana, yams, taro	7,000

Adapted from Diamond, J. *Guns, Germs and Steel.* New York: W.W. Norton & Co. 1997.

[*]Jared Diamond, The Worst Mistake in the History of the Human Race, *Discover Magazine*, pp. 64–66, 1987.

work, decreased nutrition, and increased disease. Some important issues are glossed over, but the basics are worth consideration. In particular, we can think about using modern knowledge of nutritional needs and food production to overcome shortfalls in the balance of foods available to any group of people.

There are large and diverse numbers and varieties of foods in the world. In addition, there are larger or equally diverse methods of growing, preparing, and eating any single food. Here the attempt has been to select only the most important food crops for inclusion. In some Mediterranean countries dishes that use grape leaves are common, and in parts of Africa, particularly in the south west, the leaves of cassava are eaten. However, neither of these is commonly eaten the world over, even where the crop from which they come is grown extensively. Guinea pigs are grown and eaten in some South American countries but are limited to these areas. We have concentrated on crops and animals that are raised and eaten on at least two continents.

Even with this restriction the number of plants and animals grown for food is too large to be exhaustively covered in this book. Likewise not all aspects of a food's nutritional characteristics can be given. Therefore, the authors have included what they believe from their experience to be the more common foods. Additional information about other foods is readily available from both the U.S. Department of Agriculture (USDA) and the Food and Agriculture Organization (FAO) of the United Nations. These are referenced frequently in the various chapters.

Different chapters emphasize different aspects of food production and use. For this reason not all topics or concepts are presented in all chapters but rather different aspects are emphasized in different chapters.

Data has been collected from many sources and has been checked; however, much information is subject to updating, and different sources will give varying, sometimes conflicting, data. For this reason it is suggested that the data be used to compare situations rather than provide absolutes. Also, the Internet references have been checked; however, Internet sites are changed and updated. If a problem is encountered, one should go to the parent site and proceed from there.

The authors would like to thank the following persons for helping in the preparation of this book. In Ecuador, Jenny Valencia, David Céron, and Mickey Zambrano for their help, and both Señora Aída Jiménez and Señor Octavio Tipán for allowing us to learn about their farming operation. In the Philippines, Henry Goltiano and his wife Sarah B. Goltiano for the excellent job they did and both Celedonio Derecho (Donio) and his wife Sita Derecho for allowing us to learn about their farming operation. We also wish to mention Nelson L. Cabaña, agricultural extension agent, Florante T. Sabejon of ICRAF, Sergio M. Abit, Jr., faculty, Jade Mesias, student, Ed Allan L. Alcober, faculty, Dr. Eduardo G. Apilar, ATI at LSU administrative officer and the Agricultural Training Institute all at Leyte State University (LSU), and also Alan B. Loreto of PhilRootcrops of LSU for help in gathering information about the farm and facilitating this part of the work. In the United States, Steve Murphy and Patti M. Murphy, their sons Nick and William, and Milton Murphy were all very helpful and open in providing all the information and some of the pictures we needed. All photographs, charts, and figures not otherwise noted were taken or prepared by the authors. Maps were prepared using MapLand™ software by Software Limited.

In preparing this book the following were most helpful in providing comments and suggestions as to how various chapters might be improved: Drs. Donald Troike, Esmail Hejazifar, Stephen Potthoff, Robert Beck, Roger Cortbaoui, Kenton Brubaker, Monte Anderson, Malcolm Manners, Laura Tiu, Wayne Haag, Guillermo Scaglia, Brad Miller, Harold Thirey, Dr. James W. Tallman, Weiji Wang, and Carolyn Stilwell. Their help is greatly appreciated.

The authors particularly wish to thank their wives Norma Durán and Petra Conklin for their help and support in preparing this book.

ALFRED R. CONKLIN
THOMAS STILWELL

Wilmington, Ohio
March 2007

1

REPRESENTATIVE FARMS FROM AROUND THE WORLD

> The authors have selected three very different farms from different parts of the world—the Philippines, Ecuador, and the United States—to illustrate the varieties of farming operations. The student should not take from this that these are the only types of farms found in these countries. All types of farms can be found in all countries, that is, subsistence, moderate size family, and large commercial farms occur in all countries including the United States.

All over the world food comes from farms to nonfarm peoples. However, the similarity stops there. The size, crops grown, equipment used, soil, water available, electricity availability, transportation, and sale of crops are all different. In spite of this wide variation, all of these various farm types provide food for themselves and their neighbors.

To begin studying world food three farms from three different places—the Philippines, Ecuador, and the United States—have been chosen to serve as examples of the wide variety of farms in the world. These are not necessarily representative farms in the world or the particular country but rather are used to show the diversity and range of farms. Figures 1.1 and 1.2 show the locations of the farms in the world and relative sizes of the three countries.

World Food: Production and Use. By Alfred R. Conklin, Jr. and Thomas Stilwell
Copyright © 2007 John Wiley & Sons, Inc.

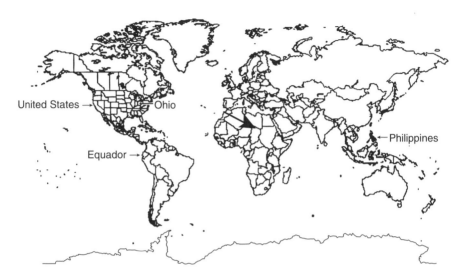

Figure 1.1. Location of the three farms.

Figure 1.2. Relative sizes of the three countries.

1.1 THE FACE OF SUBSISTENCE FARMING IN THE PHILIPPINES: DONIO

Henry Y. Goltiano*

Celedonio Derecho is 57 years old, his wife Sita is 47, and his youngest son Roger is 12, as seen in Figure 1.3. Donio, to all who know him, has been farming in *Baryo* (village) Hibunawan for 35 years. Hibunawan is a farming village about 5 km

*Agricultural Resource Management Section Chief, Agricultural Training Institute, Leyte State University, Visca, Baybay, Leyte, Philippines.

Figure 1.3. Donio, right, with Roger and Sita, left.

(1.6093 km = 1 mile) from the center of the town of Baybay, Leyte, Philippines. The road to Hibunawan is a mixture of sand, gravel, and soil that gets nasty during rainy season and dusty during the dry season (Fig. 1.4).

Farming has been Donio's life. He worked for about 3 years in an ice-cream factory in Manila, the capital of the Philippines, but when he went home to Hibunawan for a short

Figure 1.4. Road from Donio's farm.

Figure 1.5. Donio's house with Donio in front.

vacation, he was so smitten with Rosita Pancito that he decided not to go back to Manila and instead "married" Rose who was a 12-year-old girl (Donio was 22). Since marrying, Donio has been tilling the soil. He started as hired labor on neighboring farms while Sita worked on other farms, usually weeding. Donio's parents were tenant farmers, and, when they were too old and weak to work, they handed the farms they tenanted to Donio who lives with Sita in a house adjacent to the farms and about 30 m away from the road; see Figure 1.5. Transportation to the house is either by paying someone with a motorcycle to carry you or to ride in a side-car-equipped motorcycle.

Figure 1.6. Rice field being prepared with a tractor.

Donio's Fields. Donio tills one rice field and two corn fields (Fig. 1.6).

RICE FIELD. The rice field with an area of 2070 m^2 (0.207 ha, 0.511 acre) is just at the back of their house. At a market value of P100,000 to P150,000 ($1902.59 to $2853.88)[†] per hectare, the area can sell for P40,000 ($761.04) at most. Because the farm is rainfed, Donio can plant rice only twice a year.

Donio is a lessee of this farm owned by a Manila, the country's capital, resident. The owner has a caretaker who comes to the village during harvest to collect the rent. As lessee, Donio pays a fixed rent of three sacks of *tipasi* (Cebuano[§] for newly harvested rice), regardless of yield fluctuation. However, in the case of an extremely bad harvest, rent is negotiated. Rent is to be agreed upon by the lessee, the landowner, and the Department of Agrarian Reform after determining the average harvest in three cropping seasons.

MAIZE FIELDS. Donio has two maize fields. One has an area of 6580 m^2 (0.658 ha, 1.62 acres). The other has 6900 m^2 (0.690 ha, 1.70 acres). Both farms are owned by a resident of Hibunawan and could sell for about P100,000 ($1902.59).

From the smaller rice farm, Donio shares 50 percent of the harvest with the owner. His 50 percent is further divided by two because Donio is a sublessee. Thus Donio gets only 25 percent net share of the harvest. Donio gets a "fairer" deal with the bigger farm because he is the lessee. He shares the net harvest only with the owner, who gets 50 percent. The owner provides nothing other than the land.

OTHER CROPS. Aside from rice and maize, Donio has banana and cassava (*Mahinot esculenta*, a root crop presented in Chapter 3) planted in some areas around his larger corn field; see Figure 1.6. In the yard at his house he has a patch of sweet potato (*Ipomea batatas*), which he raises for *ganas*, the young leaves that are prepared as vegetable. He also has a few *gabi* (*Colocasia esculenta*), which is another root crop. It does not appear that Donio applies any improved management practices to these crops, even though his banana plants are affected by *tibak* or *bugtok*, a bacterial disease caused by a strain of *Ralstonia solanacearum*.

Rice Farming Practices. Donio plants only rice in his rice field, which is rainfed, and so he plants in January and June or when it rains. Controlling the amount of water in a rice paddy is important in obtaining a good yield. Because he is dependent on—or at the mercy of—rain, Donio cannot control the amount of water in the field, which may remain flooded when it needs to be drained or remain dry when water is needed by the crop.

Nelson L. Cabaña is the extension agent assigned in Barangay Hibunawan where Donio lives. He has been encouraging Donio and other farmers to follow new technological practices. Donio, however, does not follow these recommendations because he does not have the resources. For example, choice of seeds is the first step in any

---

[†]Using a conversion rate of P52.56 per dollar, which was the conversion rate on May, 30, 2006.

[§]Cebuano is the dominant language in Visayas, including Baybay, and Mindanao regions of the Philippines and is Donio's language. Other local names, as indicated, are also Cebuano.

Figure 1.7. Rice seedlings raised by the wet-bed method.

successful harvest, but Donio obtains his seed from his previous harvest and not from seed growers because the price for him is prohibitive.

In the Philippines rice seedlings are grown and transplanted to the rice paddy, where it grows and matures. In preparing seedlings, Donio follows the *Dapog* method (see below), which is not as costly as the wet-bed method; see Figure 1.7. In *dapog*, the seedbed can be made in the yard of the house, thus requiring less labor compared to the wet-bed method, which is prepared in the paddy, and thus requires some land preparation and higher labor cost.

Wet-Bed and *Dapog* Methods of Seedling Preparation

For both the wet-bed and *dapog* method of preparing rice seedlings, seed is placed in a sack and submerged for 48 hours in water. This allows the seed to germinate. In the wet-bed method a seedbed is prepared in a small portion of a rice paddy having a shallow covering of water. Sprouted seed is placed on the seedbed and allowed to grow for 15 to 20 days before being pulled, put into small "bundles," and carried to the field for planting. In the *dapog* method the sprouted seed is spread thickly on a dry area, or piece of concrete, which will not be part of a rice paddy. It is allowed to grow for a maximum of 10 days, during which time it must be kept constantly wet. It is rolled up much like a carpet and carried to the field and planted.

Nelson recommends that the field be harrowed two to three times before transplanting because thorough land preparation helps control weeds before transplanting. Donio harrows the field only once to lessen expenses.

Donio says that he does not have many problems with pests. It is possible that he has but just does not recognize their occurrence because he does not attend farmer's training. If he recognized the presence of pests, he may ignore them because of his

Figure 1.8. Weeding a rice paddy.

incapacity to control them. He says that the only pests that bother his rice farm are rats and golden snails, locally called *kuhol*. He does not do anything against the rats because he believes they will become nastier if attacked or killed. He just "pleads" with them not to devour all his plants but leave something for his family's needs. *Kuhol* are picked up and thrown into the road to be crushed by passing vehicles.

Kuhol was introduced in the country in the 1980s as a means of addressing protein deficiency in farm communities, but it spread so fast that they have become pests. While edible, most farmers do not eat them because of their odd smell. However, in extreme cases, when they do not have food (*viand*) farmers cook and eat the big *kuhol*.

Weeds are removed by the family by hand or using weeding machines, which Donio rents (Fig. 1.8); Donio does not hire anyone to help them. They pull weeds by hand or dig them using a *guna* (short bolo). Sometimes when there are excessive weeds, they use Rogue, a herbicide, which costs about P300 ($5.70) a quart. Another way weeds are controlled is by keeping the rice field filled with water to suppress growth. According to Nelson, while this practice works, it also curbs the growth of rice, and Donio's ability to control water is limited as noted above.

It is advised that soil analysis be done to determine what and how much plant nutrients a farm needs. However, Donio does not have the money to do this. Plant nutrients come from hay and stubble that are plowed into the soil as organic fertilizer, and Donio applies a 50-kg bag of urea (1 kg = 1000 g or approximately 2.2 lb), which he splits, applying half before transplanting and the other half at panicle[¶] initiation. The

---

[¶]The panicle is the seed-bearing part of the rice plant.

Figure 1.9. Rice drying in the sun.

recommended amount is two bags per hectare. Since the farm is only 0.207 ha, he should apply 40 kg before transplanting and about 10 kg at panicle initiation. After harvest and threshing Donio sun dries his *tipasi* by spreading it on a mat laid on the area around the house or rice paddy, see Figure 1.9.

Maize Farming Practices. Donio's corn fields are also monocropped, that is, he grows only one crop on the same land year after year. Donio plants maize three times a year: January, May, and August. The last planting is usually affected by typhoons that come during the last months of the year. Donio gets his maize seed from his previous harvest, selecting the biggest and fullest bodied ears as seed sources.

He uses only the carabao drawn plow (Fig. 1.10) and harrow (Fig. 1.11) in preparing the soil for planting. While two or more plowings and several harrowings are

Figure 1.10. Donio with his plow.

Figure 1.11. Corn field harrow (*sudlay*).

recommended, he plows and harrows once. Rosita describes the harrowing of their maize field as *sinampayu*, or cursory.

As with the rice farm, no soil analysis was done and Donio just applies one bag of urea per area about a month after planting. Before planting application of four bags of complete fertilizer is recommended. Two bags of urea 25 to 30 days after planting are also recommended. Thus, for each of his maize fields Donio needs almost three bags of complete fertilizer and about one and one-third bag of urea. Unlike in the rice field where hay and stubble are mixed with the soil during plowing, stubble and other plant parts in the maize fields are removed at harrowing using the *calcag* (maize field harrow).

To control weeds in maize, Donio and his family use weeding machines and hand weeding. They do not set a particular number of hours to weed each day. Weeding is done when their time and other activities permit. Just like the rice farm, they do not have problems with pests, except for rats, which they just "talk" to.

Harvested maize is husked, shelled, and dried in the field by spreading kernels on mats as is done with rice shown in Figure 1.8. When it is rainy, ears are hung inside the farmhouse, and husked ears are spread on the farm house elevated floor. The floor is made of bamboo slats; see Figure 1.12, so air flows freely and allows maize to dry. To shell the maize, they use a manual *banguran*, or Sheller. About 120 kg of clean grain can be shelled in 8 hours.

Daily Routine. Donio usually wakes up at 4 a.m. so that he can work during the cooler part of the day. After checking in and around the house, he moves out to his farms when there is a little streak of light. He weeds the fields, feeds the weeds to the year-old carabao he is taking care of, and allows the carabao to graze where there are grasses. He then brings the carabao to the *suba* (river) to *lunang* (wallow). After that he does other farmwork and, when done, goes home and does other work in and around the house and then relaxes. At about 4 or 5 in the afternoon, he

Figure 1.12. Lamp on slatted bamboo floor.

fetches the carabao. This routine is disrupted only when he gets the opportunity to work somewhere off the farm such as on a construction job.

The family usually eats breakfast, lunch, and dinner together, except when farm-work interferes. At times Donio cannot join the family for breakfast when he cannot return early enough from the farm. During planting, weeding, and harvest season, the family cannot have lunch together when Donio, Sita, and Nano work in other farms. They are, however, mostly together during dinner. Their usual meal consists of either a combination of fish, *ginamos* (salted fish paste), *buwad* (dried fish), and *utan* (vegetables). Sita prepares breakfast and cares for a female carabao and its 3-month-old calf. Daughter Genedina helps in the household chores during weekends when she is not in school.

For the most part, except for rice and maize grain, any purchased food is fresh every day and immediately eaten because Donio and Sita have no way to store food. People selling fish or meat or other food items will pass by the house every day making their produce easily available. There are some exceptions to this general situation. Dried fish and to a lesser extent beef, which may be stored without refrigeration, may be available. After dinner at about 6 p.m., they watch TV at Donio's brother's house, who is a neighbor.

Farm Tools. Donio owns simple farm tools; a plow (*daro*) (Fig. 1.10), harrow (*sudlay*) (Fig. 1.11), one single weeding machine, one double weeding machine (Fig. 1.8 inset), and one maize field harrow (*calcag*) (Fig. 1.11) made of iron. An iron plow like Donio's costs about P1,500 ($28.54). An iron harrow costs about P1,300 ($24.73). A single weeding machine is P350 ($6.66) and a double one is P500 ($9.51). Donio rents other farm implements like a tractor and a thresher as needed; see Figure 1.13.

Figure 1.13. Threshing machine.

Donio also has a wooden rake (*kandos*) used in mixing grain when drying, a wooden lining board (*bagis*) used in marking the rice paddy so that planters know where to plant the seedlings, and a wooden yoke (*yugo*) for his carabao. He also has a wooden mortar and pestle (*lusong* and *alho*), a maize shelling board (*banguran*), and bamboo gong (*karatong*) (Fig. 1.14) struck with bamboo stick to inform farmworkers that food is ready. Sickle (*garab*), bolo (*sundang*), and *guna* (a shorter version of the bolo) are among Donio's smaller farm tools (Fig. 1.14). A bolo is a long (∼45 cm, or 18 inches) wide knife that is an all-purpose utility farming implement.

Farm Animals. Donio has chickens (Fig. 1.15) that freely roam around the house and the neighborhood. Most of the food for the chickens comes from the surrounding

Figure 1.14. Examples of Donio's tools.

Figure 1.15. Donio's chickens foraging for food.

environment; however, he feeds them with some grain kernels and bran. The hens give the family both meat and eggs. Donio only sells them when money is urgently needed.

Aside from chickens, Donio does not own any other farm animal, although he does have a dog. He is caring for a year-old carabao owned by a neighbor, which he feeds, takes to the river to wallow (Fig. 1.16), and does other things to ensure the carabao's wellbeing. He also trains the carabao for farmwork. For his efforts, he gets three-fourths of the carabao, P6000 to 10,000 ($114 to $190), and the owner gets one-fourth if sold. Wife Sita also cares for a neighbor's carabao and its 3-month-old calf. She gets paid P100 ($1.90) a week for this task.

Figure 1.16. Carabao in a wallowing hole.

In addition to his house, Donio has a farm shed on one of their adjacent maize fields about half a kilometer away from the road. The farm shed is only one 2.5-m^2 room where harvest and some farm tools are stored, is made of indigenous materials, wood and bamboo as is their house. The house floor is raised about a meter from the ground and is made of bamboo slats. Under the floor is a space where chopped wood and coconut husks for fuel are stored. Adjoining the room is the kitchen and work area. The roof above the area is made of nipa[||] and the floor is the ground, which becomes muddy when it rains. At the front wall of the farmhouse is an overhang with a stall for resting or keeping farm tools.

The estimated farm shed cost is a P1000 ($19.02) because Donio did not buy most of the materials and did not put value on his labor. But if all are included the value of his farm shed is about P4000 ($76.10). Except for the farm shed, Donio has not made any major or permanent investment in the farms because they are not his.

Harvest and Cost of Attaining It

RICE. Donio said that he is planting a *Masipag*,[1] an inbred variety yielding an average of 120 sacks of *tipasi* (newly harvested unmilled grains) per hectare per harvest in his 2070-m^2 rice field. He is not certain whether it is really *Masipag* or not; this is what other farmers told him.

Donio's April 2006 harvest yielded 990 kg of *tipasi*, which is about average. Donio spent an estimated P6213.34 ($118.21) in producing this rice and would get a gross income of P8910 ($169.52) from selling his rice. Since he plants twice in a year, we may double these figures; however, many times, his second cropping does not do well because of typhoons that come toward the end of the year.

Donio "pays" some of his expenses in harvested grain and so does not have the full amount of the harvested 990 kg grain as "profit." Although on paper Donio may appear to make a little from rice farming, the reality is very different. When major costs are deducted from the 990 kg of harvest, he is left with 331.32 kg of harvest, which he dries and takes to the rice mill.

After drying and milling, he recovers 183.05 kg of rice. About a month after harvest, Donio gleans his field during which he is able to gather about another three sacks of *tipasi*, which will ultimately give him 82.86 kg of milled rice. Thus, overall, he makes 265.93 kg of milled rice in one cropping. If sold after milling, Donio's rice would return P5584.53 ($106.26). Since his production cost was P6213.34 ($118.21), rice farming is a losing venture for Donio, although he does not realize this. He does not sell the rice because it is used to feed his family.

MAIZE. Donio plants maize three times a year: January, May, and August. The last planting is usually affected by typhoons that come during the last months of the year. Donio's harvest varies, depending on the weather, but his most recent harvest was about 440 kg from each field, or a total of 800 kg.

[||]Nipa is a marsh plant whose leaves are ~3 to 4 cm wide and 30 to 50 cm long, which is folded over a strip of bamboo 60 cm long and "sewn" into place to form a "shingle" used to cover the roof.

After shelling, drying, and milling, similar to the situation with rice described above, he is left with 157.95 kg, which would return P3159 ($60.10), which pales against the estimated average production cost of P12,727.34 ($242.15) for both fields. As with rice Donio does not sell his maize because it is used for food for the family. Maize farming, therefore, is also a losing proposition for Donio.

MAIZE FOR SNACKS. There have been times when Donio sold maize while it was still green for *tilaub* (boiled maize with the husk still on for snacks). This is usually done when the rains are heavy and constant and thus adversely affect crop growth or when the crop is attacked by worms. Selling maize green for *tilaub* seems profitable because it minimizes labor cost. Donio stopped doing this because he was not paid in cash.

Donio's farming does not yield a profit, although it does produce food for his family. However, his farming is not an overall sustainable food production system.

1.1.1   Question of Survival

Farming is a losing proposition for Donio, but he and many others like him never bother to see whether they are making money from farming. Farming is the only thing Donio is capable of doing; it is his life. Whether gaining or losing, he has no other option.

In 2005, the daily cost of living for a Filipino family of six was P534.80 ($10.18).[2] Donio receives only P150 ($2.85) a day from his nonfarmwork and no profit from his farming. How do Donio and family survive then?

While he may not make money from farming, Donio ensures the food needs of his family by not selling his produce. The average per capita consumption of Filipinos is about 115 kg of rice a year.[3] With five members, Donio's family needs 575 kg of grains. Extra grain is shared with other family members. If they have a shortage of grain, they eat cooked banana and root crops grown on the edges of fields, see Figure 1.6.

1.1.2   Other Survival Strategies

While energy food is secured, how can Donio and family meet their other food and basic family needs such as clothing, children's education, and the like?

They accomplish this in many different ways, mostly involving working on other farms and doing nonfarmwork. Donio, Sita, and two older children (Nano and Genedina) weed on other farms, and family members get real cash when they work on other farms during planting season.

During threshing, Donio and Nano work in threshing groups. At times Donio works as an assistant in carpentry and construction in the *poblacion* (town center). Getting to town he could ride a *habal-habal* (motorcycle for hire) which costs P7.00 ($0.13), however, he walks starting at 5:30 a.m. so that his daughter can ride to the school in the *poblacion*, Baybay.

Sita makes and sells *silhig*, brooms made out of coconut midribs. She peddles the *silhig* house-to-house in Baybay. One daughter, Genedina, helps earn money for the family by doing laundry in the neighborhood for P50 ($0.95) a day when she is not in school.

Donio and many other farmers like him survive principally by *utang*, or borrowing money or goods perpetually. They become chained to those who lend to them because of two things: (1) they do not earn enough to pay what they borrow and (2) because they often cannot pay in cash and thus they pay with their produce. This is often more costly than paying with cash.

1.1.3 The Family

Donio and Sita have seven children, one died at 1 year old and one at 2 years. The living children are: Grace, 22, who is married and has one child. Renante, 19, also married, finished only grade 4, Monic, or Nano, 17, also has an elementary education. Genedina, 14, is in her first year of high school. She helps in the household chores and in the farm when not in school. Roger, 12, is in grade 5 and does not help much in the house and farm because he is in school.

1.1.4 The House

Structure. The family lives in a 5-m × 7-m (16-ft × 23-ft) one-story house made of wood and bamboo frames, bamboo floors, and nipa roof and walls; see Figure 1.5. It has three bedrooms, one for the male children, one for the female children, and one for Donio and Sita. The bedroom floors, made of bamboo slats, are raised on stilts about a meter from the ground. Under the floor is the storage area for firewood, chopped coconut husks, and other things including chickens.

Adjacent to the bedrooms are the kitchen and dining area, which occupies about a third of the house. The floor of the kitchen-dining area is the ground, which becomes muddy when it is rainy. Last year its walls and roof collapsed and were rebuilt by Donio. Below the roof of the kitchen are two baskets for the chicken to lay eggs and incubate them. Adjoining the other side of the bedrooms is a living room with a table at the center.

1.1.5 Household Appliances, Utensils, and Utilities

Donio and family live a bare existence. Their only appliance is a small, battery-operated (they do not have electricity), Avegon radio, worth P200 ($3.81), which is their major source of entertainment and news.

Their kitchen utensils include three aluminum pots of different sizes where they cook rice or maize, three aluminum kettles of different sizes, and one aluminum frying pan. They also have some ceramic and plastic plates and glasses and some ladles and forks and spoons. Their two stoves, fueled by firewood, gathered from around the neighborhood, and chopped dried coconut husks, are made of a rounded iron bar supported by a tripod of iron bars.

At night they light the house using a lamp made of bamboo, tin can, and $\frac{1}{2}$-inch-diameter pipe; see Figure 1.12. A cloth is inserted into the pipe from end to end

Figure 1.17. Water source for Donio and family.

and the pipe placed in the tin can, which contains kerosene. The cloth adsorbs the kerosene, which rises to the top by capillary action. At the top the kerosene is burned to produce light. The tin can is placed in bamboo so that it is easier and safer to handle.

For drinking water, they get the town's water supply system through a pipe and faucet (Fig. 1.17) that runs to their yard. They store drinking water in plastic water jugs in the house.

1.1.6   The Future

Asked what future he sees for himself, apathy and grief blended on Donio's face as he cheerlessly answered with a question of his own: "What future? I was a poor farmer, now I am a poor farmer; I will always be a poor farmer. The future is just the same yesterday and today." Even worse: "Who can lift us out of this poverty?" Asked what he wants his children to become he said that he does not know. "It's up to them," he said. He wants them to finish college and find a better occupation, but he cannot provide for that. When asked the same questions, Rosita said: "*Ambot. Ambot. Ambot*" (I don't know. I don't know. I don't know), her voice, broken and doleful.

1.2   THE FACE OF AN ECUADORIAN FAMILY FARM: AÍDA AND OCTAVIO

Jenny Valencia**
Photographs by David Cerón

Señora Aída Jiménez is 46 and Señor Octavio Tipán, on the right in Figure 1.18, is 48 years old. Aída grew up on a farm and only went to school for 2 years. She worked as a

**FAO Consultant

Figure 1.18. Aída Jiménez (Aída) and Octavio Tipán.

maid in Quito, the capital of Ecuador. She started helping with farming when she married Octavio to help with family expenses and meals. Octavio also grew up on a farm where he learned how to farm from his parents by working with his father and mother in the fields. His parents were able to pay for his education only through primary school (6 years), and he was unable to continue to secondary school.

1.2.1 Their Farm

The farm is in Pinantura, 30 minutes by car from Pintag. The road to the farm is shown in Figure 1.19. To get to the farm from Pintag, you need to take a bus from Pintag and get off in the place called *las minas* (the mines). From there it is a 30-minute walk to the farm. The road from Pintag is paved with stone but after *las minas* the road is unimproved. You must walk down the road, cross a ravine, and from there follow

Figure 1.19. Road from Pintag.

Figure 1.20. Aída and Octavio's farm house.

the road up and around the mountain to get to the farm entrance at the top of the ridge. No one in the family has a bicycle and so walking is the only way to get to Pintag.

Pinantura is near the Nevado Antisana, a snow-covered volcanic peak in the Andes, and from the farm you can also see Monte Tenerías, another snow-capped mountain. The location of the farmhouse on the top of the ridge gives the family a beautiful view of these mountains. The farm is also near San Alfonso and Santa Rosa, two beautiful small towns. Aída and Octavio's daughter, María Isabel, dreams about converting this spot into a tourist attraction someday.

Their farmhouse, shown in Figure 1.20, is situated on about 0.75 ha of land, on a ridge of the mountain, owned by Octavio. Another 0.5-ha field owned by Aída is about 500 m away. This field, shown in Figure 1.21, is used for maize and oats. In front of the house, across the ravine, is another field of 1 ha owned by Octavio's father. He is too old

Figure 1.21. Aída's main field.

to cultivate the property and Octavio and Aída use it for pasture for their animals. It is difficult to get there, so they only go there every two weeks or so.

The soils are excellent in this area. The topsoil is about 1 m deep with a dark black color, good organic matter content, and an ideal sandy loam texture. Even though the fields have a slope of more than 30 percent, the practices of using small raised beds and plantings of trees and shrubs reduce soil erosion and maintain soil fertility. Because the soils are derived from relatively recent volcanic ash, they are rich in minerals and are very fertile.

1.2.2 Farming Practices

The family tries not to use chemical inputs for their crops. They maintain a space for earthworm composting (Fig. 1.22) where they recycle the leaves and stems of vegetables produced for sale and kitchen scraps. In addition, they use manure from the animals (chickens, guinea pigs, rabbits, and swine) to increase soil organic matter. The manure is stockpiled and mixed with soil so that it can dry before being applied to the fields.

Purchased inputs are used for potatoes. Furadan, Maneb, 10–30–10 fertilizer and Kristalon are purchased. Furadan is a systemic insecticide to control the Andean potato weevil (*Premnotrypes vorax*). Maneb is a fungicide needed to prevent potato scab. The 10–30–10 is a granular fertilizer containing 10 percent nitrogen, 30 percent phosphate, and 10 percent potash, while Kristalon is a foliar fertilizer. The potato crop is the only one that receives chemical fertilizer.

The field closest to the house is used primarily for vegetables. This field is managed by Aída since she is in charge of family nutrition. Octavio helps her when he is at home, but Aída is the one who decides what crops to plant and whether to sell or use them. Octavio is primarily responsible for the care of the animals when he is at home.

In the garden used for vegetables Aída and Octavio have prepared raised beds like the one for lettuce shown in Figure 1.23. The vegetables are planted every 1 or 2 weeks so as to have a continuous supply. This way they have each crop in various stages of

Figure 1.22. Octavio working on a compost pit.

Figure 1.23. Lettuce growing on a raised bed.

growth and harvest. All vegetable crops are managed under an organic farming system, that is, no chemicals are used and fertility is maintained by using compost and animal manure. Drip irrigation is used during dry periods with water that comes from the same pipe as the household water. The household pays a single fee per year and the cost of water in the area is minimal. There are no water meters. Because the water is inexpensive, people in the area commonly use the same source of water for their households and for irrigation.

Aída plants crops in rotation, alternating root crops with leaf crops and medicinal herbs. Among the many crops planted for family use and sale are lettuce, Swiss chard (Fig. 1.24), beets, broccoli, carrots, white cabbage, red cabbage, cauliflower, white onions, red onions, spinach, coriander, parsley, radishes, and turnips. Other crops like celery, zucchini, garlic, cucumbers, and pumpkins have little potential for sale and are planted solely for family use.

Beehives are located around the borders of the raised vegetable beds, and Aída plants flowers such as poppies, dahlias, chrysanthemums, geraniums, and sunflowers to help with honey production. In the same field are beds of medicinal plants such as chamomile, lemon balm, mint, oregano, verbena, and garden rue.

In the field near the house there are also small plants and trees surrounding the vegetable areas serving as a windbreak and as a division between the areas. Many of these also produce fruit for family use. Among the many plants in this category are blackberry, banana, passion fruit, strawberries, and ground cherries. Fruiting trees are also in spaces farther from the vegetable beds and include avocado, tree tomato, and babaco.

Figure 1.24. Chard growing on a raised bed.

In the larger fields they have planted oats and floury maize. Oats are used for feeding the animals and none is sold. The floury maize is used for family food and some is sold. It is harvested green and used much like sweet maize in the United States (Fig. 1.25). The field located farthest from the house is currently used for pasture only. In previous years it had been used for planting potatoes and faba beans. Aída and Octavio are thinking of planting oats there in the future to feed the farm animals.

Daily Routine. Octavio is known as "Zorro" where he works as a reservoir guard for the municipal water supply company. He works 8 days at the reservoir and 8 days on the farm. Octavio is always happy when he gets to stay 8 days on the farm. While there, he works on organizing the farm, most recently changing the roof on the building where

Figure 1.25. María Isabel harvesting green maize.

Figure 1.26. Octavio working on a chicken pen.

they raise chickens and guinea pigs. He is working on his new chicken pens shown in Figure 1.26.

Aída spends her days taking care of the house and the farm. She is specifically in charge of anything having to do with caring for the vegetables. She has had training at the Antisana Foundation where she learned how to produce vegetables and small animals.

The farm provides their food needs, although they may buy a few food items, such as cooking oil and pasta, from the local market. Thus in most cases they are eating fresh food produced and prepared on their farm.

Farm Tools. A rented tractor is only occasionally used for plowing one field farthest from the house. All other operations are done by hand labor using the nine basic tools shown in Figure 1.27. Because the fields are relatively small, it is possible

Figure 1.27. Tools used on the farm.

Figure 1.28. Chickens on Aída and Octavio's farm.

for two or three persons to do planting and weeding as needed. The only time more labor is needed is in the harvest of crops such as potatoes. The weight of the harvest makes transport difficult, and several people must help sort and carry the harvest.

1.2.3 Farm Animals

Octavio is in charge of the livestock, and all of the farm animals are kept near the house. This helps in their daily care so that Aída and the children can do the chores when Octavio is away working. When he is at home, he spends most of his time working on the cages for rabbits, chickens (Fig. 1.28), and guinea pigs. His biggest project so far has been the construction of another building for the pigs. When he has time, he also works on improvements on the storage building for tools and equipment.

Aída and Octavio have started the construction of 6 pig pens. Octavio designed the pens and has done all the construction of the floors and walls for the pens and only lacks metal sheets for the roof to be complete. In spite of the lack of a finished roof, he has moved the pigs from an old building to the new one. He has plans to change the breed of pigs you see in Figure 1.29 to a more productive breed, but that takes money.

1.2.4 Farm Income

The crops and animals produced on the farm have two purposes: to provide food security for the family and for cash income. The farm provides most of their daily needs, but when they have urgent needs, for small amounts of money, they harvest some honey (Fig. 1.30) and sell it for US$5.00 each liter. More income comes from the sale of chickens, guinea pigs, rabbits, and pigs.

Every 1 or 2 weeks the family takes farm produce to the supermarket in Pintag. The market buys mostly vegetables and chickens. The chickens are delivered cleaned and in plastic sacks ready for sale.

Figure 1.29. Some of the farm pigs.

Figure 1.30. Beehives along the side of the garden.

1.2.5 Family Food

The selection of the vegetables to plant is done by Aída, who is one of the few ladies that has continued planting vegetables after her training by the Antisana Foundation. She recognizes that this training has improved the diet of her family. In this area of Ecuador most people do not have any vegetables in their diet, which is principally carbohydrates such as rice and potatoes. Since receiving this training Aída became convinced of the value of vegetables and has changed the diet of her family.

A typical evening meal is shown in Figure 1.31. It consists of fresh maize on the cob, grilled trout, boiled faba beans, and a mixed salad with vinegar and oil dressing. Although trout is an expensive food in some countries, these fish come from a local trout farm only 1 km away.

Figure 1.31. Typical family meal.

1.2.6 The Family

The couple has six children, five boys and one girl. The oldest, José, is 26 years old. He has graduated from the university with a degree in historical tourism. At the moment he is a teacher in the Colegio General Pintag, the local high school. The second eldest, Javier, is 24 years old. He works as a guard in the same water company as his father, Octavio. Maria Isabel, their third child, is 22 years old. She is a student at the Universidad Central de Quito majoring in business administration. To attend classes she must travel 2 hours each way every day. Diego, in Figure 1.32, with Javier, is 20 years old. He works with a foreigner cultivating algae in a greenhouse for food and feed. Segundo is 18 years and a student in the third year at the Colegio General Pintag where his older brother, José, works as a teacher. The youngest, David, is 10 years old and is a student in the Cosme Renelia de Pinantura. Their oldest son, José,

Figure 1.32. Two sons, Diego and Javier, in their room.

Figure 1.33. *Bodega* (barn) used for storage.

likes to collect ornamental plants. He has a corner at one end of the farmhouse where he grows these plants and has taken charge of the gardens around the house and the living fences that form a path to the orchard.

1.2.7 The House

The house, shown in Figure 1.20, is a simple concrete block construction around concrete pillars with reinforcing bars protruding from the top. This is a common practice that permits addition of a second floor when family, and money, permit. The flat roof serves as a work area to store items and to keep harvested crops away from animals. There is no snowfall or heavy monsoon rains so the flat roof is adequate. Although the house is not very large, it provides sufficient space for the children to have their rooms and for a living room and a well-equipped kitchen.

In addition to the house they have a *bodega* (barn) (Fig. 1.33) where they store various farm items including the tools and where they raise the guinea pigs.

1.2.8 Household Appliances, Utensils, and Utilities

There is electricity and even sewer service to the farm. This helps the family with the basic utilities. The total cost of water and sewer service is about US$1.60/month.

The family has a kitchen with running water and household appliances considered adequate in Ecuador. They have a gas stove, refrigerator, and microwave, as shown in Figure 1.34. In the parent's bedroom there is a television and stereo. Their son, Diego, has a large stereo set he sometimes uses at parties. The house also has a bathroom with running water.

Water for the household and irrigation comes through a supply system from the village of Pinantura. Its source is a spring (Fig. 1.35) that originates in the snow and ice covering the cinder cone of the nearby volcano Antisana. There is no treatment given to the water, but the quality is generally good. The family uses this water

Figure 1.34. Octavio and David.

without any treatment or filtration. An advantage of this water source is that it is not treated, making it well suited for organic farming.

They have no other source of water for the house or irrigation, but Octavio does not think they will have problems with shortages of water even though they plant crops all year. For now the flow of water from the spring is enough to supply them and their neighbors.

1.2.9 The Future

Aída and Octavio dream of expanding their farm. Octavio particularly likes raising hogs. His goal is to improve the farm production so that one day he can quit his job at the water company. He still needs to work since the farm cannot yet support his family with all their necessities.

Figure 1.35. Spring that provides water for Aída.

Figure 1.36. Steve and Patti Murphy.

1.3 THE FACE OF A LARGE COMMERCIAL FARM IN THE UNITED STATES: STEVE

Alfred R. Conklin, Jr. and Thomas Stilwell[‡]

Steve Murphy is 54 years old (Fig. 1.36) and has been farming all his life on a farm in Clinton County, Ohio, 5 miles southeast of the town of Wilmington. Steve grew up on the farm, which was handed down to him from his grandfather through his father and will be handed down, in the near future, to his sons. The farm is 486 ha (1200 acres) and has been in the family for about 150 years. Steve has never worked at other jobs

Figure 1.37. Road to Steve's house.

[‡]Professor of Agriculture and Chemistry and Assistant Professor of Agriculture Wilmington College Wilmington, Ohio, respectively.

Figure 1.38. Steve and Patti's home.

having obtained his livelihood solely from farming. The road leading to the farm is shown in Figure 1.37.

After completing high school, Steve went to college and earned B.S. degrees in agronomy and animal science from The Ohio State University. After college he came back to the home farm and worked for 4 years before marrying his wife, Patti (Fig. 1.36), who had graduated 4 years earlier from Wilmington High School. The family lives in the farmhouse located along the road passing the farm; see Figure 1.38.

1.3.1 Steve's Farm

Steve Murphy farms about 1300 ha (3200 acres)[§§] half of which he owns and half he rents from neighboring landowners. He pays cash to 6 landowners with areas of from 30 to 81 ha (75 to 200 acres) to rent their land for farming. Land in Clinton County sells for \$9876/ha (\$4000/acre), and thus the land he farms is valued at \$12,800,000 on an acre basis. The half he owns is thus worth \$6,400,000. A picture of one of his fields is shown in Figure 1.39.

Steve grows a number of different crops—maize, soybeans, winter wheat, barley, and hay/pasture—that are rotated between fields on a regular schedule. Maize and soybeans are planted to 486 ha (1200 acres) each in May and harvested in October and November. Figure 1.40 shows one of Steve's sons planting soybeans. Maize yields 9406.5 kg/ha [150 bushel (bu) per acre] and soybeans 3359.5 kg/ha (50 bu/acre). Barley and winter wheat are planted to 243 ha (600 acres) in the fall and barley harvested in May–June and winter wheat in June–July time frames. Barley yields 6047 kg/ha (90 bu/acre) and winter wheat 5375 kg/ha (80 bu/acre).

Fields totaling 81 ha (200 acres) are used for pasture and hay and in some cases are used for both purposes. That is, sometimes they are pastured and other times they are

---

[§§]Amounts and values are rounded in some cases.

Figure 1.39. Field on Steve's farm.

cut for hay. In total hay fields yield 9 Mg (10 tons) of hay per year (Mg = megagram or 10^6 g). Figure 1.41 shows a tractor pulling a haybine that is used to cut and condition grass for baling.

Hay and pasture fields have a mixture of grasses and legumes and are different from other crops because fields remain in this crop several years before being rotated to another crop. Fields used either for pasture or hay are changed between the two uses as need arises. Hay is cut three times a year, cured (involves mostly loss of water) in the field, and baled, compressed into large round bales or "packages" for storage. It is used as feed for the cattle during the winter months when there is no green grass pasture for the cattle to eat.

Steve also raises and sells cattle and hogs. He has 40 cattle and calves (young cattle) that are a mixture of angus and other breeds. They obtain most of their food by feeding

Figure 1.40. Planting soybeans.

Figure 1.41. Haybine for cutting and conditioning grass.

on grass growing in fields, that is, pastures. Their feed is supplemented with protein, salt, and maize and during the winter by feeding them hay harvested and baled the previous summer. Salt is supplied as blocks, which also contain minerals, that the cattle lick.

Hogs are raised in a barn, equipped with automatic feeders and waterers and with a slatted floor to allow feces to fall through into a pit, which is pumped out regularly and the manure applied to surrounding fields. Figure 1.42 shows a tractor with a tank for hauling and spreading manure. He raises two groups of 1200 hogs per year for a total of 2400 hogs. Hogs are raised on a contract basis—the young hogs and their feed is supplied by the owner and Steve contracts to raise them to reproductive age at which time the owner takes the hogs back. Thus Steve supplies the buildings, care, and labor for raising the hogs for which he is paid a set fee.

Figure 1.42. Tractor with a manure tank.

1.3.2 Other Crops

Steve's son, William, has a small garden that supplies some fresh vegetables for the family during the summer months, but this is not intended to supply all the family's vegetable needs. There are black walnut trees that produce nuts and wild blackberries growing on the farm, however, for the most part these are not used as food. He also has two dogs. Otherwise there are no other crops or animals on the farm.

1.3.3 Farming Practices

Steve practices no-tillage and minimal tillage farming. In no-till, commonly used with maize and soybeans, herbicides are used to control weeds, and seed and fertilizer are placed directly into a field containing dead weeds and plant residues remaining from the previous crop. Because crops are planted in the spring, most of the previous year's crop residue has decayed over the winter. Dead weeds and any remaining crop residue are completely gone by harvest time. Minimum tillage involves some minimal working of the soil surface just before plating.

For winter wheat and barley the only field preparation is to use a field cultivator or a disk that breaks and loosens the soil before planting. Planting is done with a planter, little or no herbicides are used, and harvesting is done using a stripper head on the combine as shown in Figure 1.43. No insecticides, other than those applied to seed, are used in crop production.

In addition to being highly mechanized, computers and GPS (global positioning satellites) play a large role in Steve's farming operation. Electronic equipment in the cab of his combine is shown in Figure 1.44. Steve uses computers to keep track of the economics of his farming operation. He has a computer in his house office (Fig. 1.45), where his records are kept (see also below). In addition every tractor has at least one computer associated with it and its operation. When planting, a computer and GPS unit connected to the steering of the tractor means that he can plant in straight

Figure 1.43. Combining barley.

Figure 1.44. Inside a combine cab showing the computer monitors.

rows across the field without steering the tractor. This also means that he knows exactly where he has planted and in the case of herbicides where they have been applied.

It is also possible for him to change rates of fertilizer or herbicide as he crosses a field, thus making a more accurate application of either or both. When harvesting, he can use sensors on the combine to keep track of the yields he is obtaining as he crosses a field. Sensors also tell him the moisture content of the grain as it is being harvested. He can electronically mark places in the field by longitude and latitude, where a particular problem, such as poor yields or increased numbers of weeds, occur.

Seeds are purchased from a local seed company and are coated with insecticides to protect the seed and young plant from root worms and maize borer. Steve also produces seed for a local company, but he never plants seed produced on his farm.

Figure 1.45. Steve's farm office.

Figure 1.46. Grain truck being filled.

Each field's soil is sampled and analyzed every other year. Fertilizer is purchased from a local supplier and is designed to fit the requirement of both the soil, adding nutrients that are not at adequate levels and the crop, to be planted. Some fertilizer is applied by spreading on the soil surface before planting while other is fertilizer applied at planting with the planter.

Harvesting is accomplished using the same combine for maize, soybeans, barley, and wheat (Fig. 1.43). For each crop a different "header" is installed. A stripper header strips the grain from the barley and wheat stalk leaving most of the stalks standing in the field. Soybean and maize headers either cut soybeans and separate beans from shells and stalks or "pick" maize ears from stalks and remove the grain from ears. All the headers use the same combine body with just the front part, the header, being changed depending on the grain to be harvested. Grain is collected in a bin on the combine and transferred to a grain wagon or truck (Fig. 1.46) when the bin is full.

Most harvested grains are immediately shipped to a local grain elevator. A grain elevator consists of several tall round buildings (silos) used to dry and store grain.

Figure 1.47. Grain bins and a grain truck being filled.

Subsequently it is shipped to markets and sold to food companies. Steve has the capacity (Fig. 1.47) to both store and dry 15,000 bu of grain. This allows him flexibility in deciding when to sell some of his crop.

Hay is cut using a haybine mower, allowed to cure in the field for a day or two, and then raked into rows and baled. Steve uses a baler that produces large round bales that are left in the field until they are needed to feed the cattle. Some farmers sell large bales, but Steve used all of his for his cattle. If there is a need, the straw, stalks of small grains, is cut using the same mower and bailed for storage. Straw may subsequently be used for bedding of animals or sold for other uses.

1.3.4 Daily Routine

Steve rises at 6:30 a.m. and eats breakfast that consist of eggs, bacon, and toast. After breakfast, at about 7:00 a.m. he checks on the hogs and cattle to make sure they have food and water and that no animals are sick. He then proceeds to do field work if weather permits. If the weather is bad, too wet, and or too cold, he may work in his shop (Fig. 1.48), maintaining his tools and farming equipment, or does paper work in his office located in the farmhouse. Steve's shop is equipped with all the tools needed to maintain and repair his equipment as needed. His office is equipped with a desk and computer that is connected to the Internet. Internet access allows him to keep up with changing weather and market conditions as well as find and order supplies.

At 11:30 a.m. Steve stops to eat lunch, which consists of a sandwich and sometimes a bowl of soup, in the farmhouse. After lunch he does field work until dinner time, which occurs between 6:00 and 7:00 p.m. During harvest, dinner may be later depending on the flow of the harvest. After dinner is family time, which may include watching television, participation in sports, or other leisure activities.

All the food eaten by the family is purchased at the local supermarket or grocery store. The store has almost all kinds of vegetables, fruits, meats, and dairy products that

Figure 1.48. Farm workshop.

may be purchased fresh, frozen, or in cans. Because these are large stores, which have operations in many states, produce may be shipped in from other locations around the United States or imported from foreign countries. In this way the shelves can contain "fresh" fruits and vegetables almost all year long.

Produce that is perishable or is likely to store better under cool or cold conditions is placed in either the refrigerator or freezer as soon as it is brought home. Because of the ability to keep food under these conditions, it is typically purchased only once a week. In addition other food items such as bread, cereals, frozen and fresh meals, and the like are available along with other needed, nonedible items used in the home.

1.3.5 Farm Tools

Steve owns five tractors, a field cultivator, a combine, a sickle mower with conditioner, hay rake, baler and sprayer. He also has a large truck that is used to transport grain to local markets. Motorized equipment, mostly tractors, is powered by diesel engines. There are also a number of electric motors that are used to supply feed to the hogs, pump water for the house and animals, fans for drying grain and to power augers used to move grain from trucks and trailers to bens and vise versa. All Steve's equipment combined is worth $750,000.

Water for the farm comes from a well that is capped and covered by a concrete block house (Fig. 1.49) .Water is pumped, using a submerged electric pump, to the house and for animals from this source. As long as there is electricity the farm is self-sufficient in terms of water.

Figure 1.49. Farm well.

In addition to the regular working tractors, Steve also restores old tractors, particularly John Deere tractors. These are sometimes used for "play," which consist of doing farming activities such as plowing, disking, and mowing with equipment no longer used in farming. This play teaches his sons about older farming methods.

1.3.6 Farm Animals

As noted above Steve raises both hogs and cattle that he sells, the hogs on a contract basis. The cattle are raised to the desired weight and sold at local markets. He has no chickens or other "farm" animals although the family does have two family dogs.

1.3.7 The House

The farm house (Fig. 1.38) is 24 m $\times$ 18 m (80 ft $\times$ 60 ft) thus having 432 m^2 (4800 ft^2) of floor space. It has two stories and four bedrooms, each of Steve's sons has his own room, two bathrooms, a family room and a living room, and a screened in porch. The house is a basic wood frame house with wooden siding and a shingled roof. The house has electricity in all rooms and running hot and cold water in the bathrooms and the kitchen.

1.3.8 Household Appliances, Utensils, and Utilities

The kitchen contains all appliances common to United States kitchens that is, stove, dishwasher, microwave, mixer, toaster, coffee maker, and the like. All appliances are electric. There is also a table and chairs in the kitchen and the family most often eats there.

1.3.9 Income

Giving a complete analysis of the expenses and profits from each of Steve's enterprises is beyond the scope of this book. However, a picture of the economics of the farming operation can be gained by looking at one enterprise.[¶¶] For the maize enterprise with a yield of 10,033 kg/ha (160 bu/acre) his variable costs are \$2.58 and his fixed costs \$1.68 per bushel. Thus the total cost of production is \$4.26 per bushel. For a discussion of fixed and variable costs see Chapter 10. The most important are the variable costs. When he has a return higher than his variable cost, Steve is making more money than he spent in cash to raise the crop. If he sells the maize for less than the variable cost, he has not recovered his cash costs to raise the crop. Because large farmers cultivate such large areas and obtain high yields, small differences in the per bushel income make a large change in total income. For 1000 acres of maize yielding 10,033.6 kg/ha (160 bu/acre) or a total of 160,000 bushels a \$0.10 difference in per bushel price results in a \$16,000 difference in total income. In the United States it requires about 304 ha (1000 acres) to support one farmer and his family.

¶¶An enterprise would be any one of the individual crops grown or types of animals raised. Thus Steve has a corn enterprise, a hog enterprise, etc.

For Steve's farming operation harvesting costs are relatively small, costing in total equipment, fuel, and the like about 10 percent of the total cost of production. Harvesting of barley is shown in Figure 1.43. For the hog operation Steve is paid a set price that provides him with a steady income of $48,000 per year.

1.3.10   The Family

Much of Steve Murphy's family, live, in and around Clinton County. However, only Steve, his wife, sons, and a brother run the farm. Steve's immediate family consists of Steve, his wife, Patti, and two sons. Steve's wife stays at home coordinating transportation and takes care of the house and their son, William. William, called Willy, the younger son, is in high school, plays football, and works on the farm. Nicholas, called Nick, has just graduated from college and is in line to take over the farming operations. The men are involved in the field work and animal farming operations along with Steve's brother, Milton. He does not live on the farm but has a home near by. Farm-work is not divided in any specific way between the men, they just do the work that needs to be done when it needs to be done. All members know how to do all jobs and so any person may be doing any job at any time. A cousin lives across the road from the farmhouse but is not involved in the farming operations.

1.3.11   The Future

When asked about the future, Steve's response is that he will retire and turn the farming operation over to his sons. He expects that he will still help in various farming operations as needed but will not be directly running the operation. He sees the farming operation going along in the same way it has. He does not see expansion of any aspects of the operation, that is, he does not see that they will increase hectares farmed, number or types of crops raised, or the number of hogs and cattle raised.

1.4   CONCLUSION: A CONCISE COMPARISON

Table 1.1 gives a comparison of a few facts and figures of the selected farmers. At first glance we can see some marked differences in the net worth of each farm and their education. The single factor of land ownership is associated with a greater investment in equipment and buildings. The farmer in the Philippines, Donio, owns no land and obviously cannot construct valuable buildings on another person's land. The fact of land ownership also changes the farmer's plans for the future. Octavio, in Ecuador, and Steve, in the United States, both are making ambitious plans for the future. Donio only seeks to continue his status quo. One of Steve's sons has graduated from college and he expects the second to do the same. He plans for them to take over the daily farmwork so that he can retire. Octavio is looking forward to building up his animal operation so the income will permit him to leave his guard job and devote full time to the farm and family. Donio has no such hopes.

TABLE 1.1. Comparison of Farmers

	Subsistence Farm (Donio)	Small Family Farm (Aída and Octavio)	Commercial Farm (Steve)
Hectares farmed	1.55	2.25	1300
Hectares owned	0	1.25	486
Hectares rented/leased	1.55	1.0	814
Value of land owned	0	US$6250	US$6,400,000
Value of equipment	US$70	US$135	US$750,000
Farmer education	4 years	6 years	16 years, BSc
Years farming	35	36	32
Source of food	Grown on farms, some purchased	Grown on farm, few purchases	All purchased from local market
Children in family	5	6	2
Children's education	4–10 years	5–16 years	11–16 years
Future	Bleak	Planning expansion	Planning retirement

Table 1.2 gives a comparison of farming operations. Donio uses the least and Steve the most inputs. Aída and Octavio are the most self-sufficient, buying little from the local marked. All of Donio's produce goes to feed the family and pay for farming operations. Aída and Octavio produce enough food to feed the family and sell for cash for other family needs. Steve sells all his produce and uses the money to purchase all the family needs. Steve's operation is all mechanized while Donio uses a mixture of hand, animal, and tractor power to carry out farming operations. Aída and Octavio use very little tractor power doing almost all farming by hand.

All three farmers are about the same age and have two to six living children. However, if we count total number of children born to each family, the most will be Donio with seven, then Octavio with six, and the least will be Steve with only two. This also follows world demographic trends as seen in Table 1.3. As family incomes rise and life expectancies increase, couples tend to have fewer children. There are more guarantees that a newborn child will grow to maturity. Farm families with greater income also tend to have less need of cheap labor. In fact, one of the reasons that a farmer such as Donio has many children is the need for labor. All of his children have small jobs that contribute to the family income. Octavio and Steve both earn enough cash income that their children can pursue the luxury of a college education. The only question for Octavio is where his children will earn their living after graduation.

Other factors that contribute to the differences in family planning are the statistics for expected life of parents and newborn children.

The two countries with highest infant mortality rates are the Philippines and Ecuador. When we review the number of children born to each of our sample farm families, we see they are similar: seven in the Philippines and six in Ecuador. The life expectancy of the parents in these two countries is also similar.

TABLE 1.2. Comparison of Farming Operations

	Subsistence Farm (Donio)	Small Family Farm (Aída and Octavio)	Commercial Farm (Steve)
Crops raised	Rice, corn, bananas	Corn, potatoes, 26 types of vegetables and herbs, 3 fruit trees	Corn, soybean, barley, hay, wheat
Source of seed/planting material	Produces himself	Purchase in local market	Purchase from commercial supplier
Equipment	Hand and animal-powered equipment	Hand equipment with little tractor use	All mechanized
Source of fertility	Plant residues left in field plus some commercial fertilizer	Plant residues, compost, manure, commercial fertilizer	Plant residues, hog manure, commercial fertilizer
Source of chemicals	Purchased in local market	Purchased in local market	Purchased from commercial supplier
Fate of produce	All consumed by family or used to pay for farming inputs	Used by family and sold in local markets	All sold to local grain and meat markets
Animals raised	Chickens, $\frac{3}{4}$ share of a buffalo	Pigs, rabbits, guinea pigs, chickens	Pigs, cattle

TABLE 1.3. Life Spans and Infant Mortalities in Three Countries

	The Philippines	Ecuador	United States
Parents expected life span	69.3 years	71.9 years	77.1 years
Mortality rate per 1000 newborns	24.2	24.5	6.4

Source: Nationmaster.com.

QUESTIONS

1. One of the differences between our three farmers is the level of their education. Discuss what importance you think this may have on their farming situation. What might this have to do with willingness to accept new farming practices?

2. Why is a $0.10 difference in the amount received per kilogram for grain important in the United States but not in Ecuador and the Philippines.

3. Comparing the situation of Donio, Aída and Octavio, and Steve, what one change in Donio's situation might significantly improve his life?

4. If Donio had access to all of Steve's equipment, what affect would it have on his farming success assuming all other things remained the same?

5. If Aida and Octavio had access to all of Steve's equipment, would it have any affect on their farming success?

6. In terms of the crops grown, considering the situation of Aída and Steve, what might Donio try to improve his diet?

7. Compare and contrast the family size, income, and education level for the three farmers, that is, Donio, Aída, and Steve.

8. What condition of the land farmed might be most important in determining the types of tools used by Donio as compared to those used by Aída?

9. Contrast the animals and the conditions under which they are raised by the three farmers.

10. Compare the future prospects for the three farmers.

REFERENCES

1. Farm Trial First Cropping Leave Farmers Speechless. Available at http://www.panaynews.com.ph/archives/1002/news6.htm.

2. Getting the Hard Facts. *IBON Facts and Figures*, **28**(24) (31 December 2005):2–9.

3. Rice Sufficiency Fight Seen to be Decided in Bedroom. *Philippine Daily Inquirer*, June 13, 2006, p. B8.

2

HUMAN NUTRITION

2.1 MEALS OF FARMERS

Donio and his family eat rice or ground maize, a little meat to flavor it, and little else as seen in Figure 2.1. The meal shown is very typical of almost all meals in the Philippines among almost all Filipinos, that is, breakfast, lunch, and dinner, from northern to southern Philippines. Water will be the most common drink at all meals.

Figure 2.2 shows a typical meal in Ecuador. It contains meat, in this case fish, fava beans, maize, and salad containing onions, tomato, and cabbage. Although not in this meal, it is also common for meals in Ecuador to contain two or more carbohydrate sources, for instance, rice and popped, baked, boiled maize, a cooking banana, and potato as part of the meal. This would be true of all meals, although eggs might be eaten for breakfast. Such a meal might also include some small amount of meat and some salad. Water or a drink mix will be the most common beverage at a meal, although tea or coffee may also be served.

Figure 2.3 shows a typical or common summer time farm meal in the United States. It consists of maize on the cob, roast beef, salsa, and ice tea. Each person will eat three ears of maize on the cob, which acts as both a vegetable and a source of carbohydrates in this meal. It would also be common for this meal to include potatoes in one form or another and some form of bread in the form of slices of bread, muffins, or biscuits.

World Food: Production and Use. By Alfred R. Conklin, Jr. and Thomas Stilwell
Copyright © 2007 John Wiley & Sons, Inc.

Figure 2.1. Donio eating a meal of fish and rice, which is a typical meal in the Philippines. (Courtesy of Henry Goltiano.)

Figure 2.2. Typical meal for Aída and Octavio consisting of corn on the cob, trout, and a salad of tomato, cabbage, and onion.

In the United States a farm meal will often be somewhat different from meals eaten by nonfarm families, having more carbohydrates and always incorporating bread in one form or another. Breakfast, the first meal of the day, will be different from lunch, the second meal of the day, eaten around noon, and dinner, the evening meal, eaten around 6:00 p.m. Breakfast may be bacon, toast, and eggs or cereal with milk. Lunch will typically be a sandwich, drink, and a desert, that is, something sweet. In Ohio, dinner* may be a meat, carbohydrates, potatoes, and two vegetables followed

---

*Names of meals are used differently by different people; some will call the noontime meal dinner and the evening meal supper, but breakfast is always breakfast.

Figure 2.3. Summer meal for Steve of corn on the cob. Each person will eat three ears, beef roast, and salsa with ice tea to drink. (Courtesy of Steve Murphy.)

by desert. All four will be in about equal size proportions, although the meat and carbohydrates may be in larger amounts in some cases, for instance, at the evening meal. The meal will be finished with a dessert, commonly pie or cake, sometimes with ice cream. The drink will also change with the meal. Breakfast will usually include juice and coffee, lunch ice tea or a soft drink, and dinner ice tea.

2.2 INTRODUCTION

All people need food but the amount and type of food varies with age. Children need enough food not only to maintain their body but also to provide energy and nutrients to build new cells. There is also evidence that children need certain nutrients that adults no longer need. Mature, nongrowing people, need only enough food to maintain their body. However, there is more to nutrition than this. One food component cannot totally substitute for another. That is, foods that primarily provide carbohydrates cannot provide sufficient amino acids for the body, no matter how much a person eats. Likewise protein-rich foods cannot provide all necessary fats for the body. All types of components, except for minerals, can be used by the body to produce energy, although not with the same efficiency.

Because of these needs, various organizations and government agencies have devised recommendations as to the types and amounts of foods people need to eat to remain healthy. Examples are recommended diets from the United States, through the Food and Drug Administration (FDA), and from the United Nations, through the Food and Agriculture Organization (FAO). The FAO dietary recommendations are based largely on calorie intake and are shown in graphical form in Figure 2.4. The FDA gives its dietary recommendations on the basis of amounts of various food groups to be eaten as shown in Figure 2.5.

Both of these systems have advantages and drawbacks. The FDA system is easy to understand and specific about the food groups to be eaten. However, it has the drawback

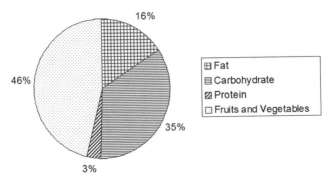

Figure 2.4. FAO (Food and Agriculture Organization of the United Nations) dietary recommendations in terms of grams. Fat, carbohydrate, and protein are calculated using data
in Dietary Recommendations in the Report of a Joint WHO/FAO, Expert Consultation on
Diet, Nutrition and the Prevention of Chronic Diseases (WHO Technical Report Series 916,
2003): Potential Impact on Consumption, Production and Trade of Selected Food Products
http://www.ifap.org/issuesifapreport-Whodiet.pdf. Table 2.2. was also used.

that in many parts of the world certain foods, for example, milk and dairy products, are
not readily available. The FAO system gives details about the various types of fats and
oils, including cholesterol, needed or recommended. Both the FDA and FAO system
fail to note the fact that certain amino acids (components of proteins) and oils are

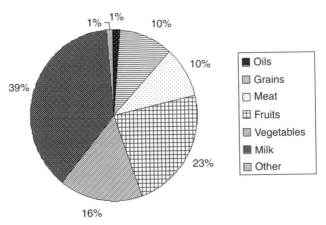

Figure 2.5. FDA (Food and Drug Administration) dietary recommendations on the bases of
grams of recommended foods. Grams are an average of selected fruits, avocado, passion
fruit, dates, guava, papayas, strawberries, and banana for vegetables, spinach, lettuce, tomatoes, carrots, peas, lima beans, sweet potato, and cauliflower. Average grams meats and
beans were converted from ounces to grams. The milk group is an average using whole milk,
cream cheese, and American cheese. Oils are given in grams, and other was converted from calories to grams using Table 2.2. All data for these conversions came from Agricultural Research
Service Nutrient Data Laboratory USDA National Nutrient Database for Standard, Release 17,
http://www.nal.usda.gov/fnic/foodcomp/Data/SR17/reports/sr17page.htm.

TABLE 2.1. General Energy and Nutrient Needs of Children and Adults

Person	Energy Calories (kcal)	Protein (g)	Carbohydrate (g)	Fat (g)	Fiber (g)
Growing child (20 kg)	1800	45–68	247–337	30–60	27–40
Nongrowing male adult (80 kg)	2900	73–109	399–544	48–97	27–40
Nongrowing female adult (60 kg)	2000	50–75	275–375	33–67	27–40

Source: Calculated from WHO Technical Report Series 916, Table 6. Available at http://WHO.int/publications/en/and data available in Bibliography.

essential in that the human body cannot make them and thus need to be part of the food intake. Both of these issues are discussed below.

People around the world obtain their food from different locally available sources, and this is sufficient as long as the total of their diet contains all the essential components and contains them in proportions that the body needs. A diet containing a mix of carbohydrate and protein-rich foods along with some fat, oil, and fiber, particularly from fruit and vegetables, is probably sufficient. This is particularly true if the diet contains mixed food sources, that is, grains, root crops, meat, vegetables, fruits, berries, and nuts. However, it is likely not to be true if the diet consists of only one food type such as meat or a single type of grain or root crop.

It is possible for people to obtain some of their needed minerals such as calcium and iron from nonfood sources, such as salt, and those minerals dissolved in drinking water and eating minerals such as calcium carbonate. However, it is generally agreed that these sources are not as readily absorbed by the body during the digestion process as are those available from food.

In more specific terms what does the human body need from food and how does it use these needed foods? It needs energy, most often provided by carbohydrates, protein to build and maintain muscle, fat and oil for both energy and cellular membranes, vitamins to augment enzyme activity, and minerals for teeth, bones, and enzyme activity. As shown in Table 2.1, the amounts of each of these will depend on age, size (body weight), level of activity, climate, and heredity.

2.3 ENERGY

Carbohydrates are the most common source of metabolic energy around the world. Four of the more common carbohydrate-rich foods are rice, maize, potato, and cassava, as shown in Figure 2.6, and are considered staple foods. Wheat, which is not shown, is also very important. These can be cooked in various ways and are often the bulk of the meal.

As seen in Table 2.1, the human body needs energy. Several significant things about this need are not generally recognized. One is that a lack of energy can substantially reduce the development of children and the health of adults. Second, all food, except minerals and fiber, can be used as a source of energy if insufficient energy is supplied by carbohydrates or fat. This means that other components of the diet will not be available to be used for other essential bodily needs such as muscle development. For this reason an appropriate intake of energy-rich foods is essential for normal growth and health.

Figure 2.6. From left to right, rice, corn, potato, and cassava—the four most widely eaten carbohydrate sources in the world.

Energy is measured in many different ways by scientists, but for foods the most common unit of measurement is the calorie. When spelled with a capital C, that is, Calorie, it represents or is understood to be a kilocalorie (the kilocalorie being 1000 times greater than the calorie). This measure of energy is very easy to understand and to measure. It is simply the amount of energy, obtained by oxidation of food [see Eq. (2.1)], needed to increase the temperature of 1 g of water (at 25°C),[†] 1°C (actual measurements are made with sophisticated scientific instruments in a laboratory). Table 2.2 gives the kilocalorie content of the three basic food components. Equations (2.1) and (2.2) show reactions by which the body obtains energy from carbohydrates. For the specific reaction given, 686 kcal/mol (kilocalories per mole) are obtained.

$$\text{Carbohydrates} \xrightarrow[\text{Enzymatic breakdown}]{\text{Metabolism}} \text{Glucose} \qquad (2.1)$$

$$\underset{\text{Glucose}}{} + \underset{\text{Oxygen}}{6O_2} \xrightarrow{\text{Metabolism}} \underset{\text{Carbon dioxide}}{6CO_2+} \quad \underset{\text{Water}}{6H_2O+} \quad \underset{686\,\text{kcal/mol}}{\text{Energy}} \qquad (2.2)$$

The body converts all carbohydrates consumed to glucose, which is combined with oxygen from air (similar to burning) to release energy. Half of this energy is lost as heat, which maintains body temperature. The other half is converted to adenosine triphosphate (ATP), which is the most important high-energy compound in the body.

TABLE 2.2. Energy Content of Basic Food Components

Component	Calories/Gram
Protein	4
Carbohydrate	4
Fats	9

[†]°C represents degree celsius where 0°C is the freezing point of water and 100°C is the boiling point of water.

In spite of this latter reaction, energy content of food is universally measured in calories, which is a measure of heat.

The main source of carbohydrates (for production of energy or calories) come from starchy foods. These are primarily from grains such as maize, rice, wheat, and root crops, such as potatoes and cassava. See Figure 2.6 for examples of starch-rich crops, although other small grains and root crops, barley, oats, millet, sweet potato, and others are also important in certain localities. Of these sources grains are generally higher in other dietary essentials, such as protein, than are root crops. Usually carbohydrate-rich foods are consumed in larger amounts, and thus considered staple foods, than other components of the diet. Generally speaking, carbohydrate-rich foods are less expensive than protein-rich foods and this also contributes to their being important foods.

Figure 2.7 shows the kilocalorie content of four major food carbohydrate sources. Note that although potato and cassava are lower in carbohydrate on a 100-g wet-weight basis, that is, they contain considerably more water than the grain; thus on a dry-weight basis all three have similar amounts of carbohydrates. Note also that grains will usually not be eaten dry and so comparing these sources of carbohydrates on a raw (wet) weight basis can be deceiving.

When comparing the energy content of foods, it is also important to also keep in mind the other components contained in that food source, that is, protein, fat, fiber, vitamins, and minerals. Some carbohydrate sources provide substantially more of these essentials than do others. Fat in the diet is another source of energy and thus can add

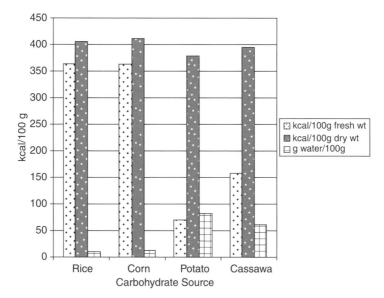

Figure 2.7. Kilocalorie content of rice, corn, potato, and cassava on both a wet (raw) weight and dry bases per 100 g of each. Water content is in terms of grams and uses the same scale as kilocalorie. (Data from Agricultural Research Service Nutrient Data Laboratory USDA National Nutrient Database for Standard Reference, Release 17, http://www.nal.usda.gov/fnic/foodcomp/Data/SR17/reports/sr17page.htm.)

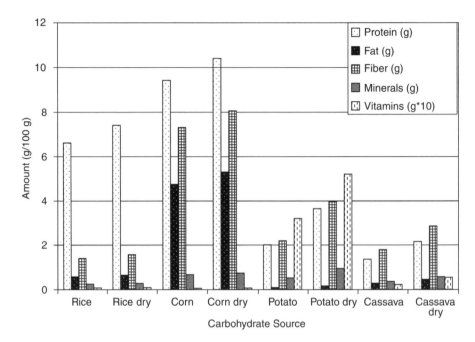

<u>Figure 2.8.</u> Protein, fat, mineral, and vitamin content of common carbohydrate sources. (Data from Agricultural Research Service Nutrient Data Laboratory USDA National Nutrient Database for Standard Reference, Release 17, http://www.nal.usda.gov/fnic/foodcomp/Data/SR17/reports/sr17page.htm.)

substantially in the required energy component of a diet. In Figure 2.8 both maize and rice are shown to have significantly more protein and fat than potato and cassava. None of the carbohydrate sources have significant fat content. Cassava is maligned because it has high carbohydrate content but low protein content. However, it is seen to have much higher vitamin content, mostly vitamin C, than do the other carbohydrate sources. Also none of these carbohydrate sources is a significant source of minerals.

In some parts of the world fats, oils, and refined foods are a significant source of energy. Fats and oils have high energy content (see Table 2.2). Thus, if high-fat foods, such as fatty meats (e.g., pork) or foods cooked in fats such as fried foods are consumed on a regular basis, their fats and oils will provide significant kilocalories in the diet. Refined foods are food items that have been significantly changed from their original form by removing some of their components. This often means that vitamins, minerals, and fiber have been removed. The resulting food is often easier to digest and has higher energy content. While as an energy source they are better, overall they are often less nutritious.

2.4 PROTEIN AND AMINO ACIDS

Proteins, which are polymers of amino acids, make up the muscle of the body and are responsible for all bodily movement including the heart and lungs, without which the

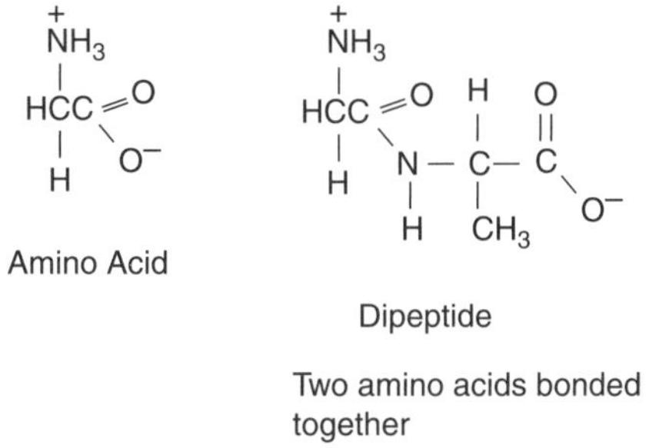

$$\overset{+}{N}H_3$$
$$HCC \overset{O}{\underset{O^-}{=}}$$
$$H$$

Amino Acid

$$\overset{+}{N}H_3$$
$$HCC \overset{O}{=} \quad N - C - C$$

Dipeptide

Two amino acids bonded
together

Figure 2.9. Chemical structure of amino acids and bonding (connection) between them used to form proteins. Two amino acids bonded together would be a dipeptide.

body cannot function. In Figure 2.9 a simple amino acid and two simple amino acids bonded together are shown. In a protein or muscle a long chain of different amino acids are bonded together in a similar way. In addition to muscle, all enzymes, which are the body's catalysts, are made of proteins. They also often contain one or more metal ions, which explains, in part, the importance of minerals in the diet. The shape and function of the protein depends upon its amino acid composition.

Two examples of major sources of high-quality protein in the diet are meats and nuts. Examples, steak and walnuts, are shown in Figure 2.10. Typically people eat far more meat than nuts. One possible exception might be peanuts, which are eaten in a number of different ways and as parts of meals in several regions in the world, particularly parts of Africa and India. Like meats, nuts are also a source of fat in the diet (see Section 2.5).

Proteins are made up of a variety of amino acids that can be classified as either nonessential or essential. Nonessential amino acids, listed in Table 2.3, are not essential

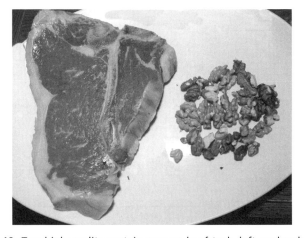

Figure 2.10. Two high-quality protein sources, beefsteak, left, and walnuts right.

TABLE 2.3. Common Nonessential Amino Acids

Name	Abbreviation	Chemical Formula
Alanine	Ala	$\overset{+}{N}H_3$ $\overset{\|}{CH_3CHC} \overset{O}{\underset{O^-}{\diagup}}$
Aspartic acid	Asp	$O \quad \overset{+}{N}H_3 \quad O$ $\overset{\diagdown}{CCH_2CHC}$ $HO \diagup \qquad \diagdown O^-$
Cysteine	Cys	$\overset{+}{N}H_3$ $\overset{\|}{HSCH_2CHC} \overset{O}{\underset{O^-}{\diagup}}$
Glutamine	Gln	$O \qquad \overset{+}{N}H_3$ $\overset{\|\|}{H_2NCCH_2CH_2CHC} \overset{O}{\underset{O^-}{\diagup}}$
Glutamic acid	Glu	$O \qquad \overset{+}{N}H_3 \quad O$ $\overset{\diagdown}{CCH_2CH_2CHC}$ $HO \diagup \qquad \qquad \diagdown O^-$
Glycine	Gly	$\overset{+}{N}H_3 \quad O$ $\overset{\|}{CH_2C} \overset{}{\underset{O^-}{\diagup}}$
Hydroxyproline	Hyp	$\overset{+}{N}H_2 \qquad O$ $CHC \overset{}{\underset{O^-}{\diagup}}$ HO
Proline	Pro	$\overset{+}{N}H_2 \qquad O$ $CHC \overset{}{\underset{O^-}{\diagup}}$
Serine	Ser	$\overset{+}{N}H_3 \quad O$ $\overset{\|}{HOCH_2CHC} \overset{}{\underset{O^-}{\diagup}}$
Tyrosine	Tyr	$HO \qquad \overset{+}{N}H_3$ $\overset{\|}{-CH_2CHC} \overset{O}{\underset{O^-}{\diagup}}$

TABLE 2.4. Common Essential Amino Acids

Name	Chemical Abbreviation	Chemical Formula
Arginine	Arg	$$\underset{\quad\;\; \overset{\displaystyle NH}{||}\qquad\qquad \overset{\displaystyle \overset{+}{N}H_3}{|}}{H_2NCNH(CH_2)_3CHC} \overset{\displaystyle O}{\underset{\displaystyle O^-}{}}$$
Histidine	His	$$CH_2\underset{|}{\overset{\overset{+}{N}H_3}{C}}HC\overset{O}{\underset{O^-}{}}$$
Isoleucine	Ile	$$CH_3CH_2\underset{\displaystyle CH_3}{\overset{}{C}}H\underset{|}{\overset{\overset{+}{N}H_3}{C}}HC\overset{O}{\underset{O^-}{}}$$
Leucine	Leu	$$(CH_3)_2CHCH_2\underset{|}{\overset{\overset{+}{N}H_3}{C}}HC\overset{O}{\underset{O^-}{}}$$
Lysine	Lys	$$H_2N(CH_2)_4\underset{|}{\overset{\overset{+}{N}H_3}{C}}HC\overset{O}{\underset{O^-}{}}$$
Methionine	Met	$$CH_3SCH_2CH_2\underset{|}{\overset{\overset{+}{N}H_3}{C}}HC\overset{O}{\underset{O^-}{}}$$
Phenylalanine	Phe	$$-CH_2\underset{|}{\overset{\overset{+}{N}H_3}{C}}HC\overset{O}{\underset{O^-}{}}$$
Threonine	Thr	$$CH_3\underset{\displaystyle OH}{\overset{}{C}}H\underset{|}{\overset{\overset{+}{N}H_3}{C}}HC\overset{O}{\underset{O^-}{}}$$
Tryptophan	Trp	$$-CH_2\underset{|}{\overset{\overset{+}{N}H_3}{C}}HC\overset{O}{\underset{O^-}{}}$$
Valine	Val	$$(CH_3)_2CH\underset{|}{\overset{\overset{+}{N}H_3}{C}}HC\overset{O}{\underset{O^-}{}}$$

because the body can make them as long as prerequisite carbon, nitrogen, and sulfur compounds are present in the diet. These amino acids range from the simplest, glycine, to more complex amino acids such as arginine.

Essential amino acids are listed in Table 2.4. They are essential because they cannot be made by the body and must be available in food in order for the body to obtain them for use in producing muscle, enzymes, and the like. This group of amino acids ranges from the simple leucine to the multifunctional arginine and the multiringed tryptophan. Both groups have amino acids with multiple nitrogen-containing groups and other functionalities such as —OH, —SH groups. These groups play key roles in determining both the shape of proteins and, in the case of some enzymes, their ability to accommodate metal ions needed for enzymatic activity of enzymes.

The best sources of protein and amino acids are meat and nuts. For example, in Figure 2.11 it is seen that on a raw-weight basis meat and nuts have about the same protein content. However, on a dry-weight basis meat has much higher protein content. This is caused, as can be seen, by meat having much higher water content than nuts. Thus, when adjusted for water content, that is, on a dry-weight basis, meats have a considerably higher protein content than do nuts. Meat is, for most people, an essential part of the diet in terms of obtaining enough protein and amino acids for good health.

Tuna and beef, as shown in Figure 2.12, have no carbohydrates or fiber while nuts do. Nuts have much higher fat content than do meats and also have a much higher vitamin content on a wet-weight basis, but tuna has about the same amount on a dry-weight basis. Because they contain little water, nuts have small changes in composition when calculated on a dry-weight basis. However, they have considerably more fat and vitamins than beef.

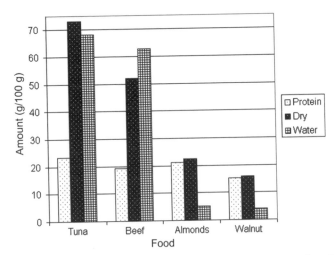

Figure 2.11. Protein content of selected meats and nuts. (Data from Agricultural Research Service Nutrient Data Laboratory USDA National Nutrient Database for Standard Reference, Release 17, http://www.nal.usda.gov/fnic/foodcomp/Data/SR17/reports/sr17page.htm.)

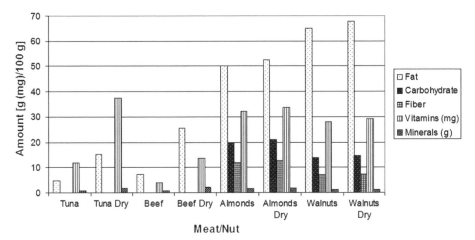

Figure 2.12. Fat, carbohydrate fiber, vitamin, and mineral content of selected meats and nuts. Note that vitamins are in micrograms (mcg) but still using the left-hand numbering. (Data from Agricultural Research Service Nutrient Data Laboratory USDA National Nutrient Database for Standard Reference, Release 17, http://www.nal.usda.gov/fnic/foodcomp/Data/SR17/reports/sr17page.htm.)

Another aspect of protein is that the body has no means to store either protein or amino acids. On the other hand the body can store fat and can convert carbohydrates to fat, which is thus a way to store carbohydrates. This creates a problem in that if the diet does not contain sufficient protein the body can only get needed amino acids from already formed muscle and thus muscle wasting will occur.

2.5 CARBOHYDRATES

Figure 2.13 shows a two-sugar unit of a larger polysaccharide. Carbohydrates are mostly made up of polysaccharides and are common in nature and food except meat as noted above. They are the chief source of energy in most diets also as noted above. Carbohydrates can be simple, as in the case of single-sugar molecules or as two-sugar molecules bonded together. They can also be complex such as the starches found in seeds, grain, and root crops. Simple sugars are readily taken up and used by the body, while complex carbohydrates must be broken down before absorption.

The glucose molecule shown in Figure 2.13 is readily taken up by the body and used to produce energy. The disaccharide, sucrose, common table sugar, must first be broken into its two constituent sugar molecules before use. Sucrose is made up of two sugars that are only different in the arrangement of their —OH and —C=O substituents. However, both types of sugars are readily metabolized in the body and thus provide a rapid source of energy.

Sugar molecules can be bonded together in different ways producing many different polymers that are significantly different from starch. Thus, although both cellulose

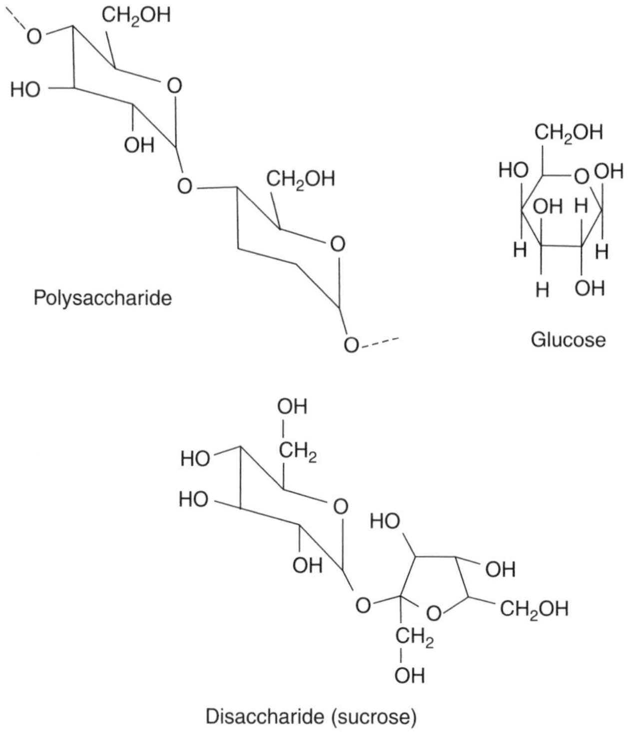

Figure 2.13. Glucose molecule, a disaccharide, and two units of a polysaccharide (not all hydrogen atoms are shown; unsatisfied bond positions are occupied by hydrogen).

and starch are polymers of sugars, because of differences in bonding, starch is digested by the body while cellulose is not. Cellulose forms the roughage, one constituent of fiber, used by the digestive tract for bulk.

Because they can be both straight chained and branched, complex carbohydrates, such as the various starches, require a number of different enzymes to break them down into their component sugar molecules that can then be used by the body. Thus, one enzyme is required to break off branches so that they are single unbranched chains. A second enzyme is used to break the chains into individual molecules of sugar. Because complex carbohydrates must be broken down before they can be absorbed, they provide a slower increase in blood sugar level and release of energy over a longer period of time.

The carbohydrate content of three major foods is given in Figure 2.14. There is a significant difference in content when presented on a dry rather than a wet or fresh basis. On a raw- or wet-weight basis, maize and rice have more carbohydrates and less water than potato and cassava. However, on a dry-weight basis all four have about the same carbohydrate content with cassava having the highest amount.

Unrefined carbohydrate sources contain dietary fibers that the refining process removes and discards. This material provides bulk that the digestive tract muscles use to maintain tone. These materials also hold water to keep the stool moist and soft.

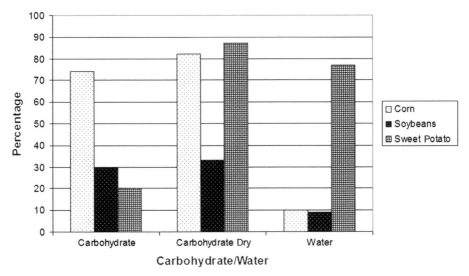

Figure 2.14. Carbohydrate and water content of corn, soybeans, and sweet potato. (Data from Agricultural Research Service Nutrient Data Laboratory USDA National Nutrient Database for Standard Reference, Release 17, http://www.nal.usda.gov/fnic/foodcomp/Data/SR17/reports/sr17page.htm.)

2.6 FATS, OILS, AND LIPIDS

Figure 2.15 shows two common sources of lipids—fats and oils in the diet, for example, beef fat and vegetable oil, in this case olive oil. Both are good sources of fats in the diet and as described below, beef fat is also a source of essential fats. Figure 2.16 gives the fat content of three common sources: almonds, avocados, and bacon, all high in this food component.

Fats, oils, and other lipids are discussed as if they occurred as free fatty acids; however, they are most often in the form of triglycerides, an example of which is shown in Figure 2.17. They are made up of three fatty acids bonded through their acid functional group to glycerol, which is also shown in Figure 2.17. Because fatty acids are the component used for energy production, they are larger than glycerol, and, because each triglyceride yields three fatty acids when broken down, they are the component that will be discussed.

The term lipid, in addition to being used for as a general term for fatty acids and triglycerides, also includes cholesterol and other water-insoluble molecules. While these other lipids may be found in fats and oils from any source, cholesterol is only found in lipids from animal sources. Common saturated and monounsaturated fatty acids are given in Table 2.5. Almost all common fatty acids have an even number of carbon atoms in their chain. For this reason most odd-numbered fatty acids have not been given. Dodecanoic, tetradecanoic, tetradecenoic, hexadecanoic, hexadecenoic,

Figure 2.15. Examples of sources of fat: beef fat (left) and olive oil (right) in the diet.

octadecanoic, and octadecenoic acids are the most common mono- and polyunsaturated fatty acids. All are water insoluble and have high boiling points. In the scientific naming system the occurrence of a double bond is indicated by a change in ending, that is, from -anoic to -enoic, while keeping the same root name.

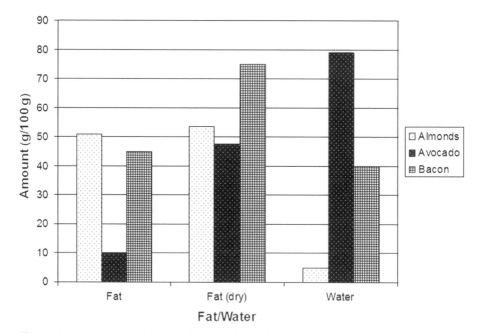

Figure 2.16. Fat content of almonds, avocado, and bacon. (Data from Agricultural Research Service Nutrient Data Laboratory USDA National Nutrient Database for Standard Reference, Release 17, http://www.nal.usda.gov/fnic/foodcomp/Data/SR17/reports/sr17page.htm.)

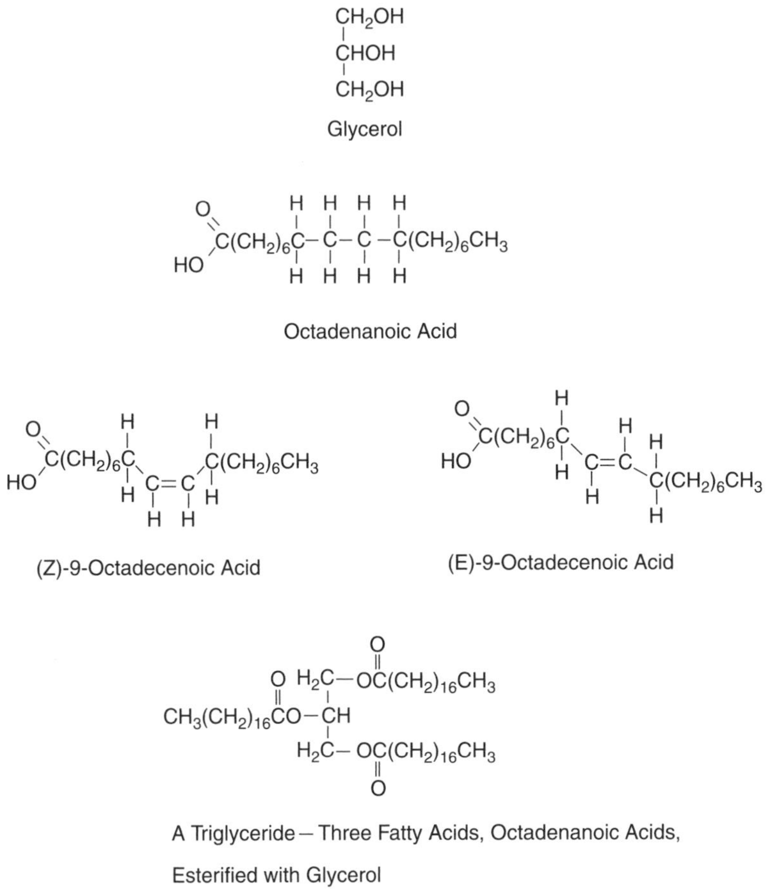

Figure 2.17. Examples of the components of lipids, glycerol, unsaturated, Z (cis), E (trans) fatty acids, and triglyceride.

Table 2.6 gives the common polyunsaturated fatty acids. As noted, they all have more than one double bond and docoshexenoic acid has six double bonds. It is common that these double bonds be Z (cis) as indicated in Figure 2.17. Increasing the double-bond content of fatty acids decreases their melting point, and thus they are liquids at room temperature and below. Double bonds can be removed by a process called hydrogenation, which results in fatty acids that remain solid at room temperature. This is advantageous when producing butter substitutes. However, trans fatty acids are produced during hydrogenation, and there is increasing data that indicates that trans fatty acids are deleterious to health and should be avoided (http://www.hsph.harvard.edu/reviews/transfats.html).

Although the body can make most fatty acids from any food source, there are two, octadecadienoic and octadectrienoic, acids that are essential fats. This means that they

TABLE 2.5. Common Saturated and Monounsaturated Fatty Acids

Systematic Name	Common Name	No. of Carbons	No. of Double Bonds
Butanoic acid	Butyric acid	4	0
Hexanoic acid	Caproic acid	6	0
Octanoic acid	Caprylic acid	8	0
Decanoic acid	Capric acid	10	0
Dodecanoic acid	Lauric acid	12	0
Tetradecanoic acid	Myristic acid	14	0
Tetradecenoic acid	Myristoleic acid	14	1
Hexadecanoic acid	Palmitic acid	16	0
Hexadecenoic acid	Paimitolic acid[a]	16	1
Heptadecanoic acid	Margaric acid	17	0
Heptadecenoic acid		17	1
Octadecanoic acid	Steric acid	18	0
Octadecenoic acid	Oleic acid[a]	18	1
Eicosanoic acid	Arachidic acid	20	0
Eicosenioc acid	Gadoleic acid	20	1
Docosanoic acid	Behenic acid	22	0
Docosenoic acid	Erucic acid[a]	22	1
Tetracosanoic acid	Lignoceric acid	24	0
Tetracosenoic acid	Nervonic acid[b]	24	1

Note: When present, all are in gram (g) quantities in typical foods.
[a]Both the cis and trans isomers are included in this name.
[b]Only the cis isomer.

TABLE 2.6. Common Polyunsaturated Fatty Acids in Food

Systematic Name	Common Name	No. of Carbons	No. of Double Bonds
Octadecadienoic acid	Linoleic acid[a]	18	2
Octadecatrienoic acid	Linolenic acid[b]	18	3
Octadecatetraenoic acid	Parinaric acid[a]	18	4
Eicosadienoic acid	None	20	2
Eicosatrienoic acid[a]	None	20	3
Eicosatetraenoic acid	Arachidonic acid[a]	20	4
Eicosapentaenoic acid	Timnodonic acid	20	5
Docosadienoic acid	Brassic acid	22	2
Docosapentaenoic acid	Clupanodonic acid	22	5
Docosahexaenoic acid	None	22	6

Note: When present, are in gram (g) quantities.
[a]These acids occur as cis and trans isomers and no distinction between the two is made in this name.
[b]Double bonds occur in different positions and no distinction is made here between the compounds with different double-bond positions.

must be eaten as part of the diet. Octadecadienoic acid is commonly found in milk fats, while in general animal fats are sources of both these essential fatty acids.

2.7 FIBER

Fiber represents all those components of food that are not digested by the body and pass through unchanged. There are two types of fiber, soluble, specifically water soluble and insoluble, and each play a different role in the diet. Insoluble fiber helps move food through the body and makes elimination easier. Soluble fiber also helps move food through the digestive system, but in addition it absorbs fat and can thus decrease fat uptake and the potential adverse effects of fat on health. Although all foods, except meats, have some of both types of fiber, grains, fruits, and vegetables vary widely in the amount of each that they have.

Fruits, barley, and oats have soluble fiber, while wheat and brown rice are noted for having insoluble fiber. Legumes, seeds, and vegetables in general are noted for having insoluble fiber. Soluble fiber includes things such as gums, mucliages, and pectins, while insoluble fiber is composed of cellulose, lignin, and some hemicelluloses. Examples of the molecular structure of two fibers is given in Figure 2.18. A balanced diet containing a mixture of fruits, vegetables, and whole grains will contain a suitable mixture of both kinds of fiber.

Generally, the more processing a food is subject to the less fiber it contains. For instance, components such as skin and core are removed from fruits and the remaining portion strained. Each step will remove some fiber and thus make the food less desirable as a fiber source. Grain as it comes from the field has outer layers that are darker, commonly brown, in color. These outer layers are often removed during processing making

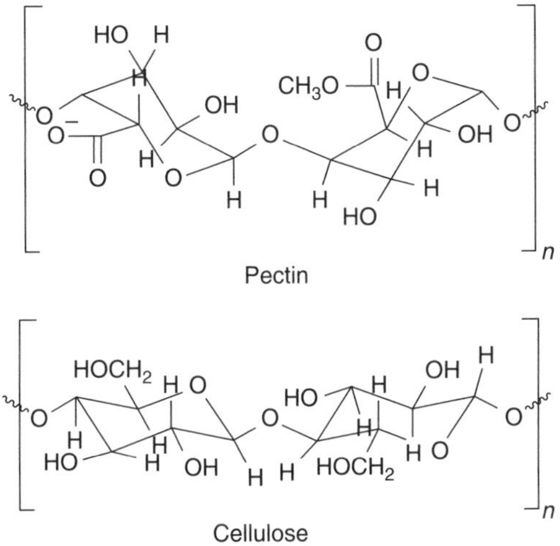

Figure 2.18. Two types of dietary fiber, pectan and cellulose.

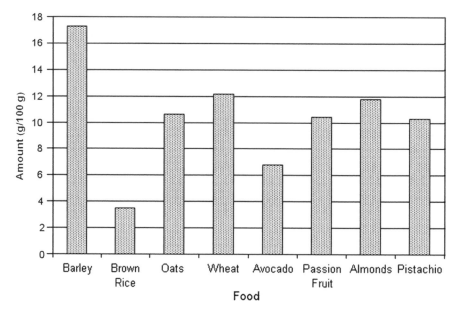

<u>Figure</u> 2.19. Fiber content of selected common foods. (Data from Agricultural Research Service Nutrient Data Laboratory USDA National Nutrient Database for Standard Reference, Release 17, http://www.nal.usda.gov/fnic/foodcomp/Data/SR17/reports/sr17page.htm.)

the grain white, for example, rice or wheat flour. Removal of this layer or layers removes much of the fiber and many vitamins and minerals.

Eating whole fresh food that has undergone a minimum of processing results in the intake of the maximum amount of fiber and the eater receives the maximum benefit from it. As seen in Figure 2.19, barley has the highest fiber content of the common grains, while passion fruit and almonds in the fruit and nut food groups have high fiber. Although brown rice has a lower fiber content, it is still important because its fiber is highly soluble. It should be noted that the outer layers of rice are high in vitamins that are lost when they are removed.

2.8 VITAMINS

Vitamins are essential components or molecules required by the body in milligram (mg, 10^{-3} g) or microgram (μg, 10^{-6} g) amounts on a daily basis to carry out essential functions. Table 2.7 lists the vitamins, their scientific and common names, and the general concentration levels in foods. In some cases as with ascorbic acid (vitamin C) and the B vitamins, they may be relatively simple molecules. In other cases such as vitamin B_{12} they may be complex structures as shown in Table 2.8. Vitamins can be divided into water soluble (Table 2.8) and fat soluble (Table 2.9). This is an important distinction in that water-soluble vitamins are readily excreted from the body and not stored, with the exception of B_{12}, while fat-soluble vitamins are stored in fat and the liver.

TABLE 2.7. Vitamins

Vitamin	Common Name	Concentration[a]
Alpha-tocopherol[b]	Vitamin E	mg
Ascorbic acid	Vitamin C	mg
Folic acid[c]	One of the B vitamins	mcg
Thiamin[c]	One of the B vitamins, i.e., B_1	mg
Riboflavin[c]	One of the B vitamins, i.e., B_2	mg
Niacin[c]	One of the B vitamins, i.e., B_3	mg
Pyridoxine[d]	One of the B vitamins, i.e., B_6	mg
Cobalamin	One of the B vitamins, i.e., B_{12}	mcg
Pantothenic acid	One of the B vitamins, i.e., B_5	mg
Phylloquinone[e]	Vitamin K	mcg
Retinol[f]	Vitamin A	IU or mcg
Cholecalciferol[g]	Vitamin D	IU or mcg

[a]mg = milligram (10^{-3} g), mcg = microgram (10^{-6} g), IU = International unit.
[b]A family of molecules, alpha being the most important.
[c]Part of what is called the B-vitamin complex.
[d]One of six forms of B_6.
[e]One of two forms.
[f]One of a number of related compounds called retinoids.
[g]One of several compounds that can be converted to vitamin D in the body.

TABLE 2.8. Water-Soluble Vitamins

Vitamin/Name[a]	Structure[b]	Function
B_1/thiamin	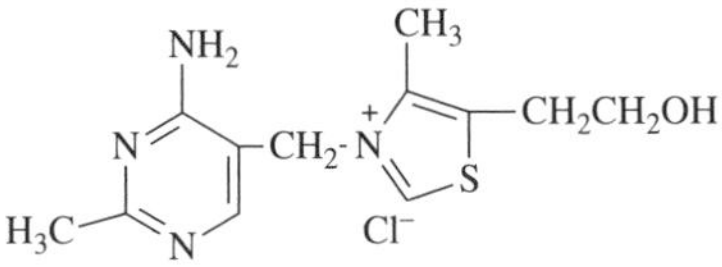	Energy metabolism
B_2/riboflavin	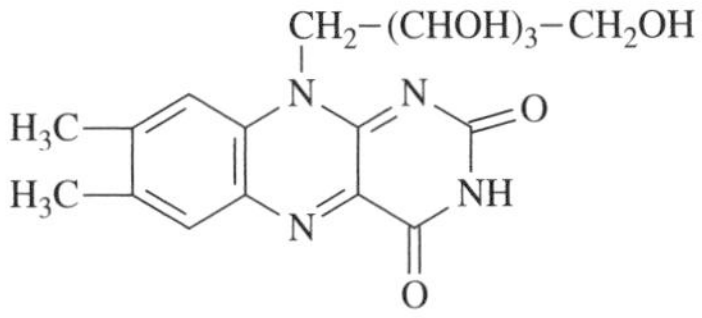	Energy metabolism/ vision
B_6/pyridoxine	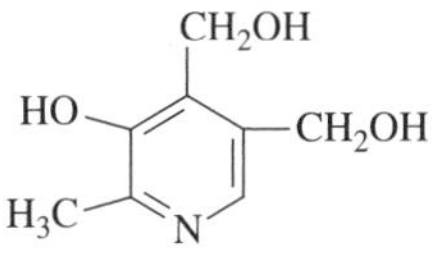	Amino acid and fatty acid metabolism
Niacin/nicotinic acid	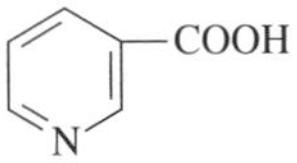	Energy metabolism

(Continued)

TABLE 2.8. *Continued*

Vitamin/Name[a]	Structure[b]	Function
Folate/folic acid		New cell development
Pantothenic acid		Energy metabolism
Biotin		Fat glycogen metabolism
C/ascorbic acid		Collagen synthesis
B$_{12}$/cyanocobalamin		New cell synthesis

[a]Vitamins often have more than one name and only a more common name is given.
[b]Not all hydrogen atoms are shown; unsatisfied bonding positions are occupied by hydrogen.

TABLE 2.9. Fat-Soluble Vitamins

Vitamin/Name[a]	Chemical Structure[b]	Function
A/retinol		Vision—health of cornea
D/calciferol[c]		Bone development/ calcium and phosphorus in blood
E/tocopherol[d]		Antioxidant
K/phylloquinone		Blood clotting

[a]Vitamins often have more than one name and only a more common name is given.
[b]Not all hydrogen atoms are shown; unsatisfied positions are occupied by hydrogen.
[c]A number of different structures form a group of compounds of similar structure and activity called vitamin D. The specific compound shown is calciferol.
[d]A number of different structures form a group of compounds of similar structure and activity called vitamin E. The specific compound shown is α-tocopherol.

While vitamins are essential to the proper functioning of the body, they can become toxic in high concentrations. For the water-soluble vitamins this is not much of a problem as excess vitamin is readily excreted. Thus, in this case there is less possibility of toxic symptoms from taking large doses of these types of vitamins. *Caution:* This is not the case for the fat-soluble vitamins. They are absorbed by the fat in the body and are not easily excreted, thus toxicities are possible when people take excess vitamin supplements. Toxic symptoms are generally not seen when vitamins are obtained solely from food.

The vitamin levels in the fruits, a nut, and berry shown in Figure 2.20 are obtained by summing all the vitamins and thus do not show the relative levels of each individual

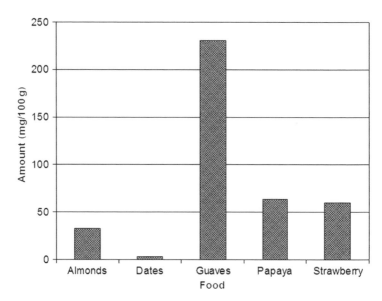

Figure 2.20. Sum of vitamin content of a nut, selected fruits, and a berry. (Data from Agricultural Research Service Nutrient Data Laboratory USDA National Nutrient Database for Standard Reference, Release 17, http://www.nal.usda.gov/fnic/foodcomp/Data/SR17/reports/sr17page.htm.)

vitamin. This is misleading because while guavas have a high total level of vitamins this is due to a very high level of vitamin C (ascorbic acid) but low levels of other vitamins, as shown in Figure 2.21. This is even more dramatically illustrated by considering that guavas contain very low or undetectable amounts of pantothenic

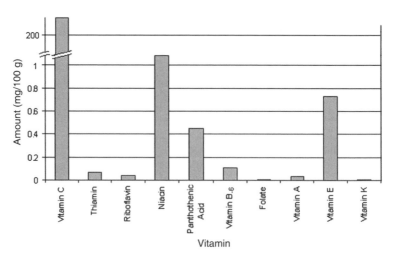

Figure 2.21. Vitamin content of guava. (Data from Agricultural Research Service Nutrient Data Laboratory USDA National Nutrient Database for Standard Reference, Release 17, http://www.nal.usda.gov/fnic/foodcomp/Data/SR17/reports/sr17page.htm.)

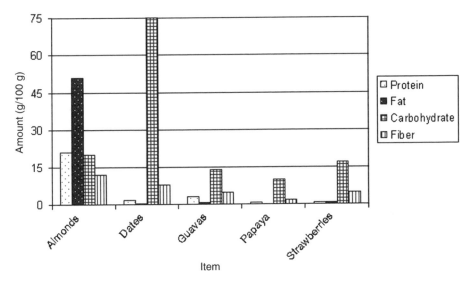

Figure 2.22. Protein, fat, carbohydrate, and fiber content of nut, selected fruits, and berries. (Data from Agricultural Research Service Nutrient Data Laboratory USDA National Nutrient Database for Standard Reference, Release 17, http://www.nal.usda.gov/fnic/foodcomp/ Data/SR17/reports/sr17page.htm.)

acid, vitamin B_6, folic acid, and vitamins E, D, and K! Vitamin B_{12} is only found in animal sources and so it not present.

In addition to vitamins, guavas are also not a particularly good source of other food elements, that is, protein, fat, carbohydrates, and fiber as seen in Figure 2.22. Of these four components guava is higher in carbohydrate and fiber than in protein and fat. In looking at almonds and fruits given in Figure 2.22 dates are seen to be very high in carbohydrates and almonds high in fat and higher in protein and fiber than the fruits. Guaves, paypaya, and strawberries have similar levels of all four components. This should reinforce the idea that a diet consisting of a variety of fruits would be best for supplying all vitamins.

2.9 MINERALS

Commonly agreed upon important minerals are given in Table 2.10. Minerals can be divided into two groups: major minerals, those required at higher levels (i.e., greater than 200 mg per day), and trace minerals, those required at lower levels (i.e., less than 200 mg per day). Sodium, potassium, calcium, and magnesium are examples of minerals required at higher levels, while iron, manganese, copper, and zinc are required in trace amounts. The common mineral form given is not necessarily that present in food and are not those commonly available in supplements.

The major minerals, particularly calcium and phosphorus, are used by the body to produce structures such as bone and teeth. Calcium is also important in maintaining

TABLE 2.10. Minerals Found in Foods in Milligrams[a] Quantities

Mineral	Chemical Formula	Ionic Form	Common Mineral Form
Calcium	Ca	Ca^{2+}	$CaCO_3$
Magnesium	Mg	Mg^{2+}	$MgCO_3$
Phosphorus[b]	P	MPO_4	$CaHPO_4$
Potassium	K	K^+	KCl
Sodium	Na	Na^+	NaCl
Zinc	Zn	Zn^{2+}	$ZnCl_2$
Copper	Cu	Cu^{2+}	$CuCl_2$
Manganese[b]	Mn	Mn^{3+}	$KMnO_4$
Iodine	I	I^-	KI
Molybdenum[b]	Mo	Mo^{4+}	MoS_2
Selenium	Se	Se^{4+}	SeO_2

[a]Except for selenium, which is found in microgram quantities (mg = milligrams = 10^{-3} gram, μg = micrograms = 10^{-6} gram.)
[b]Commonly occur in a number of different oxidation states. M stands for metal.

blood pH. Magnesium is an important constituent of bones and other components such as muscle and the liver. Sodium finds its major use in maintaining the proper fluid content of the body in addition to helping to maintain blood pH. There are also many other functions of these minerals in the body.

The trace minerals are important in the functioning of blood and enzymes. Iron is an essential component of hemoglobin, which carries oxygen throughout the body. Zinc and selenium are both essential components of numerous enzymes.

In addition to the above minerals that, except for phosphorous, occur as cations, there are three important minerals, chlorine as chloride, sulfur as sulfate, and iodine that occur primarily as anions in the body, that is, Cl^-, $SO_4^=$, and I^-. Chloride and sulfate both maintain the electrical neutrality, fluid levels, and pH in the body. Iodine on the other hand is an essential component of thyroxin, which maintains the body's base metabolism, and thus is essential to health. Because electrical neutrality is essential, it is rare to find a deficiency of one of these minerals if the cationic minerals are in sufficient supply.

Some sources of these minerals are shown in Figure 2.23, and amounts of minerals in these fruits and nuts are given in Figure 2.24. A more detailed analysis of the major and trace minerals in almonds, guavas, and strawberries is given in Figures 2.25 and 2.26. In both cases almonds are high in all minerals, except sodium and selenium, when compared to guava and strawberry. Sodium is present at low levels, 1, 2 and 1 mg, respectively, for almonds, guavas, and strawberries. Selenium is present in microgram quantities, 2.8, 6, and 4, respectively, for almonds, guavas, and strawberries. With this type of comparison it is always essential to keep in mind that a major portion of the variation between almonds and fruits occurs because of their different water contents, strawberries having higher water contents than either guavas or almonds.

Figure 2.23. Examples of fruit: (top left to right) strawberries, guava, papaya, (on bottom from left to right) olives and a nut, almond, unshelled.

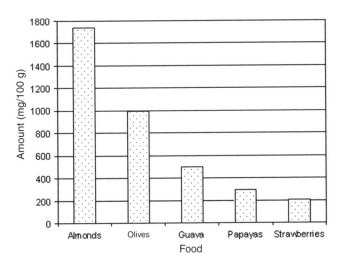

Figure 2.24. Sum of mineral content of a nut and selected fruits. (Data from Agricultural Research Service Nutrient Data Laboratory USDA National Nutrient Database for Standard Reference, Release 17, http://www.nal.usda.gov/fnic/foodcomp/Data/SR17/reports/sr17page.htm.)

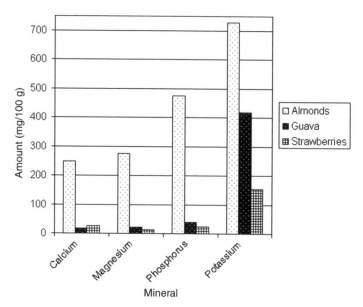

Figure 2.25. Major minerals in almonds, guava, and strawberries. Note that the level of sodium is so low, ranging from 1 to 2 mg, that it is not included in this graph. (Data from Agricultural Research Service Nutrient Data Laboratory USDA National Nutrient Database for Standard Reference, Release 17, http://www.nal.usda.gov/fnic/foodcomp/Data/SR17/reports/sr17page.htm.)

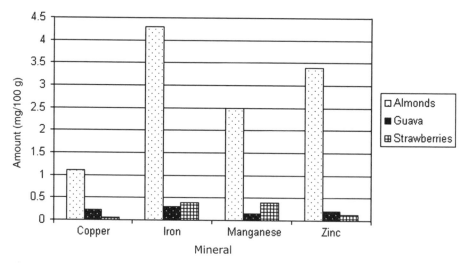

Figure 2.26. Trace minerals in almonds, guava, and strawberry. Note that the concentration of selenium is so small that it is not included in this graph. (Data from Agricultural Research Service Nutrient Data Laboratory USDA National Nutrient Database for Standard Reference, Release 17, http://www.nal.usda.gov/fnic/foodcomp/Data/SR17/reports/sr17page.htm.)

2.10 HOW THE BODY USES NUTRIENTS

All parts of the body must have a continuous supply of energy to function. Carbohydrates, both simple and complex, are the body's primary source of energy.
Complex carbohydrates are broken down by several different enzymes into simple
sugars. At this point they are either immediately used for energy or stored in the
body as lipids. When carbohydrates or sugars are not available, the body can use
either fats or even protein to produce energy.

The body's use of lipids is much more complex than that of carbohydrates in that
they are involved in both energy storage and in essential structures' cellular
components. All cell membranes contain lipids, and thus this component provides an
essential structural component of the body. Cell membranes control the flow of all components in and out of the cell, and the lipid portion is important in this control.
However, a high-lipid diet is not recommended for adults as it results in obesity and
contributes to health problems such as heart disease and diabetes. However, a
low-fat diet is not recommended for children because they need both the additional
calories and essential fats for growth.

The major use of proteins is the production of muscle, which includes the heart as
well as those muscles needed for breathing and movement. They are also used to construct other essential components such as connective tissue and enzymes. Amino acids
from protein taken in as food must be balanced to provide the amino acids needed to
produce the necessary proteins, muscles, enzymes, and so forth.

Enzymes contain accessory components necessary for function, that is, cofactors,
some of which are provided, either directly or indirectly, by vitamins and minerals.
Vitamins used by the body as accessory components for various processes are
shown in Tables 2.8 and 2.9. A lack of these important components of the diet can
lead to conditions that have been considered diseases in the past and present. Vitamins
are essential but also poisonous if present in too high a concentration. In addition some
enzymes contain minerals such as zinc, copper, and iron that are essential parts of the
mineral component of the diet. Thus enzymes are complex but essential components of
the body.

In addition to their importance in enzyme function, the minerals calcium and
phosphorus are used to build bones and teeth and are important in other structures
as well. Fiber plays no direct role in development or functioning of the body on
either the macro or cellular level; it is nevertheless extremely important to proper
function. Movement of food through the body is essential in that different food components are digested and absorbed in different portions of the digestive tract. Also it is
important to move digested food out of the body if for no other reason than to allow
room for more food intake. A mixture of soluble and insoluble fibers works best for
carrying out the important functions of fiber in the diet.

In all cases the body manages its intake of essential nutrients in such a way as to
allow the individual to grow, survive, and reproduce before death. For this reason a lack
of the proper nutrition leads to a decrease in stature and, in extreme cases, a decrease in
mental facility. A basic reality is that deprivation at any time, particularly during the

growing phases of an organism, cannot be made up at some later life stage. Thus a person's stature is fixed at the end of the growth phase, and subsequent good or even excellent nutrition will not cause the individual to grow in stature, although he or she may grow in weight even to obesity.

2.11 DIETS OF THE THREE FARMERS

In observing the meals of the three farmers, as shown in Figures 2.1, 2.2, 2.3, Donio's meal consists of fish and rice; Aída and Octavio's meal contains maize, beans, salad, and fish; and Steve's meal contains beef, maize, and salsa. Donio, Aída, and Octavio are all likely to drink water as part of the meal, while Steve will drink tea flavored with lemon and sugar. These meals represent only one meal out of a whole year and so do not represent any type of average of the different foods eaten. However, a majority of Donio's meals will consist of simply rice and fish with an occasional addition of chicken, egg, camote tops, camote tuber, taro, and cooking banana.

For Aída and Octavio the meals will vary in the meat eaten in that they will eat pork, chicken, and guinea pig meat during the year. Also they will eat meals that have maize of various kinds and in various forms, potatoes, and cooking bananas. In some cases a meal may contain all three, for example, maize, banana, and potato as a major part of the meal. They will eat various vegetables from the garden as they become available.

Steve and his family will eat a variety of meats, primarily pork, beef, and chicken, but little or no fish, goat, or lamb during the year. It is common for there to be both potatoes and bread with the meal, and many meals, including breakfast, will include both of these. Vegetables eaten will depend on the tastes of the family because all vegetables will be available in the local grocery store at all times during the year. During the summer more of a particular vegetable, for example, maize, may be eaten fresh out of the garden or from a fresh garden market.

Comparing the energy content (Figure 2.27) of the meals shown in Figures 2.1, 2.2, and 2.3, the calorie content is seen to increase from Donio to Steve's meal. This is due to the carbohydrate, meat, and fat content of the meals.

Pie graphs (Figures 2.28, 2.29, and 2.30) show the relative percentages of the four main diet components—protein, fat, carbohydrate, and fiber—in each of the diets. Donio's meal contains carbohydrate and protein with a significant amount of fat. Aída and Octavio's meal has a majority of protein with decreasing amounts of carbo-hydrates, fiber, and fat in that order. Steve's meal has a large amount of carbohydrate, mostly because each person will eat three ears of maize, with a significant amount of protein and lesser amounts of fat and fiber. No attempt has been made to estimate the amounts of butter or butter substitute applied to the maize and the increased fat and salt content this would add.

When comparing the meals, Steve eats the most (relative) amount of carbohydrate and Aída and Octavio the highest amount of protein. Aída and Octavio eat the largest amount of fat and Steve the least with Donio inbetween. Donio eats little fiber while

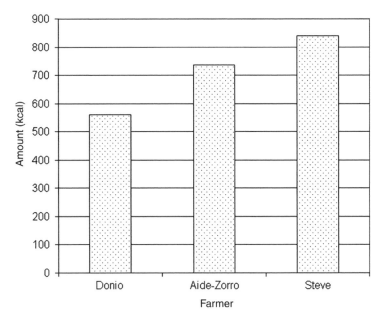

<u>Figure</u> 2.27. Energy content of meals of Donio, Aída, and Octavio and Steve. (Data from Agricultural Research Service Nutrient Data Laboratory USDA National Nutrient Database for Standard Reference, Release 17, http://www.nal.usda.gov/fnic/foodcomp/Data/SR17/reports/sr17page.htm.)

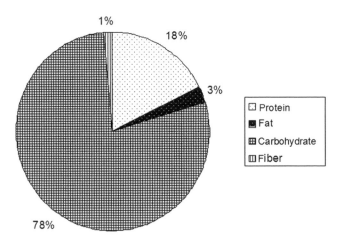

<u>Figure</u> 2.28. Protein, fat, carbohydrate, and fiber content of Donio's meal. (Data from Agricultural Research Service Nutrient Data Laboratory USDA National Nutrient Database for Standard Reference, Release 17, http://www.nal.usda.gov/fnic/foodcomp/Data/SR17/reports/sr17page.htm.)

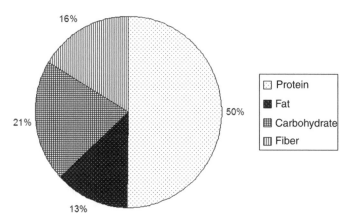

Figure 2.29. Protein, fat, carbohydrate, and fiber content of Aída and Octavio's meal. (Data from Agricultural Research Service Nutrient Data Laboratory USDA National Nutrient Database for Standard Reference, Release 17, http://www.nal.usda.gov/fnic/foodcomp/Data/SR17/reports/sr17page.htm.)

Áda and Octavio eat the most. It is important to note that there is very little variety of types of foods eaten by Donio and a great variety of foods eaten by Steve. To have a healthy diet, it is essential to eat a wide variety of foods. Thus Aída, Octavio, and Steve's meals are expected to provide a good diet.

In considering the foods farmers in an area should, or need, to be growing, it is essential to look at the diets of the local population and increase the production of foods that will enrich and improve that diet.

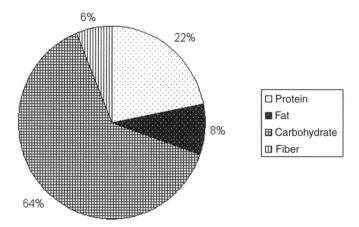

Figure 2.30. Protein, fat, carbohydrate, and fiber content of Steve's meal. (Data from Agricultural Research Service Nutrient Data Laboratory USDA National Nutrient Database for Standard eference, Release 17, http://www.nal.usda.gov/fnic/foodcomp/Data/SR17/reports/sr17page.htm.)

2.12 CONCLUSIONS

All agriculture should be viewed in terms of food production, although agriculture does produce other human necessities such as fiber, for example, cotton, flax, and silk. The human body needs energy, protein, carbohydrates, fat, vitamins, and minerals to function, grow, and reproduce. Agriculture produces the vast majority of these necessary inputs. However, it is not sufficient to have one source of food, as it is necessary to have a mixture of foods so that a balance of the necessary components is obtained. The body cannot make all the proteins, fats, vitamins it needs for proper functioning. It cannot make any of the minerals needed either. In addition providing an excess of any one dietary component will not make up for deficiencies in the intake of others.

Any change in agriculture production must keep in mind the dietary needs of the people engaged in agriculture and the people not engaged in agriculture who are dependent on the agricultural productivity of the area. Because of the need for a balanced diet, changes in agriculture production should be made in terms of providing both a more secure source and a more balanced food supply.

QUESTIONS

1. Which food sources given in this chapter have the highest and which the lowest energy contents? Considering other body needs, what might make one of these sources of energy preferable over the others? Explain.

2. Which food source supplies the most protein? Of the protein sources given in Figure 2.8, which is likely to be produced in more parts of the world and why might this be the case?

3. Of the food sources given in Figure 2.12, which has the highest and which the lowest fat content on a dry-weight basis? In considering which of these sources is the best, what specific fatty acids must be considered and why?

4. On what basis are the vitamins divided into two groups? In high concentration all vitamins are toxic. This being the case, why are some vitamins more dangerous when consumed in high concentration than others? Explain.

5. Minerals are important in the body on both a macroscale and on a cellular level. Explain and give examples of how minerals are used in both these ways.

6. Explain the importance of fiber in the diet considering that the body does not get any nutrients from it.

7. Using the Internet and the resources given, pick the three foods of those discussed that are best in overall nutritional value. Be sure to include essential components in making your decision.

8. Many proponents of certain foods promote them as the "perfect food." Describe the characteristics you would want in a perfect food.

9. When considering agricultural development, what should the nutritional considerations be, in terms of new crops or animals to be introduced into the cropping system?

10. Give the nutritional pros and cons of the meals eaten by our three farmers. What changes in their diets might you recommend?

11. Considering your recommended diet changes, what agricultural changes might you suggest?

BIBLIOGRAPHY

S. S. Gropper, J. L. Smith, and J. L. Groff, *Advanced Nutrition and Human Metabolism*, 4th ed., Wadsworth, Belmont, CA, 2004.

D. L. Nelson and M. M. Cox, *Lehninger Principles of Biochemistry*, 4th ed., W. H. Freeman, New York, 2004.

F. Sizer and E. Whitney, *Nutrition Concepts and Controversies*, 7th ed., International Thomson, New York, 1997.

E. N. Whitney and S. R. Rolfes, *Understanding Nutrition*, 10th ed., Wadsworth, Belmont, CA, 2004.

3

GRAIN CROPS

3.1 FOOD GRAINS ON THE THREE FARMS

Subsistence Farmer. Donio Derecho does not own land so he must rent or sharecrop fields. He currently has arrangements for three fields: one for rice and two for maize. The rice field is small, only 0.2 ha. Both maize fields add up to 1.3 ha. Donio can plant two rice crops a year. Most of the work is done by hand, as can be seen in Figure 3.1. The crop is rain fed so yields are irregular and crop failure is possible. The soil in the two maize fields is more sandy than the rice fields and cannot support rice. Donio can get three plantings each year, if he does a timely harvest.

Family Farmer. Octavio and his wife, Aída, plant maize for use by their livestock and some for sale in the village. Other crops are in small parcels primarily for family use. Their farming efforts are primarily designed to supply food and some sale of items to maintain their family. Their only grain crops are maize and oats, both primarily for livestock feed. They have no machinery to till the soil or harvest the crops. All farming operations are done by hand, like turning the compost shown in Figure 3.2.

World Food: Production and Use. By Alfred R. Conklin, Jr. and Thomas Stilwell
Copyright © 2007 John Wiley & Sons, Inc.

Figure 3.1. Donio and a laborer transplanting rice.

Commercial Farmer. Steve Murphy and his family cultivate about 1280 ha of crops each year. Only half is owned by Steve and his brother. The other half is rented for cash every year. Some years they have more, some years less. It all depends on the other landowners. They plant a variety of crops including maize, wheat, barley, and soybeans. A small area is planted to grass and alfalfa to make hay for their beef cattle in winter. Part of their maize crop is stored to fatten the young beef calves (steers) before selling them. The rest of their grain crops are sold for cash to a local grain elevator. To farm all the land they own and rent, they use seven large tractors, two combines, various planters, and other types of equipment. The total value of their

Figure 3.2. Octavio preparing compost.

Figure 3.3. Nick Murphy planting soybeans.

farm equipment is around $750,000. Steve's son, Nick, is shown in Figure 3.3 planting soybeans with their computer-guided tractor and zero-tillage planter.

3.1.1 Types of Food and Feed Grains

Grain crops are generally divided into two groups: those consumed by humans and those fed to animals. Within the United States, grains for animal feed are called "coarse grains." In practice, many grains are consumed by both humans and animals. Sometimes the distinction is made on the grain quality. Other times it depends solely on culture and historical practice within a particular country.

Many food and feed grains are the seeds of plants in the grass family. Examples include wheat, rice, maize, barley, sorghum, and millet. A second important class of grains includes the seeds of plants classified as legumes. Examples of these are soybeans, peanuts, lentils, beans, cowpeas, and chickpeas. Other grains of major economic importance are oil crops such as sunflower, safflower, and mustard. The world production of 12 major grain crops is shown in Figure 3.4.

Grain crops are extremely important to world trade because they are uniquely well suited for shipment. When dried and protected from insects and animals, wheat, maize, and rice can be stored without deterioration for several years. They contain high-quality starch and protein that can be easily assimilated by the human body. Rice requires only removal of the hull before being boiled. Both maize and wheat only require grinding to form flour used in many food products. The grains are durable and small enough to flow through automated loading and unloading equipment without damage. It is this durability and nutritional value that led our ancestors to select these seeds over others.[1]

3.1.2 Importance of Grain Crops in World Trade

Cereal grains are a major source of income for the United States. The United States net balance of trade in cereal grains averages $10 to 15 billion each year. All of this is

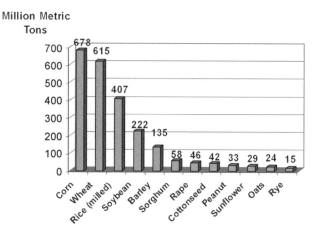

Figure 3.4. Production of major food grains. (USAID/FAS Circular WAP 12-05.)

surplus, after supplying the national market with some of the least expensive food in the world.

On an international level, wheat is the grain most widely traded. The United States is the major wheat-exporting country. It also supplies much of the coarse grain (primarily maize) that enters into world trade. Rice appears to be of lesser importance in international trade simply because most rice is consumed within the country in which it is produced. A relatively smaller proportion of rice enters international trade, when compared to maize and wheat.

3.2 MAIZE AND SORGHUM PRODUCTION

Maize

Latin Name: *Zea mays*

Other English Names: Corn, Indian maize, flint corn, mealies[2]

Sorghum

Latin Name: *Sorghum bicolor*

Other English Names: Chicken corn, guinea corn, milo, sorgo[2]

Maize is grown in most countries of the world. It is the third most widely planted crop after wheat and rice. The leading maize-producing countries are the United States, the Peoples Republic of China, and Brazil. There are large variations among the varieties of maize. Visible differences in color, grain size, and grain type form the basis for classification of the grain.

An ear of maize always has an even number of rows of seeds. This is because each pair of rows is the result of a cell dividing in two.

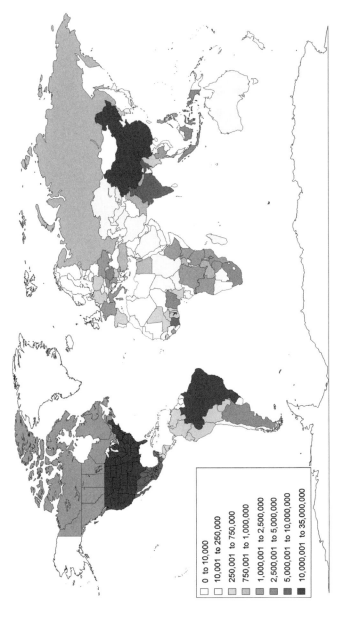

Figure 3.5. Hectares of maize in the world. (FAOSTAT, 2005.)

0 to 10,000
10,001 to 250,000
250,001 to 750,000
750,001 to 1,000,000
1,000,001 to 2,500,000
2,500,001 to 5,000,000
5,000,001 to 10,000,000
10,000,001 to 35,000,000

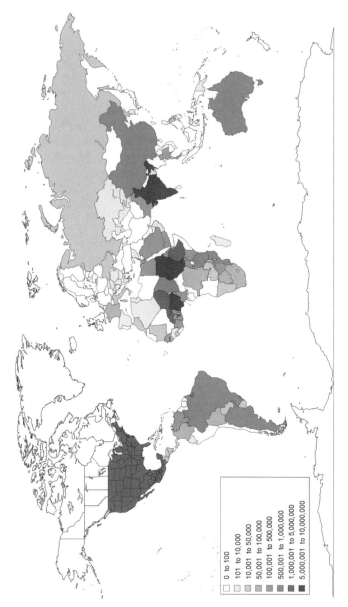

	0 to 100
	101 to 10,000
	10,001 to 50,000
	50,001 to 100,000
	100,001 to 500,000
	500,001 to 1,000,000
	1,000,001 to 5,000,000
	5,000,001 to 10,000,000

Figure 3.6. Hectares of sorghum in the world. (FAOSTAT, 2005.)

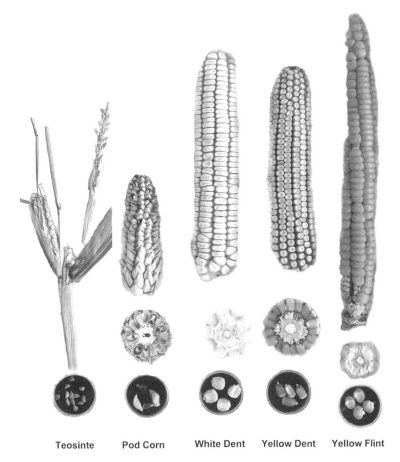

Teosinte   Pod Corn   White Dent   Yellow Dent   Yellow Flint

Figure 3.7. Ancestors and modern types of maize.

Sorghum, although appearing very similar to maize, has very different characteristics that make it well adapted to drier areas of the world. It also grows well in the areas where maize is grown, but comparison of Figures 3.5 and 3.6 show significant differences. Sorghum is a major crop in the Sahel countries of Africa, India, and Australia, while it is less widely planted than maize in the United States, China, and Brazil. There are great differences among varieties and races of sorghum, causing problems in its botanical classification. See Figure 3.7 for a history of maize types.

3.2.1 Climatic Adaptation

Maize is widely adapted to many climates and growing season lengths. Maize is planted from sea level to over 4000 m altitude. It is produced from 50° N latitude (southern Canada and Russia) to 40° S latitude (Argentina and southern Australia).[3] Some varieties are relatively insensitive to photoperiod changes, while others can only be grown in a narrow north–south range. It can tolerate climates with as little as 500 mm annual rainfall but yields best in areas with 1200 to 1500 mm rainfall.[4]

Sorghum is normally considered to be more "drought tolerant" than maize. The root system of sorghum has more fine, secondary root hairs than maize, making it especially efficient at extracting moisture in the soil. The physiology of the sorghum plant also enables it to survive heat and water stress better than maize. Although it can produce grain in very dry climates, most world production occurs in areas with annual rainfall between 250 and 1250 mm.[5]

3.2.2 Importance of Grains in Human Diets

The usage of maize in human nutrition depends on both culture and physical factors. In some parts of Latin America, other calorie sources are expensive. Animal protein is expensive and difficult to preserve. As a result, almost 84 percent of the protein consumed by lowest income groups in Mexico comes from maize. It is eaten fresh, popped, as tortilla flour, cornbread, and as a thickening agent in soups. It is estimated to provide about 70 percent of all caloric intake in rural areas of Mexico.[3]

Sorghum poses a special problem for human and animal nutrition. The seed of many varieties has a high tannin content, giving it a bitter taste and making it difficult to digest. There are varieties with very low tannin contents that are used in human diets. In the countries where sorghum is widely planted, it is a staple food, being used boiled, popped, fermented, in soups, and in various types of breads.

3.2.3 Botanical Description

Maize is a large member of the grass family, reaching heights of 2 to 5 m. It is a C4 type plant, meaning it is especially efficient at converting sunlight to starch and proteins. Plants bear both male and female flower parts, but they are separated. The male flower is the tassel, developing at the top of the stem. The ear forms from the female flower, which bears fine styles called *silk*. When pollinated, grain forms on the cob to form an ear. The ear is covered by a husk made of specialized leaves, which protect the grain from birds, insects, and fungi. The length of growing season varies from 60 days to over 13 months (time from planting to dry grain).

Maize is classified according to its grain characteristics. Some of the major grain types are shown in Figure 3.8. The major grain types recognized for trade are yellow dent, yellow flint, white dent, white flint, floury, popcorn, and sweet corn. In terms of nutrition, yellow and white grains have little difference. Dent and flint types reflect the density of packing of starch grains in the grain. Floury-type maize has a very loose starch grain packing and is easily damaged by insects. Popcorn has a more vitreous endosperm than flint types. This holds in water vapor when heated to cause an explosion when the kernel fractures (pops). Sweet corn is a mutant that does not convert sugars to starch in the grain. The result is a grain with a sweet taste when eaten in the immature stage. Many specialized types of maize have been developed to serve specific industry requirements. Examples include high-quality protein

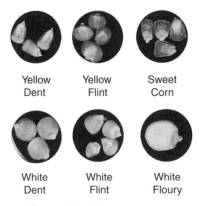

Yellow   Yellow   Sweet
Dent   Flint   Corn

White   White   White
Dent   Flint   Floury

Figure 3.8. Maize grain types.

maize (for animal feed), high-oil maize for maize oil extraction, waxy maize for industrial starch, and silage maize for animal fodder.

Sorghum appears very similar to maize in the vegetative stages. It is also a member of the grass family with both male and female flower parts in a single panicle emerging from the top of the plant as shown in Figure 3.9. Modern varieties are usually short (1 to 2 m), older varieties were much taller, sometimes growing as much as 5 m tall.

Figure 3.9. Sorghum panicle.

There are five races of sorghum in cultivation: bicolor, guinea, caudatum, kafir, and durra. Bicolor is the most widely grown type. For trade purposes, sorghum is classified by grain color, which is related to the tannin content. The official grades are yellow, white, and brown, which correspond to increasing tannin contents. The young sorghum plant contains elevated levels of hydrocyanic acid and hordenine, both toxic to animals. Farmers have developed specialized varieties suited for specific applications. For example, varieties with high levels of sucrose in the stems are crushed much like sugarcane to make molasses.

3.2.4 Origin

It is generally believed that maize originated in what is now Central America. In contrast to other domesticated crops, there is no recognizable wild maize relative. Teosinte is a wild grass that grows in various Central American countries, but the grains are borne on the "tassel" with the male flowers and appear quite different from cultivated maize. Pod corn is another maizelike plant with a male tassel and grains borne on an ear. However, each grain is enclosed in an individual husk, making it extremely difficult to remove for consumption. Maize is considered a "cultural artifact" that cannot exist in the wild and can survive only with the help of humans. It lacks a natural means of seed dispersal necessary for survival and depends on humans to harvest and plant the seed. Some representative types of maize are shown in Figure 3.7 with teosinte on the left and progressing to a yellow flint type grown in southwestern United States several hundred years ago. Maize variety selection, improvement, and hybrid development started in the United States in the 1920s. Many improved varieties of maize have been developed for tropical areas by Centro Internacional de Mejoramiento de Maíz Y Trigo (International Maize and Wheat Improvement Center) or CIMMYT in Mexico starting in 1943.

Sorghum originated in North Central or Sahelian Africa. Many wild types can be found in areas of Sudan, Egypt, Chad, Niger, and Ethiopia. It moved by ancient trade routes to India, Arabia, and China. It was first introduced to North America with the slave trade, where it was originally used to feed slaves and farm animals. Later introduction of tannin-free varieties have increased its usage in foods in the United States. Improved tropical varieties have been developed by International Crops Research Institute for the Semi-Arid Tropics (ICRISAT) in India since its establishment in 1972.

3.2.5 History

Archeological evidence has established that early Americans were using maize in 2500 BC in the Tehuacan Valley of modern Mexico. Many Mayan traditions center around maize, and it followed travelers through the Americas, becoming established throughout North and South America by the time the Spaniards arrived in the early 1500s. These explorers carried the grains with them back to Europe where it quickly spread through the rest of the world. By 1597, a document was produced in China describing maize. Though the Spanish considered maize to be a coarse grain only suited for animal consumption, other cultures quickly found it well suited for human use. The rapid growth of the maize plant, plus its ability to produce up to 1000 kernels for every seed planted, led to its rapid spread and adoption by most cultures.

Early movement of sorghum from its origin in Africa was most probably along trade routes established between East Africa and the western coast of India. Overland trade was also a means for movement into the Fertile Crescent and along the Silk Road into China. Introduction in the United States was most probably a result of the slave trade from west Africa. A syrup-type variety was the first deliberate introduction in 1857. Since then, varieties have been introduced from South Africa, India, Sudan, and China.

3.2.6 Soil Preparation and Planting

In many developed countries, maize and sorghum production is a mechanized system. One of two systems of soil preparation may be used: conventional tillage or minimum/zero tillage.

Under the conventional tillage system, the field is plowed with a moldboard plow that lifts and turns over the upper 7 inches of soil. This acts to bury weeds and loosen the soil, permitting easier root penetration by seedlings. This is an energy-intensive procedure requiring a tractor, several types of heavy equipment, and two to three trips over the same field before planting.

Minimum or zero-tillage systems reduce preplant tillage to a minimum or to none at all. Specialized planters are able to till a narrow strip or to plant in soil undisturbed since the previous year. This method reduces costs of planting but permits weeds to grow unchecked. A necessary component of the zero- or minimum tillage system is the use of chemical herbicides. When applied before, or during planting, these will kill competing weeds, permitting the crop to germinate and grow for several weeks without weed competition.

In less developed countries and in subsistence agriculture, farmers have been using minimum or zero tillage for hundreds of years. Often only a pointed stick is used to make a 2- to 3-inch hole into which two to three seeds are dropped. The hole is covered and the seeds left to germinate. True zero tillage requires use of a chemical herbicide to control weeds. When maize is a commercial crop, the additional cost of herbicides is repaid by better yields and lowered field preparation costs. Sometimes farmers will use a stick plow to till the soil before planting, but the depth is often only 3 to 5 inches. This acts to kill growing weeds and gives the crop a chance to develop before the weeds catch up. As can be seen in Figure 3.10, the tillage is sometimes less than desirable.

3.2.7 Fertilization

Maize and sorghum are nutrient-hungry crops. Each produces a heavy stalk and a large number of leaves in a short period to form nutrient-dense grains. Every kilogram of maize harvested from a field carries off 16 g of nitrogen, 7 g of phosphate, and 5 g of potash.

Figure 3.10. Bullock-drawn plow in India.

Many maize varieties are able to produce grain under a relatively low soil fertility level. Sorghum varieties have been developed in areas with low rainfall and low levels of soil fertility. Modern varieties and economics demand that grain yields per hectare be much higher than subsistence farmers in the tropics. In Ohio, a farmer will apply 180 kg of nitrogen, about 90 kg of phosphate, and 160 kg of potash fertilizer per hectare to obtain a yield of 8900 kg of maize from the same hectare.[6] Actual rates depend greatly on fertilizer cost, projected sale price of the grain, and soil test results.

Chemical fertilizer rates can be reduced as much as 50 percent by growing a legume crop the season before the crop is planted. The residual nitrogen fixed by the legume crop is available to a crop up to 3 years after the legume is plowed under. Application of animal manures also helps to reduce the need for chemical fertilizers. Under commercial growing conditions, chemical fertilizers are preferred due to their lower cost.

Commercial farmers in developing countries will often use a minimal amount of nitrogen fertilizer and supplement this with animal manures, compost, and a crop rotation. The result is a lower yield but with the benefit of less cash investment in the crop. Subsistence farmers depend almost entirely on nonchemical sources of nutrients for their crop. Often this means a longer period between plantings of maize or sorghum in the same field to permit buildup of nitrogen reserves.

3.2.8 Weeds, Insects, and Diseases

Weeds are a major problem for maize and sorghum. Commercial grain production uses heavy applications of fertilizers that also contribute to rapid weed growth. Traditional methods of weed control involve hand labor to pull or hoe weeds in a field. This is difficult even for small operations with only 1 to 2 ha fields. Large-scale commercial grain production involves planting as much as 1000 ha, which is impossible to hand weed. Mechanical weeding was practiced in the United States until the mid-1970s. A tractor with special cultivator "sweeps" would spend hours passing up and down the rows in a field stirring the soil between the rows to remove small weeds. Two to three weeding operations were required for each crop; an expensive operation that also compacted the soil. Most large commercial operations now use minimum or zero-tillage systems. Herbicide is applied at the time of planting so that only one pass through the field is needed. Under ideal conditions, this will control weeds until harvest. If there is a heavy weed infestation, another pass through the field may be needed for a second herbicide application. Witchweed (*Striga helmonthica*) is a serious pest of sorghum in the Sahel region of Africa and maize in the sub-Saharan regions of Africa. This is a parasitic weed that attaches to the roots to draw off nutrients and, eventually, kills the plant. Large areas of Sudan and Niger have suffered major losses because of this weed. Current efforts are underway to produce varieties resistant to witchweed.

Insect management is normally done by selection of a variety. Modern maize and sorghum hybrids have been genetically modified to discourage feeding by rootworms and some leaf-feeding insects. In tropical areas, planting dates are often shifted to avoid times of insect proliferation. Small commercial plantings of maize may utilize chemicals for control of the leaf-feeding European corn borer or local preventive measures may be used. One such method used in Mexico is to apply small amounts of wood ash in the whorl of the young maize plant. Many transgenic temperate varieties of maize and sorghum now utilize the *Bt* gene to discourage feeding by the corn borer without application of toxic chemicals. There has been less success incorporating this gene in tropical lines of maize and sorghum.

Disease management has been accomplished for hundreds of years by selection of the most resistant plants. The seed of these plants are replanted to form new disease-resistant, or tolerant, varieties. This is an especially effective way to put disease-resistant varieties in the hands of small farmers.

3.2.9   Harvest and Storage

Harvesting of maize and sorghum varies according to the climate and type of farm operation. The simplest is that of the commercial farmer. He takes a combine through the field that removes the ear or panicle, runs it through a sheller, then separates the grain from the cob, leaf, and stalks. Stopping when the combine grain bin is full, he unloads the shelled grain into a waiting truck. When the truck is full, it drives directly to the local grain elevator to deliver the harvested grain. In some large operations, the farmer may store the grain in his own grain bins to await better market prices or to use for livestock feed. A single hectare can be harvested in less than 30 minutes by one person driving the combine.

Harvest of maize and sorghum by a small farmer in developing countries is often done by hand, hiring day labor to twist off the ears of maize (as in Fig. 3.11) or cut off the sorghum heads and collect them in a waiting wagon. If labor is scarce, he may cut the stalks with a machete and form "shocks" (standing bunches of cut stalks) in the field. When possible, he will pick off the ears and take them to his house for shelling and later sale in the local market. Near large cities, truck drivers may pass by as the harvest progresses and buy the grain directly from the farmer in the field. This has the advantage of giving instant cash to the farmer and avoiding threshing, transport, and storage problems. Often, however, the price offered by the drivers is much below market price.

Figure 3.11. Maize harvest in Ecuador.

If the grain is stored on the farm, the farmer must construct rat-proof bins and protect it from insects and moisture. Even then, losses of 30 to 40 percent are common.

Harvest by a subsistence farmer is aimed to conserve the maize ears before birds and rodents eat the grain in the field. Usually, the cob is left with the husk intact to give some protection against insects. If the farmer is lucky enough to have a metal barrel, the grain will be stored safe from rodents. Another common method is to hang the ears from rafters in the kitchen where smoke from the cooking protects them from insects and mice. Even with these measures, losses of 40 to 50 percent are common. Varieties of sorghum are cultivated with naturally high levels of tannin to reduce losses by insects, animals, and fungus.

Stored maize and sorghum are both subject to attack by fungi that produce toxins such as aflatoxin. These can have serious, even fatal, effects on humans and animals. This is especially dangerous in humid areas.

3.2.10 Marketing

Sale of grain by commercial farmers in the United States is usually done at the time of harvest. Grain will typically move directly from the field to a grain elevator, where it is delivered to a buyer. Payment is made later that month for the grain delivered. If the grain has not dried to around 15 percent moisture, there may be some discount, or drying charge. In many cases, the operator of the grain elevator will have sold the grain on a "futures" contract several months before the harvest. This gives the operator a sure market for the grain still to be purchased and guarantees the price he will receive. It also gives him a price to be paid to farmers.

Often large commercial farmers will sign a contract to deliver their grain directly to a beef or poultry farmer for a guaranteed price. In this case, they bypass the intermediary and sell directly to the end user.

Small commercial farmers often have little choice in their marketing of harvested grain. If they have a farm truck, they will carry the grain to a central market, where it can be sold at the going rate. Without a truck, the small farmer must wait for a commercial truck driver to pass by his farm. Often the drivers conspire to offer minimal prices for grain, sometimes letting each driver buy only from a restricted area (a monopoly). This means the farmers must sell their grain for whatever price is offered. No other truckers will dare to buy their crop. There is risk for the truck drivers also. They must travel over poor roads and pass internal customs posts to reach the market. It is sometimes necessary to pay taxes and bribes to local officials, thus raising their cost of operation.

By definition, subsistence farmers have little or no surplus to sell. When there is a surplus, it may be only a few kilos. These small quantities are usually sold by the farmer at the weekly market. He will bundle up his maize or sorghum to sell, walk to a bus stop or catch a truck at a main road (sometimes several miles) where he must pay for a ride to the market and to transport his grain. If it is a good day, he will sell the grain, buy a few supplies, and have money left over to travel back home. Sometimes there are no trucks passing through his area to buy the grain because the roads are poor or nonexistent. Sometimes only a footpath connects the farm to a road several miles away.

3.2.11 Postharvest Processing

Postharvest processing depends on the end user. If the grain is to be fed to livestock, it may not be processed at all. Most animals can consume the grain directly. For most efficient digestion, however, the grain should be at least cracked or broken. In some cases it will be ground into a coarse powder to permit mixing with other supplements to form a more complete ration.

For human consumption in industrial countries, maize may go through several stages of processing. The most common type of processing is wet milling. This combines physical and chemical treatments to separate the grain into four basic components: starch, germ, fiber, and protein. The starch is then processed to form products such as adhesives, batteries, crayons, paper, plywood, and chewing gum. Oil obtained from this process goes into production of various products, including mayonnaise, salad dressing, cooking oil, printing ink, and soups. Ethanol production utilizing the starch extract is becoming an important motor fuel.

The Masa process is a cooking process done under alkaline conditions to produce specialty foods such as tortillas.[7] Other grades of fine, medium, and coarse flour are produced for specific products.

Dry milling is used primarily to process grains going to animal feed. Other foods made from dry milled maize include grits, maize chips, snack foods, polenta, and maize flour.

In the developing countries of South America, Africa, and Asia, a large portion of the maize is white. Although it varies from one country to another, many countries use 50 to 80 percent of the white maize production for human consumption.[8] Processing of maize in countries of South America is divided between large millers and small business processors. A significant amount of processing is often done on the farm or in small business settings. Much of the maize produced in the Andean region of South America goes to the fresh produce market, where it is consumed as roasted ears. In Mexico, steamed maize meal is sold wrapped in maize husks as tamales.

In African countries, maize is prepared in four main types of dishes: beverages, porridges, dumplings, and baked or fried products. A typical beverage is maize beer, known as *Kenkey* and *Mahewu* in South Africa. Porridges have a water content of about 90 percent and are soups or have a gruel consistency. One example is *Ogi* consumed in Nigeria. Dumplings only have 65 to 80 percent moisture and are often formed into rough balls. Flour is mixed with water to form balls like *Eko* (Nigeria) or *Tô* (Mali).[9] Baked or fried products include the fermented leavened bread consumed as *Injera* in Ethiopia or *Lakiri* in Ghana.

Uses of sorghum grain vary with the country and culture. Industrial processing methods closely follow those of maize. Methods of processing in homes or shops of developing countries seem to be more varied. Sugary or high-lysine sorghum grain is used in the immature stage in India and Ethiopia as a roasted product. Many cultures in Asia and Africa simply boil the dry grains to be eaten similar to rice or bulgur. In some cases, the grain is pearled (polished to remove the hull and tannin) and then boiled. Popping varieties, similar to maize, have been developed and are popular in India. Fermentation is practiced in most countries where sorghum is grown. Partial fermentation is used to form a locally preferred food, or fermentation that is more complete will produce beer. Bread can be produced from the milled grain. Mixing yeast with the dough will produce a type of loaf bread, while flat breads are made without yeast. Nearly all countries that produce sorghum use the grain in some type of porridge. This is made from flour that is boiled to form a stiff porridge.

3.2.12   Genetically Modified Maize and Sorghum

Maize has been extensively modified by natural and human selection. The latest modifications have been direct changes in the genetic makeup of the maize plant to give it specific characteristics to make its cultivation easier and more profitable. A more complete discussion of these changes will be presented in Chapter 12.

Sorghum has also been modified by nature and humans, but to a somewhat lesser extent. It has also been the subject of changes in the genetic makeup, but to a lesser degree than maize.

3.3   WHEAT AND BARLEY PRODUCTION

Wheat

Latin Name: *Triticum aestivum*

Other English Names: Bread wheat, common wheat, durum wheat, English wheat, German wheat, red wheat, soft wheat, spring wheat, white wheat, winter wheat[10]

Barley

Latin Name: *Hordeum vulgare*

Other English Names: Barleycorn, two-rowed barley, six-rowed barley, naked barley, Scotch barley[10]

Wheat is grown in countries with one mild or temperate season. There is little production in the humid tropics. In terms of total grain production, it is second only to maize (Fig. 3.4). The leading country in terms of hectares planted is India, closely followed by Russia, China, and the United States (Fig. 3.12). There is little production in the humid countries of Africa, Indonesia, Colombia, and Venezuela. In India, wheat is primarily a Kharif (winter) crop. Wheat is classified according to its grain characteristics and planting season and is cultivated primarily for human consumption.

Barley is generally planted in those areas where climatic conditions do not permit a good crop of wheat. This may be due to lower rainfall, shorter growing season, or less fertile soils. There is little production in the humid tropics. The leading country in terms of hectares planted is Russia, with 9,662,000 ha planted in 2005. Other countries with significant areas planted to barley include Ukraine, Canada, Australia, Turkey, and Spain (Fig. 3.13). There is little or no production in the humid countries of central Africa, Indonesia, and the Philippines. Barley is classified both by grain type and by planting season. It is planted primarily for animal feed and industrial use (brewing).

3.3.1 Climatic Adaptation

Wheat is cultivated in areas with 350- to 600-mm annual rainfall.[11] Bread wheat types are planted in areas with 450 mm or more rainfall, while other types are found in areas with less rainfall. Wheat is also grown under irrigation where rainfall is irregular or inadequate. Wheat is classified as either winter or spring type. Winter wheat is planted in the fall and requires *vernalization* in order to complete its growth cycle. After planting in the fall, winter wheat plants germinate and form a rosette of leaves around the plant with the growing point remaining below ground level. Only after exposure to a period of low temperatures (usually 32 to 50°F) will the plants send up a stem and flower to form seeds in early summer. Spring wheats do not have a vernalization requirement and so are planted in the spring to produce a crop for harvest in summer. In both cases, wheat gives the best yields in relatively cool temperatures during flowering. When planted in humid tropical conditions most wheat varieties will produce poorly and are subject to damaging disease problems.

Barley can be grown in drier climates than wheat, often with only 200- to 350-mm annual rainfall. Under very low rainfall conditions, barley yields may be as low as 500 kg/ha. Similar to wheat, barley types may be winter (sown in the fall) or spring (sown in spring). Winter types require vernalization before they can complete their life cycle to produce seed. Barley is widely adapted, being grown from inside the Arctic Circle to tropical latitudes. Barley is also a cool-season crop. It does not grow well in hot, humid conditions.

3.3.2 Importance in Human Diets

Wheat is a part of all western diets consumed as bread and many forms of pastries and pastas. Wheat provides 20 percent of the food supply for all developing countries.[12] In the Middle East and North Africa area it comprises 44.3 percent of the total food supply. Each pound of grain contains about 1500 calories.

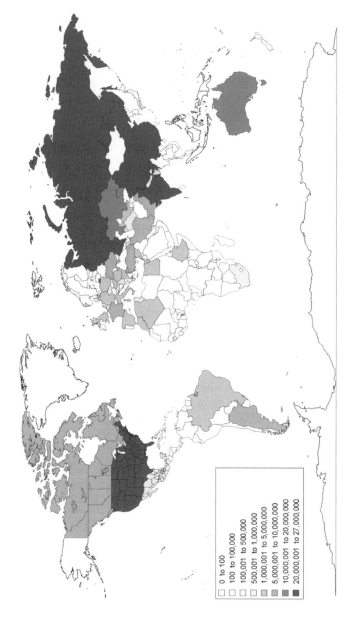

Figure 3.12. Hectares of wheat in the world. (FAOSTAT, 2005.)

0 to 100
100 to 100,000
100,001 to 500,000
500,001 to 1,000,000
1,000,001 to 5,000,000
5,000,001 to 10,000,000
10,000,001 to 20,000,000
20,000,001 to 27,000,000

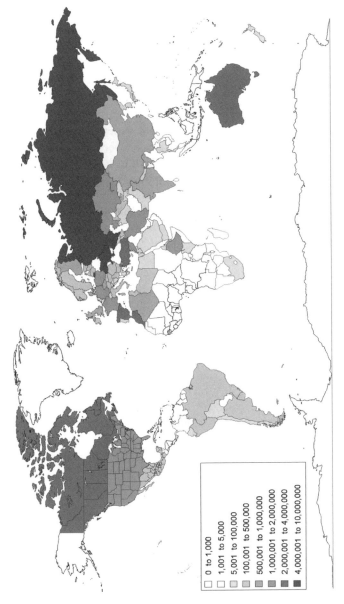

Figure 3.13. Hectares of barley in the world.

Barley is utilized much less for human nutrition in the United States. Where it does enter the human food chain, the "naked" or hulless types are preferred. The primary use for barley in most countries is for animal feed. It contributes a high level of carbohydrates plus fiber. The second most important use for barley is in the brewing industry. Specific varieties have been developed for use in malting. In the highlands of Russia and Afghanistan barley is mixed with wheat to make leavened or unleavened bread.

3.3.3 Botanical Description

Both wheat and barley are members of the grass family. Though some varieties may reach 200 cm in height, most modern commercial varieties are 100 to 120 cm tall. Modern wheat and barley varieties are a cross between traditional varieties and dwarf varieties, resulting in a compact plant able to remain upright under high fertility conditions. The grain is formed on a terminal spike. The root system consists of fine roots that go as much as 100 cm deep, an advantage in semiarid areas of the world. The seedling forms a crown of prostrate leaves until the proper day length occurs or vernalization requirements have been met. Then a central stem forms to bear the spike and the seeds as shown in Figures 3.14 and 3.15.

3.3.4 Origin

Both wheat and barley are believed to have their origin in the area known as the Fertile Crescent (Fig. 3.16). This is an area where significant numbers of wild, related species are still found growing today.

Figure 3.14. Spike of hard red spring wheat.

Two Row   Six Row

Figure 3.15. Types of barley.

Wheat is believed to have been domesticated before 8000 BC with direct descendents of modern wheat being cultivated around 7000 BC. Some of the ancestors of
modern wheat still grown in regions of the world are einkorn, emmer, spelt, and
durum. Only durum remains as a major type of cultivated wheat today. Modern
bread wheat is probably a result of crosses among several of the ancestors.

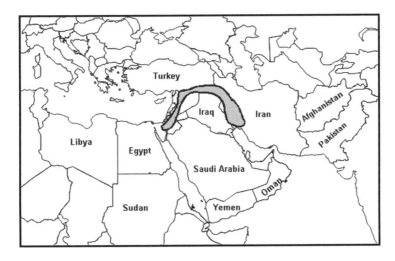

Figure 3.16. Fertile crescent.

It is estimated that barley was domesticated before 8500 BC with harvest of wild species as early as 10,000 BC. There is one two-row "wild type" of barley growing in this area considered to be the "true" ancestor of modern two- and six-rowed types of barley. The evolution from two- to six-rowed barley is considered to be a result of mutations of only three genes, a relatively simple change in plant evolutionary terms.

3.3.5   History

Early movement of bread wheat probably was in a mixture with emmer carried by hunters or migrating tribes. It has been found in archeological sites in Syria dated at 7000 BC, Turkey dated at 5000 BC, England dated at 3500 BC, and China dated at 3500 BC. Wheat was introduced to North America by Christopher Columbus on his second trip around 1493 and later by Spanish soldiers in Mexico. It found a favorable climate, and production increased to the extent that surplus grain was being exported from Mexico in 1735. Colonists in New England had problems with wheat crop failures after its introduction in 1620 but within a few years were producing enough to trade with other colonists. As new immigrants came to North America, they brought seeds of wheat from their homeland. The result was a large variety of wheat types from around the world being cultivated in the Midwest. Most of these wheat varieties were soft red winter and soft white winter types. This was due to two principal factors: (1) Soft wheat varieties yielded more grain per hectare than hard types and (2) milling technology had been developed to handle soft wheat types.

Barley is believed to have been domesticated earlier than wheat, around 8500 BC. Tools found in archeological sites dated at around 10,000 BC could have been used to cut heads of grasses, such as barley. Wooden sickles have been found at one site dated at 9000 BC. Evidence has been found that barley was cultivated in Spain in 5000 BC, England in 3000 BC, and China in 7000 BC. Barley was probably preferred to wheat before 100 BC when the practice of making bread with yeast was introduced in Greece and Rome. Movement of barley around the world in modern times parallels that of wheat. It was first introduced in North America by Christopher Columbus, with the main uses being for malting and livestock feed. Barley is better suited than wheat to areas with less rainfall, and so it was produced for grain in the drier western regions of the United States. It is interesting to note that before prohibition was enacted in 1918 much of the barley grain went into the malting industry. In 1919 the total barley production dropped but quickly increased in following years when farmers realized that it was a high-quality livestock feed.

In about 1305, Edward I of England decreed that one inch should be the measure of three barleycorns, and English shoe sizing began; thus a child's shoe that measured 13 barleycorns became a size 13.

3.3.6 Soil Preparation and Planting

Wheat and barley have similar growth requirements. Typically, soil is prepared by plowing and disking to form a uniform seedbed. With small grains such as wheat and barley, the soil must be tilled until it has no clods greater than the size of the seed. The method used depends on the degree of mechanization of the farmer. In highly mechanized farms, there may be several implements used to break down the soil into small fragments. A first will typically be with a plow that turns the topsoil over and buries weeds. A second and third pass will normally be with a disk to break up any large clods of soil. Sometimes another pass over the field will be made with another implement to further reduce the size of unbroken clods. A mechanical planter will then be used to plant the grains in rows about 20 cm apart, dropping as many as one seed per centimeter in rows. The actual seeding rate will vary with variety, distance between rows, rainfall, and the like. In practice, wheat and barley tend to compensate for over- or underseeding. If there are too few seedlings for the light, water, and nutrients available, the plants will form more tillers (stems) and grain heads. Some mechanized farmers in the United States have reduced tillage to near zero with no-till planters. These machines will open a thin slit in untilled soil and drop the seed and starter fertilizer. This has the advantage of reducing the number of tillage operations but requires application of herbicides to kill weeds.

In nonmechanized conditions, the soil is tilled repeatedly to form a smooth seedbed as seen in Figure 3.17. The seeds are broadcast by hand to obtain a uniform distribution over the field. The field is then plowed with a stick plow to lightly cover the seeds. Fertilizer is usually broadcast during tillage operations to be incorporated into the soil. The resulting stand of plants is random, sometimes tending to form rows at the spacing of the last pass with the plow. Some farmers may also use animal-drawn mechanical planters. This has the advantage of planting in rows to facilitate hand weeding.

Figure 3.17. Smoothing the seedbed for wheat in India.

3.3.7 Fertilization

Fertilization must at least supply an amount equal to the nutrients removed in the grain plus an allowance for losses in the field. For a harvest of 4400 kg wheat grain taken from 1 ha, 90 kg of nitrogen, 43 kg phosphate, and 27 kg of potash fertilizer are removed.[13] This amount must be replaced before planting just to supply the crop with needed nutrients and avoid depletion of soil fertility. The source of the nutrients for commercial farmers is usually chemical fertilizers, though this is often supplemented by manure from farm animals. The amount of fertilizer needed by wheat depends also on the type of grain being produced. Durum wheat is marketed by protein content, which is affected by nitrogen levels in the soil. The farmer must carefully balance the soil nutrient status against desired yield levels to obtain a quality product and maximize profit.

Fertility levels in developing countries are often less than desirable, which negatively affects grain yields. Chemical fertilizers are frequently expensive. This means they are reserved for high-value cash crops such as vegetables. Fertilization of wheat is often done only with animal manure, which is available in limited quantities. Even though the grain is the most important commercial product of most wheat and barley fields, the straw is used as fodder for animals during the dry season. The net effect is that all of the plant above ground is removed with all the nutrients it contains. This further reduces soil fertility since these residues are not left to decompose and replenish the soil.

Fertilization of barley is a problem because it is grown in areas of low rainfall. With lower rainfall comes the risk of irregular rainfall patterns and the risk of low yields. As a result, barley is even less likely to receive adequate fertilization for maximum yields. In commercial farming enterprises, barley is usually given a minimal application of fertilizer because of this risk. In small or subsistence farms, it often receives no fertilization.

3.3.8 Weeds, Insects, and Diseases

Weed control in wheat and barley is more difficult than maize or other crops planted in wide rows. When wheat and barley are planted in very narrow rows, tractors or animals cannot move between the rows for cultivation operations. When they are planted by broadcasting the seed, the plants are randomly spaced and weed control is even more difficult. While the plants are in the rosette growth stage, it is possible to carry out some field operations such as herbicide application or hand weeding. As soon as the stalks begin to form, any field operations will injure plants and reduce yield.

Many insect species feed on wheat and barley plants in the field. They can affect plants at all growth stages such as seedling, at tillering, and stem elongation. In most cases the damage is not severe enough to warrant application of insecticides. By using specific cultural practices, it is possible to avoid damage by specific insects. Pupae of the Hessian fly will attack the stem of wheat plants causing stunted growth and death of the seedlings. The most common control practice in the U.S. Midwest is to delay planting until after the "fly-free date" in the fall. Crop rotation, resistant

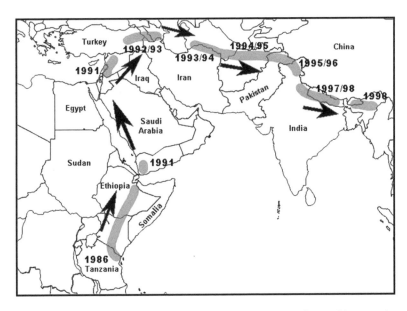

Figure 3.18. Movement of YR9 virulent race of wheat rust from Africa to Asia.

varieties, and plowing under straw after harvest help control this insect without use of expensive insecticides.

Insects cause extensive damage in stored grains of both wheat and barley. Typically, if only 5 to 10 percent of grains in a storage container may be damaged by insects, the entire lot is either destroyed or fed to animals.[14] Because these grains are destined for human consumption, they cannot contain residual chemicals. Some fumigation chemicals are used to clean storage bins. In some developing countries certain low-toxicity chemicals are used to prevent insect damage. The best control practices are to dry grain to 12.5 percent moisture and thoroughly clean grain bins before filling each year.

Diseases are a major problem for both wheat and barley. Leaf rust is especially virulent, resulting in 80 to 90 percent loss under favorable weather conditions. Resistant varieties have been developed, but the fungus is genetically variable and readily mutates to form new races that overcome resistance. Once a new race is able to infect resistant varieties, the spores follow wind currents to infect new areas. A worldwide network of wheat pathologists has been established to identify and trace movement of these virulent races. An example of such a movement from Africa to Asia is shown in Figure 3.18. Movement over the 12-year period followed the prevailing wind patterns. The rate of mutation of the fungus is such that new races develop and spread every 10 to 15 years. The development of new, resistant wheat and barley varieties is a constant battle.

3.3.9 Harvest and Storage

Mechanical harvesting of wheat and barley grain must wait until the grain has dried to around 15 percent moisture. If the grain is too moist, there is danger of incomplete

Figure 3.19. Small wheat thresher in India.

threshing and the need to artificially dry the grain after harvest. If the grain is below 10 percent moisture, there is a risk of cracking the grains as they move through the harvester, reducing their quality for sale. Mechanical harvesting of both wheat and barley is done with a combine harvester. Most are self-propelled, moving through the field cutting the dry stalks and feeding the spikes through a series of rollers and sieves to remove the grain.

Manual harvesting of wheat and barley is a laborious procedure involving large amounts of labor. The grain will typically be cut with a hand sickle, tied in bundles, and later carried to a central threshing area. It is best to cut the grain-bearing stalks with around 17 to 20 percent moisture. This helps avoid loss of grains due to shattering of the heads. The threshing area may be a concrete floor or merely a smooth, hard earth area prepared for threshing. If a motorized thresher is available (Fig. 3.19), the grain heads are fed through the machine to separate grain from straw. When using manual threshing, the grains will be separated from the stalks by beating on a log, with flails, driving over the piles of straw with a tractor, or by repeatedly driving oxen over the straw pile. After the grain is separated from the straw, it will be winnowed to separate the grain from the chaff (bits of leaves and straw), as shown in Figure 3.20.

Figure 3.20. Winnowing wheat in India.

Figure 3.21. Wheat storage: Traditional and improved.

Storage is a critical problem for wheat and barley. Grain that is too moist can develop mold, destroying its usefulness. Some fungi produce toxins, such as aflatoxins, especially dangerous to humans. Insects also damage grains when the larvae eat the starchy interior and contaminate the rest with webs and fecal material. Rodents are also a constant threat. What they do not eat is contaminated and unfit for human use.

Large commercial producers often deliver their grain directly to a local grain elevator. The elevator has large concrete towers for storage of grains and to load grain directly into railcars for shipment. Some farmers will have large metal bins for temporary storage. This is to hold the harvest in the hope of better prices later in the year. Small farmers in the developing world save much of their wheat and barley harvest for family use. Traditional storage methods include large baked clay pots, metal barrels, and simple adobe clay bins. Storage made of hardened clay by a small farmer is seen in Figure 3.21. The containers attempt to block entry of insects and rodents into the stored grain. By making an airtight seal, extensive damage by insects can be avoided. An improved model made of metal suitable for small farmers in India is shown on the right in Figure 3.21.

3.3.10 Marketing

Wheat is the grain most widely traded in the world. Over 600 million tons were traded between countries in 2005.[15] The United States dominates the international wheat market, contributing about 25.4 million tons in 2005, with Australia, Canada, and Argentina being close competitors.[16] International marketing of wheat and barley is made possible by a strictly controlled system of classification. The most widely used is that of the U.S. Department of Agriculture (USDA). The eight USDA classes of wheat are shown in Table 3.1. Within each class, there are five grades of quality from 1 (best) to 5 (worst). Wheat for human consumption is normally limited to U.S. No. 1 grade. Lower grades are used for nonfood products or livestock feed. Each class represents a type of wheat useful for specific products. Barley is only classed

TABLE 3.1. USDA Classes of Wheat and Barley

Class	End Products
Wheat	
Durum wheat	Macaroni, spaghetti, pasta
Hard red spring wheat	Bread, rolls, pizza, breakfast cereals
Hard red winter wheat	
Hard white wheat	
Soft red winter wheat	Biscuits, cakes, crackers, cookies
Soft white wheat	
Unclassed wheat	
Mixed wheat	
Barley	
Malting barley	Fermented beverages
Barley	Livestock feed

Source: http://www.gipsa.usda.gov.

as malting or just barley. Both wheat and barley must be classed before they enter into international trade.

Wheat and barley trading within the United States is handled through the Chicago Board of Trade. It works much like the stock market. Buyers and sellers work through agents to purchase and sell quantities of specific grades to be delivered at some date. Usually delivery is within a month for immediate needs. However, there is an active futures market for delivery of grain as much as 2 years in the future. This permits manufacturers to assure a supply of wheat at a known price.

3.3.11　Postharvest Processing

Processing of wheat involves the production of flour. Consumable products are then made from the flour. In contrast to wet and dry milling processes used for maize, nearly all wheat is dry milled. The four main steps in dry milling of wheat are cleaning, tempering, fractionating, and separating. The final use determines the type of wheat grain to be used. Durum is primarily used for pasta since it has little leavening power. Hard red winter and hard red spring types are mixed with soft red winter for various types of breads. Soft white is desirable for cakes and pastries. Samples of the USDA grain types are shown in Figure 3.22.

Cleaning consists of removing any remaining dirt clods, stones, stems, leaf fragments, and nonwheat seeds. The most common methods utilize screens and streams of air to separate these impurities on the basis of mass and size. Tempering is the addition of moisture to the grain to make it more resistant to breakage. This makes it easier to gradually mill the grain into smaller pieces. At each stage of the milling process the grain passes through rollers. This enables the miller to fractionate, or split, the endosperm, bran, and germ from the flour. The final step is separating where the individual fractions are separated by sieves, air blasts, or gravity into uniform fractions. Each product has specific uses, the most valuable being flour for baking.

Figure 3.22. USDA wheat types.

In most of the world, wheat mills are small, family-owned machines. These produce mostly whole-wheat flour for home consumption. In subsistence farms, the farm wife may use a hand grinder to form coarse flour for baking. The flour produced from these processes has the benefit of retaining all the nutrients in the grain. A disadvantage is that some fractions easily spoil and the flour has a short shelf life.

> One family of four can live 25 years off the bread produced by one hectare of wheat.

Barley processing for livestock feed only requires a coarse milling operation to produce cracked barley. This enhances digestion by animals. Malting barley is first partially germinated for 4 to 5 days. The germinated seeds are then killed by heat to form malt. The dried malt is stored for later use in fermenting vats to produce alcoholic beverages. Barley is useful for this process because the hulls remain on the grain and facilitate filtration.

3.4 RICE PRODUCTION

Latin Name: *Oryza sativa*

Other English Names: Asian rice, common rice, cultivated rice, lowland rice, paddy rice, upland rice

3.4.1 Climatic Adaptation

Rice is cultivated in over 100 countries on all continents, except Antarctica. Over 91 percent of the rice in the world is grown and consumed in Eurasia.[17] Four basic ecosystems describe cultivated rice. Irrigated rice is the stereotypical "rice paddy" crop grown in standing water. This accounts for about 55 percent of the total area of rice grown in the world and around 75 percent of the total grain production.[18] Rain-fed lowland and upland ecosystems are nonirrigated and produce less yield per hectare than irrigated rice. Flood-prone ecosystems are areas subject to uncontrolled flooding that cover the fields with as much as 4 m of water for months at a time. Varieties of rice adapted to this ecosystem are able to elongate so that the leaves and grain float on the water surface.

Referring to Figure 3.23, it is easy to see that the largest rice-producing countries are in Asia. Rice is a tropical crop, needing warm temperatures to grow and produce grain. Depending on the specific variety, it takes from 90 to 200 days to produce a crop. Rice does not tolerate freezing temperatures. It generally is grown in areas with at least 1500 mm rainfall. Irrigated rice requires 7 to 9 million liters of water per hectare.[19] It is generally grown only in areas with a mean annual temperature above 21°C and high rainfall to supply the water needed by the crop.

3.4.2   Importance in Human Diets

Rice is the most important grain in Asian diets, providing 35 to 80 percent of the total calories consumed. With reference to Figure 3.35, rice has the highest carbohydrate content of the cereal grains but is the lowest in percent protein, fat, and fiber. Rice is consumed as a milled grain. After the husk is removed, the brown grain is milled, or polished, to remove the brownish bran layer from the seed. Typically, only 55 to 65 percent of the harvested weight is recovered as milled, white rice. A disadvantage of milling is that minerals, vitamins, and fiber contents are reduced.

Most rice is eaten as a boiled dish. There are many variations on its preparation. Depending on the type of rice, it may be boiled to form a sticky mass or steamed to give a light, fluffy single-grain food. In many parts of Asia, rice is considered an essential part of every meal. Manufactured rice products include rice cakes, noodles, food stabilizers, puddings, crackers, bread, puffed rice, popped rice, and alcoholic beverages.

3.4.3   Botanical Description

Rice is a member of the grass family. There are two cultivated species: *Oryza sativa* and *Oryza glaberrima*. The most widely planted specie is *O. sativa*, with *O. glaberrima* being commercially produced only in parts of Africa. Over thousands of years, two strains of *O. sativa* have evolved; japonica and indica. Some scientists consider a third class, javanica, as an intermediate strain between japonica and indica. Modern rice varieties have been formed from crosses of many strains, and, in practice, it has become difficult to easily identify a particular variety as belonging to a particular strain. For example, NERICA or New Rice for Africa released by West Africa Rice Development Association, Côte d'Ivoire, rice varieties are a cross between African and Asian rice varieties having good production under upland, or dry, conditions.

Rice is classified as an annual grass, meaning the plant dies after forming seed. In practice, some tropical rice-producing areas practice ratooning. This involves cutting off the above-ground part of the mature plant, letting the field dry for a month, and flooding the field again to permit new sprouts to grow from the base of the harvested plants. Up to three harvests can be obtained from a single planting.

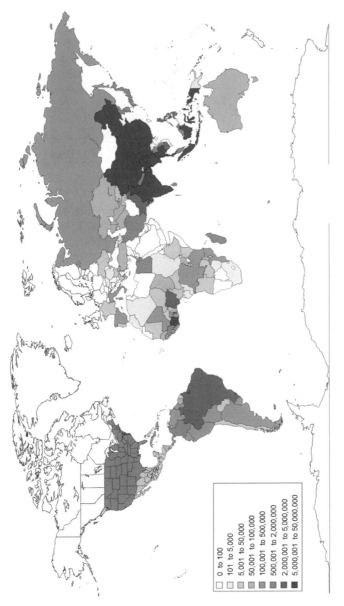

Figure 3.23. Hectares of rice in the world. (FAOSTAT, 2005.)

0 to 100
101 to 5,000
5,001 to 50,000
50,001 to 100,000
100,001 to 500,000
500,001 to 2,000,000
2,000,001 to 5,000,000
5,000,001 to 50,000,000

Figure 3.24. Rice nearing harvest.

The grain is formed on a loose panicle containing both male and female flower parts as shown in Figure 3.24. The stems of some varieties that grow in flood-prone areas reach up to 5 m in length, but more common irrigated and upland varieties seldom grow over 100 cm tall. The hull of the rice kernel is tightly attached to the seed and must be removed before consumption.

3.4.4 Origin

The specific origin of rice has been difficult to identify. It appears to have developed in an area stretching from south of the Himalaya mountains through Myanmar, Thailand,

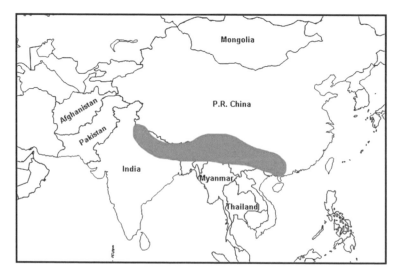

Figure 3.25. Areas of rice origin.

Laos, and Vietnam to China as shown in Figure 3.25. A more specific location is difficult to define because of the relative lack of archeological research in these areas. The best guess is that it developed around 6000 BC, spread to the Yangtze River valley in China by 5000 BC, the Indus valley of Pakistan by 2300 BC, and to Japan by 300 BC. It reached the middle east and Greece by 300 BC.

One problem, which complicates the origin of rice, is the presence of ancestors of *O. sativa* in the Americas, Africa, and Asia. It is generally agreed that the direct ancestor of modern rice was a wild species in the Pangean supercontinent, which later drifted apart to form modern India, Africa, Australia, and South America. Wild forms of rice found on these continents bear some resemblance to each other but are the result of many wild hybridizations to form present day species.[20]

3.4.5 History

Early movement of rice into other parts of the world followed wet, flood-prone river valleys. This had the advantage of planting a crop in soil that was "renewed" each year by floodwaters carrying fertile soil downstream. We are fortunate to have written records from China recording some of the innovations in early rice cultivation summarized in Table 3.2. Early introductions of rice into the Americas were by merchants traveling from Africa and Asia to Europe. The very first varieties were probably Indica types with ones introduced after 1900 being mostly Japonica types. Most of the Europeans had little practical knowledge of rice cultivation. As a result, rice in North America is classified by grain type not by ecotype. The official grain types recognized by the USDA are long grain, medium grain, and short-grain as shown in Figure 3.26.

Because rice is so important to Asian cultures, an international institute was established in the Philippines to develop rice production varieties and practices. The International Rice Research Institute (IRRI) was established in 1960 as a collaborative effort by the Ford and Rockefeller foundations and the government of the Philippines. Early work showed the need for improved varieties and accompanying cultural practices. The introduction of a semidwarf variety (IR-8) resulted in a plant that could grow under high soil fertility conditions without lodging (falling over). The same semidwarf characteristics were widely adopted in most rice-producing countries. This variety, with others following it, was the key that started the "Green Revolution." Countries

TABLE 3.2. Chinese Innovations in Rice Culture

Approximate Year	Practice
1500 BC	Use of water buffalo
1122 BC	Use of hoe introduced
600–500 BC	Controlled flooding of rice paddies
400 BC	First use of iron plow
AD 155	First use of transplanting
AD 900	Foot pedal pumps invented
AD 1000	Harrows and rollers introduced

Figure 3.26. USDA rice grain types.

that had previously only imported rice to feed their population became rice exporters. Average yield levels rose and the quality of rice on the markets improved.

3.4.6   Soil Preparation and Planting

Upland rice is similar to wheat and barley with respect to soil preparation and planting operations. Soil preparation and planting of irrigated rice differ greatly from other cereal grains. Irrigated rice needs periods of inundation or flooding to produce maximum yields. To keep water on the field, paddies are created with small earthen levees to contain water. In most parts of the world these levees (or bunds) are semi-permanent field boundaries and remain in place for years.

The first step is to flood and "puddle" the soil in a paddy. Techniques vary depending on the equipment or animals available. Frequently, oxen are used with a plow or leveling board, as shown in Figure 3.27. There are three basic goals in puddling. The first is to level the soil within the enclosed paddy. The second is to form a layer of soil below the surface that prevents excessive loss of water. The third goal is to break up any clods of soil to give a uniform, smooth transplant bed. Puddling also serves to control weeds.

Planting normally is first done in a nursery near the field. The plants are raised until they reach a height of 15 to 20 cm when they are transplanted to the open field. This is commonly a manual operation requiring large amounts of labor. Mechanical transplanters exist but are not widely used.

The purpose of transplanting is to raise the plants until they are large enough to grow and overtake any weeds existing in the field. It also permits raising plants in a small area while the main field is cleared of the previous crop and prepared for transplant. With carefully planned farm operations, it is possible to raise three crops of rice per year on one field. In temperate climates, the transplant nurseries are covered to

Figure 3.27. Puddling a rice paddy in Dominican Republic.

prevent cold damage, and transplanting to the field is done when the weather is warm enough for the crop to grow normally.

Mechanized soil preparation and direct seeding methods are used to produce rice in developed countries. Laser-guided leveling equipment is used every 7 to 8 years to level a field. Initial tillage is done with a tractor and plow or disk while the paddy is dry. Plastic levees are placed around the field before flooding. Direct seeding is done because of the high cost of labor involved with transplanting. Seeding may be done with a conventional grain drill before flooding, or presoaked seed may be broadcast by airplane on a flooded paddy.

3.4.7 Irrigation

Because rice yields are higher when irrigated, it is raised in paddies whenever possible. Irrigated rice is the most demanding of labor since even pump-driven irrigation requires someone to supervise the pump. Many irrigated rice fields are irrigated manually, using equipment thousands of years old, as in Figure 3.28. Upland rice can yield as much as

Figure 3.28. Rice irrigation by Dhenki in India.

irrigated rice when adequate water is supplied, but the problem is one of supplying enough water for the crop at critical stages. In most areas, upland rice is rain fed.

The level of water maintained in a paddy is not constant but varies depending on the stage of crop development. If direct seeded, only 2 to 3 cm of water is sufficient to keep the soil moist and permit emergence of seedlings. Transplanted plants can tolerate higher water levels, 10 to 12 cm being ideal. Water levels may be reduced to 4 to 5 cm when the plants start to tiller (forming additional stems). This ensures an adequate oxygen supply near the growing point of the plant. As the crop nears flowering, an adequate water supply is key to maximizing yields. Any water stress at this stage will directly reduce yields. Keeping water levels at least 15 cm deep also aids with weed control. The paddy is drained 15 to 20 days before harvest so that the soil is not muddy. This facilitates the use of combine harvesters and manual harvest of the crop.

3.4.8   Fertilization

The most limiting nutrient in rice production is nitrogen. Because paddy rice is grown in standing water, the soil around the roots is under anaerobic conditions. This creates different problems for application of chemical nitrogen fertilizers. When nitrate forms of nitrogen are applied to soils under anaerobic conditions, soil bacteria transform this into the gaseous form of nitrogen that escapes into the atmosphere. As much as 70 percent of nitrate fertilizer may be lost to this denitrification process in rice paddies. Remaining nitrate forms are also subject to leaching down through the soil profile. This increases environmental contamination and reduces nutrients available for plant use. To reduce denitrification problems, ammonium forms of nitrogen are applied to rice paddies. Upland rice is not grown under water and so does not have the problem of anaerobic denitrification losses.

As much as 70 percent of the nitrogen fertilizers are applied before planting and flooding. Later applications are done by inserting the fertilizer below the flooded soil to prevent volatilization losses. Upland rice can receive applications of fertilizer after the seed is planted. In some areas, application of midseason fertilizer is done by airplane. Other systems use the irrigation water to apply any remaining nitrogen.

3.4.9   Weeds, Insects, and Diseases

Certain weeds cause major problems in flooded rice paddies. One that causes problems in the United States is the so-called wild rice, *Zinzania aquatica*. While it is not a member of the *Oryza* genus, it is one of the related families found in North America. Because it grows in standing water, it can compete with cultivated rice. Herbicides are less effective against this species since any chemical that can kill *Zinzania* can also kill cultivated rice. Active research is underway to develop allelopathic characteristics in rice varieties. Allelopathy is the characteristic of a plant that discourages, or prevents, growth of competing plants of other species nearby. The means by which this is accomplished varies among plants, but seems to involve release of organic compounds that prevent germination of plants belonging to other species.

Rice blast is the serious disease facing most rice producers in the world. It is caused by the fungus *Pyricularia grisea*. It stops development of the grain in the floret,

Figure 3.29. Manual rice harvest.

resulting in empty panicles. Spores of the fungus remain in the soil for several years to serve as a source of infection for new crops. Resistant varieties are being developed.

3.4.10 Harvest and Storage

Rice is harvested when grain moisture drops to 15 to 18 percent. Harvesting grain with higher moisture content will cause problems of mold and insect losses. As shown in Figure 3.29, much of the rice in the world is harvested by hand. The plants are cut, bundled, and transported to a central threshing area.

Large areas of rice production have rice combines to harvest the crop. These are similar to wheat combines but with track propulsion to get through damp areas of the field. A rice combine being moved to another field in the Dominican Republic is shown in Figure 3.30.

Figure 3.30. Rice combine between fields.

Disposal of the straw is a problem in some areas of the world. Rice straw has a high content of silica crystals, making it somewhat resistant to decomposition. Dry straw is not palatable to most livestock. Many farmers resort to burning dry straw in the field. This has the advantage of destroying insect eggs and disease spores on the straw. A significant disadvantage is the loss of nutrients in the straw and the loss of potential organic matter additions to the soil. The Connelly–Areias–Chandler Rice Straw Burning Reduction Act of 1991 in California specified that burning of rice straw be phased out as an air pollution measure. After 2001, burning has been permitted only for approved disease control situations. As a result, there has been a surge in alternate uses of rice straw, such as erosion barriers and home construction.

3.4.11   Marketing

Most rice is consumed in the country of production. Even though China produces large amounts of rice, it sells relatively little on the international market. The major exporting countries are Thailand, Vietnam, India, United States, and Pakistan. Many African countries are net importers of rice.

In addition to the USDA classes, rice is classed by its cooking and food characteristics. One extremely important classification is the degree to which the endosperm starch is gelatinized during cooking. Types that cook into fluffy, individual grains are classed as nonsticky rice. Indica varieties have this characteristic. This is the type most commonly raised in the United States. Sticky rice has a slightly gelatinous consistency. Japonica varieties normally form a sticky, or mushy, mass when cooked. These types are very easy to overcook. Glutinous, or waxy, rice cooks differently than the previous two types. It absorbs less than half the amount of water during cooking but forms a gelatinous mass. This is used primarily for sweet desserts in Southeast Asia.

Another classification, unique to rice, is that of aroma. All rice has some aroma released upon cooking. Specific varieties have a slightly perfumed aroma that is highly prized. One of the best known is the basmati cultivar grown in India and Pakistan. Although it yields less per hectare, the higher price for this variety makes it profitable for farmers. Other varieties have been developed in the United States with similar aromas, one being Texmati.

3.4.12   Postharvest Processing

A 100-kg amount of rough, or paddy, rice goes through a dehulling process to remove the outermost covering. Dehulling is normally done by machine, but traditional methods can also be found in small farmer's homes. An example is the pounder in an Indian village home shown in Figure 3.31. This yields about 80 kg of brown rice and 20 kg of hulls. The hulls are used as fuel in heating plants, as mulch, and as a mild abrasive. The next milling, or polishing, stage removes some of the outer seed coat, leaving the familiar white rice. About 11 kg of by-products are removed in this step. Of the 69 kg left, only 55 kg passes screening for the market. The remaining 14 kg consists of broken and undersized grains used for brewing and animal feed.

Figure 3.31. Rice pounder in India.

3.5 SOYBEAN PRODUCTION

Latin Name: *Glycine max*
Other English Names: Soya bean, Japan pea, soja bean, soy pea[21]

Soybean is a relative newcomer on the world grain scene. Though it has been cultivated for thousands of years, it did not reach worldwide prominence until the 1900s. It is currently the fourth ranking grain after wheat, rice, and maize (Fig. 3.4). It differs from these three grains being a legume instead of a member of the grass family. It is not commonly consumed as a grain but used as a raw material for processed foods and many nonfood items.

3.5.1 Climatic Adaptation

Soybean is an annual crop, maturing in 80 to 120 days after planting. It requires a minimum soil temperature of 6°C to germinate and 16°C for flowering. These temperatures are easily met in most temperate climate zones similar to the United States, but areas further north, such as Canada, Europe, or Russia, are not suitable for optimum yields. Ideal temperatures for growth range from 20 to 24°C. This makes the U.S. Midwest a nearly ideal growing region. A similar climate is found in central and southern Brazil, another major soybean producer shown in Figure 3.32.

3.5.2 Importance in Human Diets

Uses of soybean for human food without industrial processing are limited mostly to Asian societies. In Japan, the average soybean consumption is about 70 g/day in the form of tofu, miso, natto, and other products.[22] Nutrition programs are in place in Africa to encourage the use of soybean products such as *tufu* and *dawadawa*. There is little published information available on direct soybean consumption in other countries.

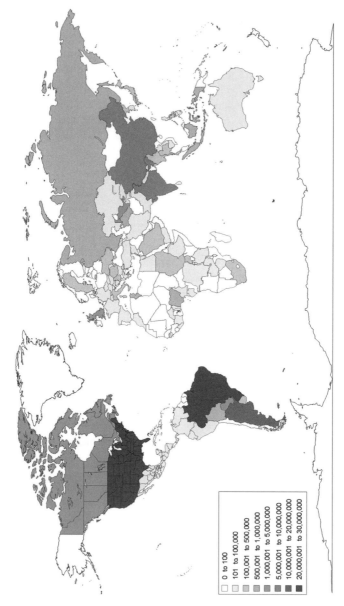

Figure 3.32. Hectares of soybean in the world. (FAOSTAT, 2005.)

Some products familiar in the United States are soymilk, tofu, edamame, and tempeh. Edamame soybeans are used green, similar to green peas, in salads and cooked foods. Soymilk is the liquid portion remaining after the beans are soaked in water, cooked, ground to a paste and the solids removed by filtering through a fine cloth. Tofu is the curdled soymilk produced in a process similar to cheese making. Tempeh is a fermented soybean food that can be fried and eaten or included as an ingredient in soups or other cooked dishes.

The soybean possesses antinutrients that may explain its slow spread to home prepared foods. Trypsin inhibitors act to reduce the effectiveness of trypsin, an enzyme needed for digestion of proteins. Heating or toasting the beans destroys these trypsin inhibitors. Lectins formed in soybeans are proteins that bind to carbohydrates. In sufficient concentrations, they can cause death. Heating destroys lectins. Isoflavones in soybeans have been reported to have biological activities similar to estrogens in humans. There is some research to indicate that these compounds may have anticarcinogenic effects. Other compounds tend to produce flatulence, reduce phosphorus utilization in the body, and allergic reactions in some individuals.

3.5.3 Botanical Description

Soybean is a legume, like peas, peanuts, and lentils. Legumes are plants that form a symbiotic relationship with bacteria of the *Rhizobium* genus. These bacteria infect the roots of a legume to form small nodules. Within the nodule, bacteria convert atmospheric nitrogen into a form that plants can use. As these bacteria die off, the nitrogen-containing compounds are released to become available to plants for growth and development. Typically, soybeans will "fix" enough nitrogen for the soybean crop and leave a surplus for a subsequent crop.

Soybean plants have a trifoliate leaf and a plant height around 50 to 70 cm. Soybean is highly sensitive to changes in day length. Varieties developed at a specific latitude cannot be planted more than 100 miles north or south of that latitude without a significant reduction in yield. Soybean varieties are classified in maturity groups numbered from 000 to 9, with 000 being those adapted to the short days of Canada and 9 being varieties adapted to southern Florida.

In addition to day-length sensitivity, soybean varieties are classified as determinate or indeterminate. Determinate varieties start to flower when 80 percent of the vegetative growth has occurred. Flowering occurs over a 2- to 3-week period. All vegetative growth stops when flowering begins. Indeterminate varieties continue vegetative growth during flowering. The flowering period may continue for 5 to 6 weeks. Normally, most varieties in maturity groups 000 to 4 are indeterminate and maturity groups 5 to 9 are determinate.

3.5.4 Origin

Soybean originated in central China (Fig. 3.33) where the wild predecessor still grows. Early cultivation of soybean was probably for grain. The earliest historical mention of soybean was about 2700 BC, when Emperor Sheng-Nung named five sacred plants— soybeans, rice, wheat, barley, and millet. However, some scientists dispute this record, insisting that the first verifiable mention was during the Chou period

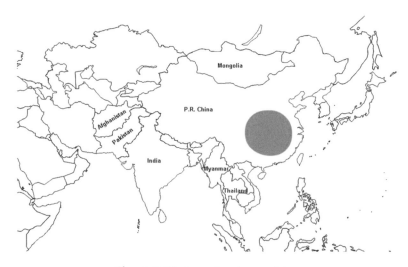

Figure 3.33. Area of soybean origin.

(1027–221 BC).[21] At some point, the use of soybean as an oil source was discovered. The first mention of soybean oil was in 1895 when Japan started importing soybean oil cake for use as a fertilizer. Soybean was introduced in Europe as early as 1712, but the conditions for cultivation were not ideal and it remained a botanical curiosity.

The ancestor of soybean, *Glycine soja*, is a small, vinelike plant with small seeds that shatter from the pods easily. The typical oil content is only 9.8 percent compared to modern varieties containing 21 percent oil. Modern varieties are stiff-stemmed plants that hold the seeds in the pods for some time after maturity. Although the two species are genetically similar, the characteristics of *Glycine max* have been greatly modified by human selection over the centuries.

3.5.5 History

Commercial uses of soybeans remained confined to a few Asian countries until 1905 when Japan withdrew from Korea at the end of the Russo-Japanese war. During the war Japan had imported soybean cake for fertilizer, but their withdrawal left a surplus on the market. In order for the Japanese to maintain their influence over Manchuria, they started export of soybean oil for use in soap and mixed livestock feeds. Prior to World War II, soybean was primarily a forage crop in the United States. Starting in 1930, the United States increased soybean production to replace the vegetable oils previously imported from Asia. During World War II, soybean production increased 246 percent to replace imported fats and oils. This emphasis on soybean made the United States a world leader in production and research. This leadership is now being challenged by Brazil and Argentina. Brazil has developed varieties well adapted to its *cerrado* area and encouraged farmers to plant large areas to this crop. With mechanization and an efficient market infrastructure, Brazil may soon surpass the United States as the world's largest soybean producer.

3.5.6   Soil Preparation and Planting

Production of edamame and some tofu varieties are manually cultivated in parts of Asia. Soybean production in the rest of the world tends to be a completely mechanized system for oil grain. Soil preparation and planting are done with equipment similar to that used for maize and sorghum. Under conventional tillage systems, a field will be plowed or disked to break up clods and leave a friable, weed-free seedbed. Minimum, or zero-tillage, systems permit planting without tillage or other soil preparation.

Soybeans are planted in rows spaced from 7 to 30 inches apart. Wide-row spacing permit tractors to pass through the field without damaging plants. When planted in narrow rows, all fertilizers and chemicals are applied at planting to reduce plant damage from tractor tires. In some parts of the United States, soybeans are planted after harvest of wheat or barley in early summer. This permits the harvest of two crops in the same year. In more tropical climates this is routine practice.

3.5.7   Fertilization

Because soybean is a leguminous crop, it does not require a nitrogen fertilizer to yield well. In fact, application of nitrogen fertilizer can have the effect of reducing or preventing nitrogen fixation by the symbiotic *Rhizobium* bacteria. Every kilogram of grain harvested will remove 16 g of phosphate and 23 g of potash. With a good yielding crop of 1500 kg/ha, at least 24 kg of phosphate and 34 kg of potash will be removed. These nutrients must be replaced to maintain adequate levels of soil fertility.

In addition to chemical or organic fertilizers, it is sometimes necessary to inoculate seeds with *Rhizobium*. Commercial preparations of *Rhizobium* were first sold in the United States in 1905. If a field has not been planted to soybeans for 5 to 7 years, it is recommended to inoculate the seeds with *Rhizobium* before planting.

3.5.8   Weeds, Insects, and Diseases

Until the 1990s, weeds were a serious problem with soybean. Because the crop grows relatively slowly, many weeds were able to outgrow the soybean in early stages and rob it of nutrients and moisture. Available herbicides carried risks of damage to the crop. In the spring of 1995, the U.S. Environmental Protection Agency (EPA) approved the planting of herbicide-tolerant soybeans. In practice, farmers could spray a growing crop with herbicide and kill all weeds. This meant that several tillage operations before planting and a herbicide application after planting could be eliminated. Since the introduction of this trait, plantings of herbicide-tolerant soybeans have captured about 65 percent of the market in the United States.

Various diseases can cause serious yield losses in soybean. Most strike in the seedling stage. Phytophthora root rot, pythium root rot, and rhizoctonia root rot are the most common diseases affecting seedlings. Some varieties have moderate resistance to some of these diseases. A combination of varietal resistance and seed treatment with a fungicide work best to prevent losses to these diseases.

Some problem insects include aphids, spider mites, and leaf beetles. Normally, these do not cause serious yield losses. Nematodes are microscopic worms that can cause serious damage to soybeans. They infest the root system and can kill significant portions of a field. The worms form cysts in the soil that can persist for several years. The best control measure seems to be crop rotation and prevention of nematode spread from infested fields.

3.5.9 Harvest and Storage

Harvesting of most soybean types is done by machine. The combine cuts the grain stalks, separates the grain from leaves and dirt, and stores it in a small bin on the top of the machine. A self-propelled combine is shown in Figure 3.34. The same combine harvester used for maize and wheat can be used for soybeans with internal adjustments and a change of a few screens. A grain moisture of 13.5 percent is considered ideal. Higher levels result in storage losses from fungi and costs for mechanical drying. Lower moisture levels can result in splitting of the beans during harvesting and losses caused by shattering of the pod before harvest.

Handling of soybeans in the field and transport to storage is similar to maize and sorghum. The relatively high oil content of soybean grains makes it especially susceptible to oxidation and fermentation. This makes low moisture content and avoidance of heat important.

Edamame soybeans are eaten as a fresh vegetable and are harvested before the grain matures and dries. The stage of harvest is critical. Harvesting too early results in smaller grains and lower yields. Harvesting a few days too late gives hard grains and a low sugar content. The entire plant is harvested and sold in bunches for home use. For canned or frozen edamame, the beans are stripped from the pods.

3.5.10 Marketing

Soybeans are classed according to their biochemical characteristics, color, size, and quality. Commercial channels recognize yellow soybean and mixed soybean, yellow

Figure 3.34. Combine in soybeans.

being the preferred type. Discolorations in the seedcoat usually indicate damage during harvesting or a disease on the grain. As with other grains, the weight per bushel is a major factor in determining the grade (No. 1, 2, 3, or 4). The best soybean grade (U.S. No. 1) has a weight of 56 lb/bu (70 kg/hl). It also has minimal heat damage, foreign material, and split grains.

Soybeans intended for specific markets may have other requirements. For example, tofu soybeans sold in Japan go through a rigorous selection process to remove blemished seeds, misshapen seeds, and small or large seeds, The processed seeds are packaged and sold for home preparation of tofu, so that appearance is of great importance. Soybeans intended for industrial processing have fewer restrictions.

3.5.11 Postharvest Processing

Most of the soybeans harvested are intended for industrial processing. Three main methods are used to extract oil from soybeans: distillation, expression, and solvent extraction.

Distillation involves passing steam through cracked soybeans to vaporize oils. As this steam is cooled, it releases oils of different densities in liquid form. Expression simply means pressing the grains after they have been cooked. Solvent extraction removes oil from grain by mixing cracked grains with a nontoxic fat solvent such as hexane at low temperatures. The solvent is then evaporated and recycled to extract more grain.

Soybean oil intended for human consumption will be further processed to remove impurities, objectionable odors, and colors. Depending on the end use, the oil may be hydrogenated to convert it to a semisolid and reduce oxidation. This gives it temperature stability and reduces breakdown before consumption. Common end products of soybean oil are margarine and cooking oil.

The soybean meal remaining after oil removal is also used for human consumption. A common product is textured soy protein (TSP) used in vegetarian dishes and cooking ingredients such as Hamburger Helper. Processing of the meal also gives tofu and soymilk. Soybean cake is another product used in animal nutrition. It has a high protein content and makes up a significant portion of cattle and poultry rations.

Nonfood products using soybean as the main ingredient are too many to mention in the space available. They range from paint and crayons to furniture and shampoo. Traditional and genetically modified varieties of soybeans are being released for specific food and industrial applications such as low saturated fat, high sucrose, or high protein types.

One hectare of soybeans can produce 205,920 crayons.

3.6 OTHER GRAINS

Many other grains are used by humans for food, fiber, and industrial applications. Their omission is merely due to space limitations not importance in the human diet. Grain legumes are especially important in some cultures as a source of plant protein.

3.6.1 Peanut (*Arachis hypogaea*)

Peanut is widely consumed around the world as a roasted seed and is a source of oil for cooking. A more complete description of this crop is covered in Chapter 6.

3.6.2 Lentil (*Lens culinaris*)

Lentil is a small seeded legume with high protein content, often included in vegetarian diets. It is widely consumed in Middle Eastern and Asian countries. It grows best in semiarid areas with a dry period at harvest time. Being a cool season crop, it is planted in the summer in the United States and Canada and in winter in India and Pakistan. The form consumed by humans is mostly as a cooked seed. Simply boiling the dry seed produces a nutritious food. The grain may also be ground into flour for specialty foods. Lentil flour also has uses in the textile and printing industries.[23]

3.6.3 Chickpea (*Cicer arietinum*)

Chickpea has a similar distribution as lentil, with the addition of semiarid parts of Latin America. It is also known as Garbanzo bean and is a high protein staple food in many diets. Two main types are grown: kabuli and desi. The kabuli types have larger seeds and grow well under irrigation. Desi types grow well under rainfed conditions but the smaller seed size is less desirable on the market. It is prepared by boiling the seeds or by grinding the dry seeds to form flour. There are few industrial, nonfood uses.

3.6.4 Faba (*Vicia faba*)

Faba beans are consumed as a fresh vegetable or in the dry bean form. In the Middle East it is a winter crop, being tolerant of cool temperatures. The high protein content makes it an important part of many Asian and Latin American diets. The beans may be roasted, boiled, or eaten fresh. In Egypt the dry beans are ground into a flour and deep fried with spices to form falafel. Faba has higher concentrations of hemagglutinins than other legumes and can cause intestinal reactions in humans. These proteins are destroyed during cooking.[24]

3.6.5 Field Beans (*Phaseolus vulgaris*)

This is a large and varied class of grain legumes. It includes French beans, kidney beans, wax beans, snap beans, and string beans. The vegetable types are normally eaten in their green form as a boiled vegetable. Dried beans have about 22 percent protein content. There are large variations in seed size and coloration. Their amino acid makeup complements that of maize, which may explain why they are commonly grown in association in Latin America.[25] They are normally grown as a summer crop. Harvesting is done by hand and using grain combines.

3.6.6 Oats (*Avena sativa*)

Oats are primarily a cool season forage crop with the grain being used for horses. Some specialty types (hulless oats) are used for human consumption. It is a member of the grass family, as are maize and sorghum. It has declined in importance for animal feed in the United States since the 1940s when tractors replaced horses for farming operations. Oats grain is still fed to young livestock and horses. It has received increased attention for human nutrition after clinical studies indicated that the water-soluble fibers in oat bran inhibit formation of cholesterol in the human body.[26]

3.6.7 Rye (*Secale cereale*)

Rye is believed to be the first member of the grass family cultivated by early humans. It was replaced by barley, then wheat. It remains second to wheat as a grain for human consumption.[27] It is used primarily for bread making, though ergot on the grain may be a problem. It is postulated that ergot poisoning was a cause of the hallucinations of young girls involved in the Salem witch trials. It is a cool season cereal crop, grown from the most northern climates to subtropical areas.

3.6.8 Millet

Several crops come under the general name of millet.[28] Pearl millet (*Pennisetum glaucum*) appears similar to maize and sorghum in the seedling stages but bears a cattail-like seed head. Foxtail millet (*Setaria italica*) is named after the large, bushy seed head similar to a miniature foxtail. Japanese millet (*Echinochloa frumentaceae*) is grown mostly as a forage grass in the United States but is a minor grain crop in Japan and Australia.

3.6.9 Sunflower (*Helianthus annuus*)

Sunflower is raised primarily as an oil crop. It is one of the few crops cultivated in the world that originated in North America.[29] From its introduction in Spain, it spread through Europe to Russia where oil content was increased from 28 to 50 percent. Over 80 percent of the crop in the United States is raised for the oil market. Most oil goes for the cooking market, with high-oleic-acid oil being preferred for its stability and resistance to oxidation. Problems limiting cultivation of sunflower are diseases and bird damage to the mature heads.

3.6.10 Safflower (*Carthamus tinctorius*)

Safflower is a broadleaf plant adapted to dry areas with warm summers. It was originally cultivated for the flowers used in preparation of dyes and colors. Today it is raised primarily for the oil, which is considered to be "high-quality" edible oil. A secondary market in the United States is for birdseed.[30] Safflower is best suited to mechanical harvesting. The leaves and seed head have prickly spines that make hand harvesting a problem.

3.6.11   Mustard (*Brassica* sp.)

Several mustard species are believed to have originated in Europe. It is a cool season crop. It spread to Asia, Africa, and the Middle East where the oil pressed from the seeds is important in cooking. Three main types are grown in North America: yellow, brown, and oriental.[31] Yellow mustard varieties are primarily cultivated for the powder and condiment market. The Brown and oriental mustard varieties are used more for oil. A variant of mustard, canola, was developed by the Canadian Canola Association. It has less than 2 percent erucic acid content, the pungent flavor of typical Indian cuisine. Canola now ranks second to soybeans as the leading source of vegetable oil.[32]

3.6.12   Cowpea (*Vigna unguiculata*)

Cowpea is a leguminous grain crop that originated in Africa where it is still widely grown. It is also grown in Latin America, Southeast Asia, and the United States. Depending on the country, it is known as southern pea, blackeye pea, or crowder pea.[33] The grain is highly nutritious, having over 20 percent protein content. It is an especially valuable source of lysine and tryptophan in contrast to cereal grains. Because of a low content of methionine and cystine, it is best used as a dietary supplement to complement cereals and animal proteins. It grows best in warm conditions and does not tolerate frost. Production has increased significantly in most of the world over the last 25 years but has declined in the United States.

3.6.13   Pigeon Pea (*Cajanus cajan*)

Pigeon pea is also a leguminous grain crop that probably originated on the Indian sub-continent.[34] It is widely grown in tropical climates ranging from subtropical to humid tropics. The plants are usually grown as annual crops, though many varieties are short-lived perennials. It has various names depending on the variety and country where it is grown. Some of the more common include Congo pea, gandul, dal and red gram. Though mainly grown for the grain, it is also used for organic matter additions to soil, a shade crop for vanilla, and even as a host for the insect that produces lac. The seeds are very nutritious, having up to 22 percent protein. It is susceptible to many diseases and insects, making its cultivation difficult.

3.7   NUTRIENT CONTENT OF CROPS

As can be seen in Figure 3.35 the cereal grains are good sources of carbohydrates and some types of protein but sometimes lack fiber and fat.[36] Soybean, in contrast, is lower in carbohydrates but much higher in protein and fat. The high oil content of this crop makes it a favorite for industrial processing, while the cereal grains are more suited to direct human consumption.

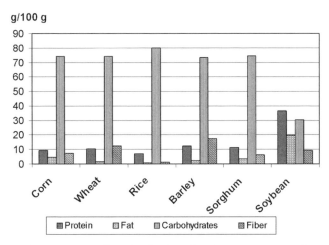

Figure 3.35. Nutrient content of major food grains. (From USDA Nutrient Database for Standard Reference, Release 18.)

3.8 CONCLUSIONS

Cereal grains provide a major portion of our daily diets. They are grown in climates ranging from temperate to humid tropical. With proper management cereal grains can provide significant amounts of food to growing populations. Minor changes in production practices can provide large increases in food production in many parts of the world. There are possibilities of major benefits from transgenic modification of cereal grains.

QUESTIONS

1. Maize, wheat, and rice are members of which botanical family?
2. In which area of the world did maize originate?
3. Where did wheat and barley originate?
4. What is winnowing?
5. What are the four basic ecosystems of cultivated rice?
6. Where did soybean originate?
7. What is a legume crop?
8. Give at least three products containing soybean.
9. What is a food product derived from safflower?
10. What food product comes from wheat?

REFERENCES

1. J. Diamond, *Guns, Germs, and Steel: The Fates of Human Societies*, W. W. Norton, New York, 1997.

2. Available at: http://www.plantnames.unimelt.edu.au/Sorting/Frontpage.html.

3. The Maize Page, Iowa State University, 2006. Available at: http://maize.agron.iastate.edu/general.html#article.

4. *Zea mays.* Available at: http://www.fao.org/ag/AGP/AGPC/doc/Gbase/DATA/Pf000342.htm.

5. Sorghum Bicolor, Center for New Crops and Plant Products, Purdue University, 2006. Available at: http://www.hort.purdue.edu/newcrope/duke_energy/Sorghum_bicolor.html.

6. P. Thomison, P. Lipps, R. Hammond, R. Mullen, and B. Eisley, Corn Production, *Ohio Agronomy Guide*, Bulletin 475–05, Ohio State University, Columbus, 2005.

7. Iowa Corn Growers Association, The Corn Refining Process, 2006. Available at: http://www.iowacorn.org/cornuse/cornuse_7.html.

8. *White Maize: A Traditional Food Grain in Developing Countries*, FAO and CIMMYT, Rome, 1997.

9. *Utilization and Processing of Maize, IITA Research Guide*, Vol. 35, Ibadan, Nigeria, 1997.

10. Available at: http//www.hort.purdue.edu/newcrop/default.html.

11. *Fertilizer Use by Crop in the Syrian Arab Republic*, FAO, Rome, 2003.

12. CGIAR Research & Impact, Consultative Group on International Agricultural Research, Wheat. Available at: http://www.cgiar.org/impact/research/wheat.html.

13. G. Schwab, *Grain Analysis: A Fertilizer Management Tool*, Washington State University Extension Service, 2002. Available at: http://cariety.wsu.edu/updates/2002_Nutrient Removal.htm.

14. U. K. Baloch, *Wheat: Post Harvest Operations*, FAO, Rome, 1988. Available at: http://www.fao.org/inpho/content/compend/text/ch06-02.htm#P715_38104.

15. *Food Outlook*, FAO, No. 4, December 2005.

16. FAPRI Agricultural Outlook—2005, Food and Agricultural Policy Research Institute, Iowa, 2006. Available at: http://www.fapri.iastate.edu/outlook2005/.

17. Available at: http:/www.cgiar.org/impact/research/rice.html.

18. *Basic Facts about Rice*, International Rice Research Institute, Philippines, 2005. Available at: http://www.irri.org/about/faq.asp.

19. Available at: http://www.fao.org/ag/AGP/AGPC/doc/GBASE/data/Pf000274.htm.

20. K. Kiple and K. C. Ornelas, *The Cambridge World History of Food*, Cambridge University Press, Cambridge, England, 2005.

21. W. Shurtleff and A. Aoyagi, The Soybean Plant: Botany, Nomenclature, Taxonomy, Domestication, and Dissemination, unpublished manuscript, Soyfoods Center, 2004. Available at: http://www.thesoydailyclub.com/SFC/historys&s31.asp.

22. *Consensus Document on Compositional Consideration for New Varieties of Soybean: Key Food and Feed Nutrients and Anti-Nutrients*, Organization for Economic Co-operation and Development, JT00117705, Paris, 2001.

23. Available at: http://www.hort.purdue.edu/newcrop/cropfactsheets/lentil.html.

24. Available at: http://www.hort.purdue.edu/newcrop/cropfactsheets/fababean.html.

25. Available at: http://www.hort.purdue.edu/newcrop/articles/ji-beans.html.

26. Available at: http://www.agron.iastate.edu/courses/agron212/Readings/Oat_wheat_history.htm.

27. Available at: http://www.hort.purdue.edu/newcrop/duke_energy/Secale_cereale.html.

28. Available at: http://www.hort.purdue.edu/newcrop/afcm/millet.html.

29. Available at: http://www.hort.purdue.edu/newcrop/afcm/sunflower.html.

30. Available at: http://www.hort.purdue.edu/newcrop/afcm/safflower.html.

31. Available at: http://www.hort.purdue.edu/newcrop/afcm/mustard.html.

32. Available at: http://www.hort.purdue.edu/newcrop/ncnu02/v5-122.html.

33. D. W. Davis, E. A. Oelke, E. S. Oplinger, J. D. Doll, C. V. Hanson, and D. H. Putnam, *Alternative Field Crops Manual*, University of Wisconsin Extension, University of Minnesota, Center for Alternative Plant & Animal Production, 1991. Available at: http//www.hort.purdue.edu/newcrop/afcm/cowpea.html.

34. J. A. Duke, 1983. Handbook of Energy Crops, unpublished. Available at: http://www.hort.purdue.edu/newcrop/duke_energy/Cajanus_cajun.html.

BIBLIOGRAPHY

G. Acquaah, *Principles of Crop Production: Theory, Techniques, and Technology*, Pearson Prentice-Hall Education, Upper Saddle River, NJ, 2002.

An Assessment of Race Ug99 in Kenya and Ethiopia and the Potential for Impact in Neighboring Regions and Beyond, CIMMYT, Mexico, 2005.

Composition of Foods: Raw, Processed, Prepared, USDA National Nutrient Database for standard References, Release 18, Beltsville, MD, Aug. 2005.

A. E. Desjardins and S. A. McCarthy, *Milho, Makka, and Yu Mai: Early Journeys of Zea mays to Asia*, USDA, National Agricultural Library Website, 2004. Available at: http://www.nal.usda.gov/research/maize.

Grain Inspection, Packers & Stockyards Administration, USDA, Beltsville, MD, 2005. Available at: http://www.gipsa.usda.gov.

Grain: World Markets and Trade, Foreign Agricultural Service Circular PG 12-05, December, 2005.

Grassland and Pasture/Crop Systems, FAO, Rome, 2006. Available at: http://www.fao.org/ag/AGP/AGPC/doc/pasture/pasture.htm.

J. C. Ho, S. Kresovich, and K. R. Lamkey, Extent and Distribution of Genetic Variation in U.S. Maize: Historically Important Lines and Their Open-Pollinated Dent and Flint Progenitors, *Crop Sci.*, **45**, 1891–1900, 2005.

T. Hymowitz and W. R. Shurtleff, Debunking Soybean Myths and Legends in the Historical and Popular Literature, *Crop Sci.*, **45**, 473–476, 2006.

C. W. Smith, *Crop Production: Evolution, History, and Technology*, Wiley, New York, 1996.

The World Sorghum and Millet Economies: Facts, Trends and Outlook, FAO and ICRISAT, Rome, 1996.

4

VEGETABLES

4.1 VEGETABLE PRODUCTION IN EACH FARMING SYSTEM

Donio has no garden and produces little in the way of what would be called vegetables. He does grow *camote* (boniato), which is planted around the fields, as shown in Chapter 5. It is also called sweet potato, and its leaves are also harvested and used as a vegetable in the diet.

Aída and Octavio have a large garden (Fig. 4.1) used to grow a large variety of vegetables. It is intensively cultivated and kept productive by the addition of manure and compost. In addition, it is irrigated so that vegetable production is continuous throughout the year. Of the 21 vegetables grown (Table 4.1) 62 percent are used both in the home and are sold in the market, and the remaining 38 percent are only used in the home. The vegetables are grown in raised beds formed in the garden. The garden is equipped with drip irrigation (Chapter 9 for further information about drip irrigation), which allows them to grow vegetables all year around. Water for irrigation, supplied by the local water system (Chapter 9), is taken from a holding tank, filtered, and fed into the drip irrigation system.

Steve does not grow any vegetables, although his son does have a small vegetable garden that supplies a small amount of vegetables for the family during the summer.

World Food: Production and Use. By Alfred R. Conklin, Jr. and Thomas Stilwell
Copyright © 2007 John Wiley & Sons, Inc.

Figure 4.1. Aíde and Zorro standing in furrow between raised garden beds.

TABLE 4.1. Vegetables Grown by Aída and Octavio in Garden Near House

Local, English, and Scientific Names	Used in Home	Sold in Market
Lechuga, lettuce, *Lactuca sativa*	X	X
Apio, celery, *Apium graveolens*	X	
Acelga, chard, *Beta vulgaris*	X	X
Zucchini, zucchini, *Cucurbita pepo*	X	
Remolacha, beet, *Beta vulgaris*	X	
Brocoli, broccoli, *Brassica oleracea*	X	X
Zanaboria, carrots, *Daucus carota*	X	X
Col blanca, white cabbage, *Brassica* spp.	X	X
Col morada, red cabbage, *Brassica* spp.	X	X
Ajo, garlic, *Allium sativum*	X	
Cebolla blanca, onion white, *Allium cepa*	X	X
Cebolla colorada, onion red, *Allium cepa*	X	X
Pepinillo, cucumber, *Cucumis sativus*	X	
Espinaca, spinach, *Spinacia oleracea*	X	X
Coilantro, cilantro, *Coriandrum sativum*	X	X
Perajil, parsley, *Petroselinum crispum*	X	X
Calabaza, pumpkin, *Cucurbita maxima*	X	
Rabano, radish, *Raphanus sativs*	X	X
Papa nabo, rutabaga, *Brassica* spp.	X	X
Nabo, turnip, *Brassica rapa*	X	

Except for specialized farming operations or specialized farms, it is typical for a farm family to have a garden with vegetables for family needs. The area devoted to vegetables is usually small, that is, less than 0.1 ha, and is intensively cultivated, often by the farmer's wife. In many countries one of the wife's roles is tending the garden, and this is not a minor role as indicated by the fact that one Kenyan farmer complained that his wife would not let him in her garden! The vegetables grown are locally adapted and easily grown by the person tending the garden and are common to the local diet and palate.

4.2 TYPES OF VEGETABLES AND THEIR USES

Any attempt to discuss vegetables is complex for many reasons. First, in some cases vegetables that look similar may come from very different plants. Thus, the leaves of both lettuce and spinach are eaten in salads, but lettuce belongs to the *latuca* and spinach the *spionacia* genus. Second, plants belonging to the same family may look very different and thus be interpreted as being different kinds of vegetables. Thus cabbage, which has a stalk and forms a tight head made up of leaves that are eaten, and broccoli, which forms a green flower head that is eaten, both belong to the same genus *Brassica*.

Third, different parts of plants are eaten. The leaf of some plants, leafy vegetables, is eaten. In some cases other parts of the plant may also be eaten. Both the belowground portion of beets and their leaves are eaten as vegetables. A fourth instance is where the mature grain, such as maize is eaten, while the immature seed or seed and pod are also eaten such as with green beans and some types of peas.

In this chapter vegetables are combined into groups that are likely to be seen as similar by the nonagronomist or nonhorticulturalist. In some cases the groupings are based on the part of the plant used as a vegetable. In other cases those used in a similar fashion as food are grouped together. In addition, as dictated by local custom, other parts of plants may be eaten, for instance, pumpkin leaves and the flowers of many types of plants including banana and squash are used as food.

Vegetables are used in three ways as food: as a main dish in a meal, as a garnish, and to add color or flavor to various dishes. They also provide vitamins, minerals, and fiber to the diet. Although they are an essential part of the human diet, vegetables do not and cannot supply a large amount of any major dietary needs, that is, calories, protein, carbohydrate, and fats. They do provide additional vitamins, minerals, and in some cases significant amounts of fiber, which may be lacking or in low concentration in other foods.

In many cases vegetables are grown in gardens and worked by hand using shovels, rakes, hoes, and the like or using small powered equipment such as rototillers. Manual work also includes such things as spraying to control insects and harvesting. Because gardens are near the house and produce is used in daily meals, produce typically goes directly from the garden to the table during the growing season. In Ohio excess garden production may be canned or frozen for use during winter months. In the Philippines and Ecuador gardens can be planted year around, although in Ecuador, and some areas of the Philippines, irrigation will be necessary to have production during dry periods.

Figure 4.2. Salad containing different types of vegetables, that is, lettuce, tomato, pepper, cucumber, broccoli, onion, and snow peas.

In Figure 4.2 vegetables used as part of the meal are shown combined in a salad. In addition to salads, vegetables are prepared using local or family recipes and eaten as part of the meal. On the whole the Filipino farmer eats less vegetables than does the Ecuadorian or U.S. farmer. This, however, is true only of the people in the

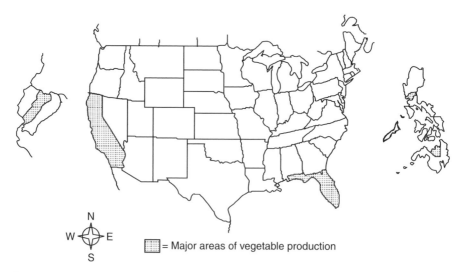

Figure 4.3. Major vegetable-producing regions of the Philippines Ecuador, and the United States. The countries are drawn to approximately the same scale.

TABLE 4.2. World Production of Selected Vegetables for 2005 (ha)

Vegetable	Philippines	Ecuador	United States
Asparagus	1,600	402	21,850
Beans, green	3,000	17,533	15,870
Cabbages	7,500	1,593	87,000
Carrot	0	4,185	40,010
Cantaloupe and other melons	1,300	1,426	45,500
Cucumber and gherkins	1,500	339	69,150
Eggplant	21,000	0	2,150
Lettuce	0	1,584	131,320
Onion and shallots	0	11,995	55,200
Pea, green	9,000	8,290	85,590
Pumpkin, squash, gourds	8,500	1,717	40,190
Tomato	17,000	7,595	166,670
Watermelon	5,500	3,010	55,200

Source: Abstracted from http://faostat.fao.org/faostat.

middle or Visayas region of the Philippines, other areas, particularly in the north, eat more vegetables and indeed have a local vegetable dish called *pinakbit*, which is eaten everyday.

Figure 4.3 shows the major vegetable-producing areas of the Philippines, Ecuador, and the United States. These are the major commercial areas of vegetable production and all production is mechanized. Most operations are carried out using tractors, tractor-mounted sprayers and harvesters, and the like. Significant vegetable production occurs in other areas of all the countries. In the United States vegetables are grown on farms in smaller quantities, and this production is also available locally and regionally. Table 4.2 shows the area, in hectares, devoted to selected vegetable production in each country (vegetables not grown in at least two of the countries are not shown).

4.3 IMPORTANCE OF VEGETABLES IN WORLD TRADE

Vegetables represent a small portion of the world's total food production. Less than 2 percent of food produced is vegetables. Partly this reflects the fact the vegetables are often grown in small areas and gardens for family consumption and often sold in small quantities locally. Because of this, these vegetables do not get counted in compiling world vegetable production. However, vegetables are important in diets around the world as part of peoples food intake and in terms of nutrition they contribute to the diet.

In world terms only 30 countries are major producers of vegetables, and these represent only 16 percent of the world's countries, not including territories, colonies, and

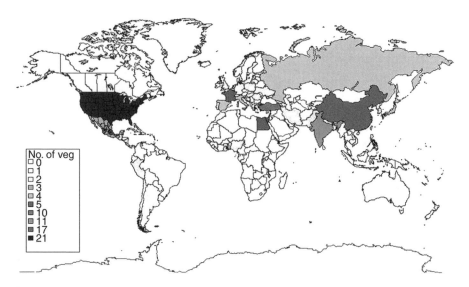

Figure 4.4. Major vegetable-producing countries of the world (the number of different types of vegetable crops produced is shown in the key at the bottom of the map (associated numbers are the numbers of vegetable crops produced). (Data used to construct this map was taken from FAOSTAT http://faostat.fao.org/default.aspx.)

dependencies.[1] Keep in mind that small countries may provide significant amounts of vegetables to local areas but do not produce enough to be counted on a world basis and that small amounts of specialty vegetables are often grown in local areas. The distribution of the major vegetable of the Philippine, Ecuador, and the United States is shown in Figure 4.3. Figure 4.4 shows the major vegetable-producing areas of the world today. The number next to the shaded block is the number of different vegetables produced.

4.4 HISTORY

Because vegetables are general highly perishable and they have been adapted and bred into plants that look nothing like their progenitors, identifying their origins is quite difficult and open to controversy. In Figure 4.5 the areas of the world where various vegetables are thought to have originated is indicated. However, as can be seen several vegetables are thought to have originated in several locations. These areas are particularly rich in historical or archeological records, and this may lead to the assumption that they are the general origins rather than being the true, specific, origin points of the particular vegetables mentioned.

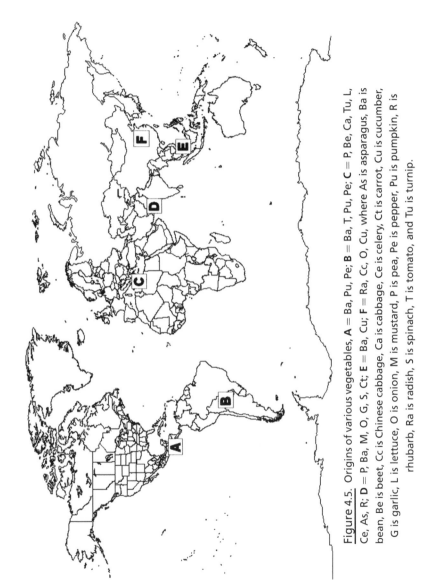

Figure 4.5. Origins of various vegetables, **A** = Ba, Pu, Pe; **B** = Ba, T, Pu, Pe; **C** = P, Be, Ca, Tu, L, Ce, As, R; **D** = P, Ba, M, O, G, S, Ct; **E** = Ba, Cu; **F** = Ra, Cc, O, Cu, where As is asparagus, Ba is bean, Be is beet, Cc is Chinese cabbage, Ca is cabbage, Ce is celery, Ct is carrot, Cu is cucumber, G is garlic, L is lettuce, O is onion, M is mustard, P is pea, Pe is pepper, Pu is pumpkin, R is rhubarb, Ra is radish, S is spinach, T is tomato, and Tu is turnip.

4.5 CULTIVATION

Cultivation of vegetables varies depending on the type of plant that produces the vegetable, the part of the plant used as a vegetable, environmental conditions, and soils. Lettuce, immature grains from maize, immature pods and grain from peas and beans, the stems of some plants, and flower heads, such as broccoli, can be grown on most well-drained soils that have sufficient moisture from rain or irrigation. However, the below-ground portion of other plants, such as carrots and beets, generally do better in light, sandy, or high organic matter soils as long as they have sufficient water.

The type of vegetable grown in an area will depend largely on the local climate, which will depend on both altitude and temperature, particularly if irrigation is used, and soil type. As noted in Tables 4.3 to 4.8, different vegetable crops are suited to different climatic zones. Keep in mind that plant breeders are constantly breeding crops that are adapted to climates that they are not currently adapted to, and so new developments in crop breeds may mean that crops are found outside the range given in these tables.

Temperature is particularly important in vegetable production. Some vegetables such as pepper and eggplant do particularly well in hot climates, while others such as lettuce, cabbage, and pea do best in cooler climates. This is offset by altitude because high-altitude regions of tropical countries will have climates suitable for cold-climate crops. In some cases warm to hot days with cool nights may be best for some vegetables.

TABLE 4.3. Suitable Climate, Preparation, and Scientific and Common Names of Selected Leafy Vegetables

Vegetable	Suitable Climate/Köppen Zone	Preparation	Name
Chinese cabbage	Tropical to temperate/A to D	Boiled	*Brassica*
Chard	Temperate/C and D	Fresh and cooked	*Beta*
Cabbage	Temperate to subartic/C to E	Fresh and cooked	*Brassica*
Collard	Tropical to temperate/A to E	Cooked	*Brassica*
Endive	Tropical to temperate/A to D	Fresh and cooked	*Escarole*
Kale	Subtropical to temperate/B to D	Cooked	*Brassica*
Lettuce	Temperate to subartic/D to E	Fresh and cooked	*Latuca*
Mustard	Subtropical to temperate/C and D	Cooked	*Brassica*
Spinach	Subtropical to temperate/C and D	Fresh and cooked	*Spinacia*

TABLE 4.4. Suitable Climate, Preparation for Eating, and Scientific and Common Names of Selected Immature Seeds

Vegetable	Suitable Climate/Köppen Zones	Preparation	Name
Pea	Temperate to subartic/C to E	Fresh or cooked	Pisum
Pea, edible podded	Temperate to subartic/C to E	Fresh or cooked	Pisum
Beans, bush	Tropical to temperate/A to D	Cooked	*Phaseolus*
Bean wing	Tropical to temperate/A to D	Cooked	*Phaseolus*
Corn	Tropical to temperate/A to D	Cooked	*Zea*

TABLE 4.5. Suitable Climate, Preparation for Eating, and Scientific Names of Melons, Squash, and Cucumber

Vegetable/Fruit	Suitable Climate/Köppen Zones	Preparation	Name
Cucumber	Tropical to temperate/A to D	Fresh	*Cucumis*
Cantaloupe	Subtropical to temperate/B to D	Fresh	*Cucumis*
Pumpkin	Subtropical to temperate/B to D	Cooked	*Curcurbita*
Squash, summer	Tropical to temperate/A to D	Cooked	*Cucurbita*
Squash, winter	Subtropical to temperate/B to D	Cooked	*Cururbita*
Watermelon	Subtropical to temperate/A to D	Fresh	*Citrullus*

TABLE 4.6. Suitable Climate, Preparation, and Scientific Name of Petiole and Stem Vegetables

Vegetable	Suitable Climate/Köppen Zones	Preparation	Name
Asparagus	Temperate	Cooked	*Asparagus*
Bamboo shoots	Tropical to subtropical/A and B	Cooked	*Phyllostachys*
Celery	Subtropical to temperate/B to D	Fresh and cooked	*Apium*
Rhubarb	Temperate[a]/C and D	Cooked	*Rheum*

[a]Must be cooked and drained to remove oxalic acid.

TABLE 4.7. Suitable Climate, Preparation, and Scientific Name of Selected Root Vegetables

Vegetable	Suitable Climate/Köppen Zone	Preparation	Name
Beet	Temperate/B to D	Cooked	*Beta*
Carrot	Tropical to subartic/A to E	Fresh and cooked	*Daucus*
Onion	Tropical to subartic/A to E	Fresh and cooked	*Allium*
Radish	Tropical to subartic/A to E	Fresh	*Raphanus*
Turnip	Temperate to subartic/B to E	Fresh and cooked	*Brassica*

TABLE 4.8. Suitable Climate, Preparation for Eating, and Scientific Names of Other Selected Vegetables

Vegetable	Suitable Climate/Köppen Zones	Preparation	Name
Artichoke	Temperate/B to D	Cooked	*Cynara*
Broccoli	Temperate/B to D	Fresh and cooked	*Brassica*
Brussels sprouts	Temperate/B to D	Cooked	*Brassica*
Cauliflower	Temperate/C and D	Cooked	*Brassica*
Eggplant	Temperate to tropical/A to D	Cooked	*Solanum*
Peppers	Tropical to temperate/A to D	Fresh and cooked	*Capsicum*
Tomato	Tropical[a] to temperate/A to D	Fresh and cooked	*Lycopersicon*

[a]Although grown in the humid tropics, tomatoes do not grow well under these conditions. High temperatures will prevent flowering and fruit development.

In regions where rainfall is insufficient for vegetable production, irrigation is essential. In some locations where rainfall is sufficient but does not come at times essential for vegetable production, irrigation is practiced. Vegetables are generally considered high value crops and so irrigation may be practiced even when it is not deemed necessary for other crops. Also more expensive irrigation methods, such as drip irrigation, may be used both to promote better vegetable production and to conserve water.

Although specific soil types are better suited to production of a particular vegetable, it is not true that soil type absolutely determines what can be grown. Typically, all types of vegetables grow well in most soils, as long as they have adequate water, nutrients, and temperature and are not compacted. Clayey soils hold nutrients well but are hard to work and are easily compacted and so are not favored by farmers, although they may be highly productive. Loamy soils, that is, sandy loams, clay loams, and silt loams (Chapter 9), are preferred agricultural soils throughout the world. Sandy and organic soils are well suited for all types of root vegetables because they allow easy development of the root portion of the vegetable. However, they do not hold water or nutrients well and so require addition of both in lager quantities than other soils.

Organic soils are well suited and very productive when used for vegetable production. However, because of their unique characteristics, such as nutrient and water-holding capacities, must be managed differently from mineral soils.

People all over the world cultivate vegetables in small family gardens or in areas smaller than those used for field crops. Cultivation of these gardens is extremely varied. The soil may be worked with Maddox, shovel, rototiller, or other tool. The ground may be left flat, hilled, or worked to produce ridges on which seed is planted; see Figure 4.1. Another popular method is called raised beds where boards, 5 cm by 15 cm, are placed on edge to form a "box" 90 cm wide and 300 cm long is made and filled with soil, as shown in Figure 4.6. Organic matter, mulch, and manure can be added and worked into

Figure 4.6. Raised beds, "boxes," surrounded by planks. Center bed is planted to broccoli. Plastic hoops are used to hold plastic covers in cold weather.

Figure 4.7. Tray for starting vegetables and on the left a plug that can be planted.

the soil before planting. Planting is then done inside the box. This method leaves a nice walkway between beds and beds can be covered if need be.

After initial working of the soil, it is often raked and seed planted. However, in many places, including tropical climates, it is common to start vegetable plants, particularly tomatoes and peppers, in separate containers and transplant them to the garden when the climate allows. Plants may be removed from soil for transplanting or the plant plus the soil it is growing in is removed from the starting chamber and the whole thing, called a plug (Fig. 4.7), is planted resulting in minimal disturbance to the plant.

Transplanting is a particularly common practice where the farmer wants to have plants produce more quickly and so starts vegetable plants in the house or greenhouse or purchases them from a commercial supplier. In this way plants can be started when it is too cold outside for them to survive, and they can be transplanted into the garden when danger of cold weather is past. The plants start producing faster and have a longer time to produce during the growing season.

Starting plants in containers in a screen house or otherwise away from fields is also practiced in subtropical and tropical climates. The advantages for this approach are varied. It may allow farmers to start plants before rains come, to start plants before fields are ready to plant, or to protect young plants from insects or other pests.

Produce is harvested by hand as it becomes ripe and used immediately or, if there is excess, stored, canned, or frozen for later use.

There are large farms specializing in the production of one or several vegetable crops for the commercial market; see Figure 4.8. In all cases the soil is worked, plowed, and cultivated to produce a fine seedbed. Vegetable seeds may be directly

Figure 4.8. Commercial field of broccoli growing in Ecuador.

planted in the field, using a planter, or started in seedbeds in green or screen houses and then transplanted to the field as young seedlings or plugs as described above. There are several different kinds of planters that can be used for planting seedlings. Irrigation of vegetable crops is common even in areas where total rainfall is adequate but may be lacking at critical points during the growing season.

   Mulching (Fig. 4.9) using any of a variety of common mulching agents, that is, black plastic, paper, crop straw, or other residue is common in vegetable production. Mulching is a general term that means covering the soil between plants in the row and between rows. Mulching accomplishes two functions, weed and soil moisture

Figure 4.9. Black plastic being used as mulch for strawberries in raised beds.

control. Mulch can prevent weed seed germination or, if germination occurs, plants cannot survive because the mulch blocks sunlight. Mulch also slows or prevents evaporation of water from the soil surface, thus saving it for plant use. Irrigation with mulching is an effective way to maximize the use of irrigation water. Mulching will also change the soil temperature depending on the type of mulch used, and this can be used to advantage to increase vegetable production.

If mulching is not done, weed control will be essential during the growing season either by mechanically removing weeds or the use of herbicides. Weeds can be removed mechanically by either or both mechanical or hand weeding. There are a limited number of herbicides that can be used on vegetable crops, and those that can be are used with caution.

Harvesting is generally done by hand, although there has been much development work designed to produce both machines and crops that are suitable to mechanical harvesting. For smaller areas hand harvesting or picking is done; however, on multihectare areas mechanical harvesting will be used if possible. Once harvested, the edible portion is washed and, frequently, stored in cool conditions until transported to market, canned, or frozen. Under the proper temperature and moisture conditions, which will vary from crop to crop, fresh vegetables can be stored for a significant period of time without spoilage.

Vegetables are generally higher value crops, and so additional expenses incurred in field preparation mulching and irrigation and harvesting are generally justified.

4.6 PROTECTION

Vegetables often require protection from insects. Chemicals, either natural or organic, such as various bacillus preparations and pyrethrins* and synthetic chemicals such as malathion and the various carbamates are called insecticides and are used to control insects. These are sprayed or dusted on vegetable plants, as per label instructions,[†] as evidence of insect infestation is noticed. Care must be exercised to cease insecticide application prior to harvesting. The recommended time between the last application and harvest is usually indicated on the insecticide container. In addition vegetables must be washed thoroughly before being cooked and eaten.

In the case of starting plants in greenhouses, it is important to make sure that insect and disease-free media is used to start plants because plant roots are susceptible to attach by various soil-borne insect and disease organisms.

Growing a different crop in a field each growing season is another way to control insects and disease organisms. This is called crop rotation and has been an effective strategy for improved crop production.

---

*Organic insecticides, although natural, are not necessarily less toxic than synthetic chemical insecticides!

[†]Insecticides are generally limited in the number of vegetables on which they are allowed to be used. Indiscriminate use is dangerous.

Figure 4.10. Shard growing in Aíde and Zoro's garden in Ecuador. (Courtesy of David Cerón.)

4.7 LEAFY VEGETABLES

Leafy vegetables are those vegetables where it is the leaf that is eaten. They are a common food around the world and are eaten fresh or cooked, often boiled, as part of meals. Of the common leafy vegetables listed, more than 50 percent are from the genus *Brassica* as can be seen in Table 4.3. Note that chard (Fig. 4.10) is in the genus *Beta* and so is related to beets, the tops of which are also eaten as leafy vegetable.

4.7.1 Climatic Adaptation

Leafy vegetables are grown and adapted to a wide range of climatic conditions, as seen in Table 4.3. They are all suited to temperate climatic zones, although some do particularly well in subartic zones, for example, cabbage, and others in tropical zones, for example, Chinese cabbage; see Figure 4.11. Leafy vegetables are grown in rows and sometimes, particularly in home gardens, in raised beds. They need fertile soil and respond well to both organic matter additions to soil and to chemical fertilization. Many of the leafy vegetables do better in cooler climates and so are not commonly grown in tropical climates.

4.7.2 Importance in Human Nutrition

As seen in Figure 4.12 leafy vegetables tend to be moderate in protein and low in fat and so are not a particularly good source of these nutrients. They are highest in carbohydrate relative to the other nutrients and are high in fiber relative to protein and fat and so can be an important source of this dietary component. Kale, however, has moderate levels of carbohydrate when compared to the other leafy vegetables and is thus less valuable as a source of this nutrient.

Figure 4.11. Type of Chinese cabbage.

4.7.3 Harvesting, Preparation, and Consumption

Leafy vegetables can be picked and eaten at almost any stage of development. However, when old, leafy vegetables can become bitter or fibrous. After picking leaves and washing they are then ready to eat. A mixture of leafy vegetables, different kinds of lettuce (Fig. 4.13), spinach (Fig. 4.14), and the like along with tomatoes, peppers, and other vegetables are often eaten, in combination as salads (Fig. 4.2),

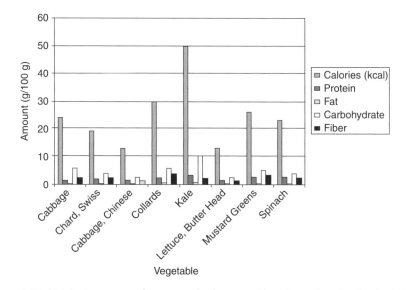

Figure 4.12. Nutrient content of common leafy vegetables. From data in the Agricultural Research Service Nutrient Data Laboratory USDA National Nutrient Database for Standard Reference, Release 17, http://www.nal.usda.gov/fnic/foodcomp/Data/SR17/reports/sr17page.htm.

Figure 4.13. Lettuce in a garden. (Courtesy of Amy Halvorsen.)

with meals, or as meals in themselves. Leafy vegetables may also be incorporated into sandwiches and other dishes.

4.7.4   World Production of Leafy Vegetables

The world's major producers of leafy vegetables are China and the United States with China producing 2 to 30 times more than the United States. Other significant producers are Spain, Italy, India, Japan, Turkey, and France. These countries produce 3 to 15 times more leafy vegetables than the rest of the world combined, representing from

Figure 4.14. Spinach a common leafy vegetable. (Courtesy of Amy Halvorsen.)

82 to 94 percent of the world production. These estimations are made on the basis the production of lettuce and spinach.[§,2]

4.8 IMMATURE SEEDS AS VEGETABLES

Peas, beans (Fig. 4.15), and maize are commonly grown to maturity and the seed used as food. Just as commonly, these crops are not allowed to mature, and the immature seed and sometimes the seeds and associated pod, as with snow peas and beans, is eaten. Peas and beans are eaten both cooked and fresh; however, maize is almost always cooked by either boiling or roasting before eating.

4.8.1 Climatic Adaptation

Immature seeds used as vegetables are grown in a variety of climates; see Table 4.4. Peas require cool climates, either as cool days and nights or warm days with cool nights. In contrast, beans and maize require warm climates and grow even in tropical climates. They will also grow in cooler climates as long as the growing season is long enough to allow for grain development. Note here that the grain is eaten in the immature stage, and so the growing season does not have to be long enough to produce mature seeds.

4.8.2 Importance in Human Nutrition

Figure 4.16 shows the nutrient content of beans, peas, and maize. Note that the scale in this graph is different from the other vegetables because both peas and maize have a

Figure 4.15. Green beans, left side, and peas, right side, at stage to be picked as vegetable.

[§]Calculated from FAOSTAT (http://faostat.fao.org) data for 2004 and 2005 on the basis of hectares.

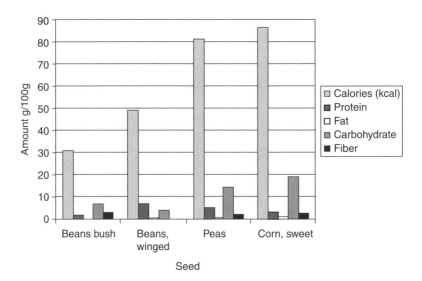

Figure 4.16. Nutrient content of immature seeds used as vegetables. (From data in USDA National Nutrient Database for Standard Reference, Release 17. http://www.nal.usda.gov/fnic/foodcomp/Data/SR17/reports/sr17page.htm.)

much higher calorie content than most other vegetables. Also the carbohydrate content is higher, and the protein content is somewhat higher than other vegetables. The fat content runs between a low of 0.12 percent for bush beans and a high of 1.18 percent for maize.

4.8.3   Soil Preparation and Fertilization

Soil is prepared as for any other crop and seed planted directly in the soil. Seedlings are never used in pea, bean, and maize production. All three crops respond well to addition of organic matter and chemical fertilizer to soil. However, both peas and beans are legumes and so can fix their own nitrogen and thus do not and should not be fertilized with nitrogen-containing fertilizers. However, they respond well to both phosphorus and potassium fertilization, depending on the fertility level of the soil. In addition, legumes are generally sensitive to soil pH, and thus the pH of soil should be near 7 for optimum production.

Maize is very different in its soil fertility requirements. It responds well to both organic and chemical fertilizer additions to soil. Maize cannot fix its own nitrogen, and so nitrogen must be one of the added fertilizer elements. Nitrogen is included in the fertilizer applications and is sometimes added after the maize is growing as what is called a side dress application. Maize is not very sensitive to acid soil pHs; however, it is always best to keep soil pH near 7 for optimum crop growth and maximum cropping flexibility.

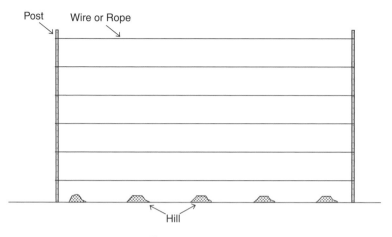

Figure 4.17. Trellis.

4.8.4 Production Practices

Many pea and some bean varieties grow as vines and so are often supported on trellises of various designs; see Figure 4.17. There are also varieties that grow as bushes and thus do not need trellises. Both are planted in rows in both family gardens and in commercial fields. Peas can be thickly sown in an area such that the developing plants support each other such that no trellises are needed.

4.8.5 Harvesting, Preparation, and Consumption

Beans and peas are harvested by picking the immature pea pods or beans before they begin to mature. As with leafy vegetables peas and beans can be harvested at any stage in development. Thus very young peas and beans may be picked and eaten fresh or cooked. However, it is common to allow the beans to develop to the point that beans (seed) are visible under the outer covering.

Maize is very different in that the ears need to fully develop kernels but need to be picked before the sugar begins to be converted to starch in the maturing process. This would apply to all varieties of maize, and all varieties are eaten in the immature stage as sweet corn. Supersweet varieties are different in that not all the sugar changes to starch and thus are sweet even when mature. However, some varieties, in addition to the supersweet, of maize have been specifically bred to be sweet and retain their sweetness for an extended period of time during development, thus making them superior as sweet corn and for use as a vegetable.

The stage of development at which maize is harvested depends on local customs and taste. In some places maize is picked when the kernels are small and just filled out. In other places the maize is left until the kernels are fully developed but before they start to become very hard.

Maize is commonly prepared in three ways (except for popcorn as describe below). The kernels may be cut off the cob and eaten as a vegetable in this way. Kernels are removed from cobs for both canning and for freezing, although whole ears can be frozen. Cobs and kernels may be boiled with or without shucks or the cob plus kernels may be roasted on a grill, frequently a charcoal grill, although open wood fires are also commonly used. In many countries roasted maize is eaten as a snack as much as a part of main meal. Popcorn is allowed to mature and is picked and the kernels removed from the cob. Heating the kernels in either air or with oil causes the kernels to explode into a fluffy "ball," after which it is eaten as a snack frequently with the addition of butter and salt.

4.8.6   World Production of Immature Seed Vegetables

The major producers of immature seeds used as vegetables are India, China, United States, France, United Kingdom, Nigeria, Hungary, Peru, Mexico, Morocco, and Turkey. The top producer of bush beans and sweet corn is the United States while the top producer of green peas is India. These 10 countries produce 2 to 18 percent more immature seeds used as vegetables than the rest of the world. This represents from 68 to 94 percent of the world's production of these vegetables. This estimate is based on bush bean, maize, and green pea production.[¶,2]

4.9   MELONS, SQUASH, AND CUCUMBER

4.9.1   Climatic Adaptation

Melons, squash, and cucumber (Fig. 4.18) grow on vines that lie on the ground or are supported on trellises. They grow in a variety of climates and, although cucumber and squash are adapted to most climates, cantaloupe, pumpkin, and watermelon are more commonly grown in subtropical and temperate climates as seen in Table 4.5.

Figure 4.18. Examples of, from left to right, melon, cucumber, and two types of squash.

[¶]Calculated from FAOSTAT (http://faostat.fao.org) data for 2004 and 2005 on the basis of hectares.

All are generally oblong or round in shape and the outside, the rind, may be yellow as with pumpkin or green as with zucchini. The interior may be solid with seeds imbedded in the flesh or hollow with seeds on the inside. Seeds can be eaten in the fresh young vegetable, such as with cucumber and zucchini, or separated, cooked, and eaten as with pumpkin seeds.

4.9.2 Importance in Human Nutrition

These vegetables are higher in calories and carbohydrates than other nutrients, as shown in Figure 4.19. Protein content is second highest, although still low, followed by fiber. Fat content is very low ranging from 0.1 to 0.18 percent.

4.9.3 Soil Preparation and Cultivation

For melon and cucumber production soil is prepared as for other vegetables. Organic matter and fertilizer is be added to soil and soil pH adjusted as needed. Depending on the planting method, soil may be left flat, worked into ridges, raised beds, or hills. Regardless of the method of preparation, mulch is commonly used to control weeds, water, and where vines are on the ground, to protect the developing melons or cucumbers.

Cultivation of cucumbers and melons is different from other crops because these crops typically grow on vines that spread over the ground. Most vegetable crops grow on upright plants, and soil can be planted in rows and cultivated between rows

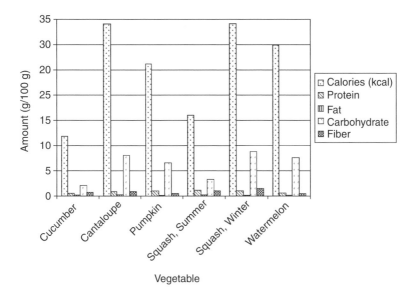

Figure 4.19. Nutrient values of melons, squash, and cucumber. (From data in USDA National Nutrient Database for Standard Reference, Release 17. http://www.nal.usda.gov/fnic/foodcomp/Data/SR17/reports/sr17page.htm.)

to remove weeds. This also makes spraying for insects easer since both the tops and lower areas of the plant can be sprayed. In the case of most melons the vines grow on the ground and spread out over the whole area planted. This makes weeding and spraying hard, thus, mulch, either plastic or straw, placed on the area between plants alleviates the need to weed.

Another way to avoid covering wide areas with vines, plants such as cucumber can be grown on trellises (Fig. 4.17). Here the plants are trained to grow vertically on a series of wires or ropes placed between posts. In this way the area between rows is clear and can be cultivated to remove weeds and spraying is made easier.

Often melons and cucumbers are planted in hills rather than as individual plants in rows. In this case soil is hilled up and several seeds planted in the hill. The multiple plants then grow out from the hill or up on the trellis as the case may be. The areas between hills may be covered by mulch to control weeds and moisture.

4.9.4 Harvesting, Preparation, and Consumption

Harvesting is done by hand picking the vegetables as they become ripe or in the case of cucumbers when they reach the desired size. Once harvested cucumbers are washed and can either be pealed and the inner flesh eaten or they can be eaten with the peal on. Melons are also washed and the interior flesh eaten. Often melons are cooled before eating. Squash are washed and all parts eaten. Cucumber, squashes, and pumpkin are used as a main dish vegetable and are combined with other vegetables when cooking. Cantaloupe and watermelon are eaten as fruits, in salads, and as snacks.

4.9.5 World Production of Melons and Cucumber

The major melon- and cucumber-producing regions of the world are China and Iran, with China being by far the largest, producing over 50 percent of the world's melons and cucumbers. Other important producers are Turkey, the United States, the Russian Federation, Egypt, Ukraine, and Spain, which produce between 1 and 4 percent of the world's production. These latter countries produce large quantities of only one or two of these crops, while China produces large quantities of all.[‖,2]

4.10 PETIOLE[**] AND STEM VEGETABLES

4.10.1 Climatic Adaptation

Stem vegetables are those where it is primarily the stem of the plant that is eaten rather than some other part. While celery is eaten fresh, the rest are cooked before eating. Vegetables such as asparagus, celery, and rhubarb are all adapted to temperate climates

---

[‖]Calculated from FAOSTAT (http://faostat.fao.org) data for 2004 and 2005 on the basis of hectares of cucumber, gherkin, watermelon, and pumpkin, squash and gourds.

[**]The petiole is the stalklike portion of the leaf that connects it to the main stem of the plant.

while bamboo is adapted primarily to tropical climates, although it can be found in sub-tropical climates, as shown in Table 4.6.

4.10.2 Importance in Human Nutrition

As seen in Figure 4.20, bamboo shoots contain the most calories and carbohydrates, while asparagus and rhubarb have about the same calorie content with asparagus a little less carbohydrate than rhubarb. All have significant fiber contents and insignificant fat content, which is between 0.1 and 0.3 percent, being highest for bamboo and lowest for asparagus.

4.10.3 Soil Preparation

Soil preparation for petiole and stem vegetables is variable because of the different growth characteristics of these plants. Asparagus, rhubarb, and bamboo grow from previously established roots and so are not planted every year, but the edible part is harvested from established, long-term, root systems. Celery on the other hand is grown from seed much as other crops in this chapter. As with other crops application of organic matter, fertilizer, and maintaining soil pH are important in obtaining maximum production.

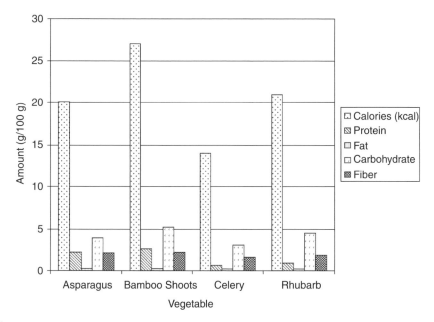

Figure 4.20. Nutrient content of petiole and stem vegetables. (From data in USDA National Nutrient Database for Standard Reference, Release 17. http://www.nal.usda.gov/fnic/food-comp/Data/SR17/reports/sr17page.htm.)

Figure 4.21. Picking young asparagus.

4.10.4   Harvesting, Preparation, and Consumption

Young sprouts of asparagus (Fig. 4.21) and bamboo are harvested for eating; however, neither is edible when the plant becomes older. On the other hand the mature stems of both rhubarb and celery are harvested and used as food. While mature stems of asparagus and bamboo are never eaten, immature stems of rhubarb and celery are never eaten. Rubarb stems are usually red in color (Fig. 4.22) before they are picked and used as food.

Both asparagus and celery are eaten in their original form, although asparagus is always cooked before eating while celery is most often eaten raw. Asparagus can be bleached white before cooking and eating. Banboo shoots are commonly sliced into

Figure 4.22. Rhubarb plant with stem ready for cooking shown in inset.

thin pieces and added to various dishes. Rhubarb is frequently cooked into a souplike mixture that is sweetened to taste before eating.

4.10.5 World Production

The major world producer (on the basis of asparagus hectares) is China, producing about 12 percent of the world's production. The other top producers—Peru, the United States, Germany, and Mexico—each produces around 1 percent of the world's production.[‡,2] Based on asparagus China is by far the world's largest producer of stem crops, producing nearly eight times more than any of the other top producers.

There are other petiole and stem vegetables, however, only small hectares are planted even in large countries such as the United States, Canada, and Asia. It is very common to find these grown in gardens and on small hectarages for home and local consumption. In spite of this, these vegetables are commonly found in the markets of many countries and are used extensively as food, snacks, and as additions to many dishes. Thus world use of these vegetables is not reflected in the large commercial-scale production records.

4.11 ROOT VEGETABLES

4.11.1 Climatic Adaptation

Here we define root vegetables as those crops where the part of the plant that is eaten grows below ground,[§§] for example, radish and beet; see Figures 4.23 and 4.24. These vegetables are grown in a wide range of climates, that is, Köppen zones A to E in most cases, as seen in Table 4.7, and are thus common food sources throughout the world.

4.11.2 Importance in Human Nutrition

As can be seen in Figure 4.25 all the root vegetables have small amounts of protein and very small amounts of fat. Although the figure does not show it, all have some fat varying between 0.008 and 0.28 g/100 g of fresh raw root. Calorie and carbohydrate are the most abundant components of these vegetables, and fiber content is higher than either protein or fat content, making these vegetables an excellent source of fiber in the diet.

4.11.3 Cultivation

Because the edible portion grows below ground, these crops grow best in sandy or high organic matter soil. Thus large additions of organic matter to soil can greatly increase

---

[‡]Calculated from FAOSTAT (http://faostat.fao.org) data for 2004 and 2005 on the basis of hectares of asparagus.

[§§]In a strict agronomic or botanical sense these may belong to different groups of plants and the edible portion a different plant part.

Figure 4.23. Red radishes in a garden. (Courtesy of Amy Halvorsen.)

productivity even in sandy soils. These crops also respond to fertilization and generally require significant amounts of potassium for optimum, high-quality, root production.

They are grown in rows on large commercial farms and in raised beds in gardens, from seed that can be planted using seed planters or by hand. These types of vegetables

Figure 4.24. Young beets.

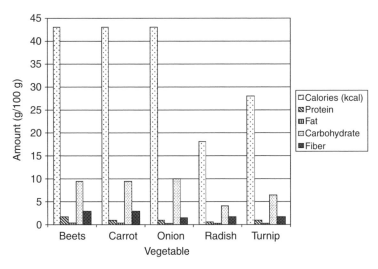

Figure 4.25. Nutrient content of root vegetables. (From data in USDA National Nutrient Database for Standard Reference, Release 17. http://www.nal.usda.gov/fnic/foodcomp/Data/SR17/reports/sr17page.htm.)

are almost never grown from seedlings. Weeding is necessary for optimum production. Plants are allowed to grow until the below-ground portion is the desired size or it is near the end of the growing season. However, many of these crops can and are harvested and eaten at any time during the growing season. Young onions, beets, and carrots are commonly eaten as soon as the below-ground portion is large enough to be used as food. Plants are harvested by pulling, although they may also be dug or plowed up before separation from soil. Equipment similar to that used for harvesting potatoes (Chapter 4) can be used to harvest any root crop, although equipment must be designed for the particular root and its characteristics.

4.11.4 Preparation and Consumption

After harvesting, root vegetables are washed to remove adhering soil, and, if washed properly, they can be stored for significant periods of time without spoiling. Removal of tops at harvest reduces the loss of water and thus extends the time these vegetables can be stored without becoming unsuitable as a food. The outer skin is frequently removed from root crops, and this makes sure that adhering soil is also removed. However, the skin is edible and does not have to be removed before eating. In some cases the skin may come loose during cooking and thus decrease the esthetic value of the food; however, the nutritional value will be higher with the skin included than if it is removed.

These are called vegetables because they are eaten as adjuncts to the main protein and carbohydrate source of the meal. Another way to look at this is that they are eaten in smaller quantities than meat or carbohydrates. Although they are eaten alone as their own separate dish, they are also frequently mixed in other dishes and cooked with them. This would be the case of cooking carrots with meat, for example.

4.11.5   World Production of Root Vegetables

The major producers of root vegetables are China, United States, Russian Federation, Poland, Japan, India, Turkey, and Pakistan. These countries produce from 1.3 to 6 times more root vegetables as do the rest of the countries in the world. China alone produces 41 percent of the world carrot production. The Russian Federation produces nearly 8 percent of the world's carrots, while the other top produces—the United States, Poland, and Ukraine—produce about 3 percent of the world's production each.[¶,2]

4.12   OTHER VEGETABLES

This group of vegetables is highly variable and hard to classify into any one type of growth or edible portion used. Artichoke looks like a head but is made of individual leaves, the fleshy part of which is eaten. Although broccoli, brussel sprouts, and cauliflower are all from the same family, they are all different in characteristics. The edible portion of both broccoli (Fig. 4.26) and cauliflower is a flower head, however, broccoli is green with discernable flower stems while cauliflower is white with no discernable flower structures in the head. On the other hand, brussels sprouts look like tiny cabbage heads growing on the stem of the plant. Most eggplant is dark purple

Figure 4.26. Broccoli plant showing head, edible portion, ready for harvest.

¶Calculated from FAOSTAT (http://faostat.fao.org) data for 2004 and 2005 on the basis of hectares of carrots.

Figure 4.27. Eggplant showing cross section with seeds visible.

(Fig. 4.27), with a white inner flesh with seeds. Peppers have a thick "skin" surrounding a hollow space containing seeds, while tomatoes have flesh throughout with seeds. Although one might think of peppers as being green or red (Fig. 4.28), and tomatoes as being red, both have been bred to have many different colors from green to red, yellow, and more different colors are being developed.

4.12.1 Climatic Adaptation

Artichoke is adapted to cool climates or at least climates with cool nights. Broccoli, brussel sprouts, cauliflower, and tomatoes are adapted to temperate climates, although they are also grown in tropical climates (Table 4.8). Tomatoes do not produce well when the temperature is too high, and when the temperature is high the tomatoes produced are generally small. Eggplant and pepper are adapted to subtropical and tropical climates, although they can also be grown in other climates.

4.12.2 Importance in Human Nutrition

As seen in Figure 4.29, of this group of vegetables, artichoke, broccoli, and brussels sprouts are highest in calories, and artichoke and brussels sprouts are highest in carbohydrates. The remaining vegetables have about the same amounts of calories and carbohydrates. All have low levels of protein and fiber. Fat content varies from 0.1 to 0.4 percent for all these vegetables.

Figure 4.28. Young peppers.

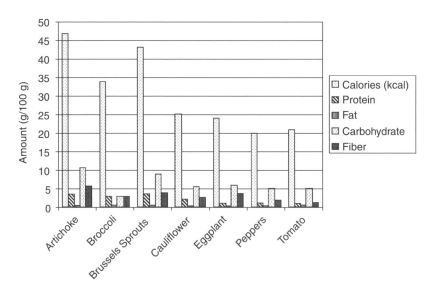

Figure 4.29. Nutrient values of various other vegetables. (From data in USDA National Nutrient Database for Standard Reference, Release 17. http://www.nal.usda.gov/fnic/foodcomp/Data/SR17/reports/sr17page.htm.)

4.12.3 Cultivation

All these crops are grown in rows with standard cultivation methods. Soil is worked to produce a fine seedbed and seeds planted in rows or in raised beds. In this group it is common for seeds to be started in the greenhouse, screen house, or in seedbeds before transplanting the seedlings to the field or garden. This is true for all these plants but is particularly common for both tomatoes and peppers. The most common, having been practiced longer than for other crops, is tomatoes, which are almost never grown from seeds in the field or garden.

These vegetables respond to organic matter and fertilizer applications to soil. Tomatoes are particularly sensitive to fertilization in that they will produce spectacular plants when fertilized with nitrogen and in some cases of excess nitrogen fertilization will produce large plants but no tomatoes. Tomatoes are also sensitive to calcium levels in soil, and thus the pH of soil and its calcium levels are important to control.

Most of these plants grow on stems or stalks that carry the heads or have the eaten portion on stems. The plants are strong enough to support the vegetable and grow upright. The exception is tomato, which is not strong enough to grow upright, and thus it is commonly grown attached to some support, such as tying to stakes (Fig. 4.30) or on a trellis, so as to obtain vertical growth. This provides for easier harvesting and keeps the tomato off the ground and less likely to become damaged. If no support is provided, tomato plants eventually bend over and grow along the ground.

Figure 4.30. Tomatoes held up by stakes.

This growth habit does not decrease tomato production and may be preferred in some large-scale production situations.

4.12.4   Harvesting, Preparation, and Consumption

Broccoli, brussel sprouts, and cauliflower are harvested before biological maturity, usually when the head is maximum size but before actual flower formation. Peppers can be harvested at any stage or can be allowed to mature and turn red or the color of the particular pepper. Tomatoes are usually harvested after turning red or the ripe color in the case of other colored tomatoes, although green tomatoes can be cooked and eaten. Tomato plants, with unripe tomatoes attached, can be pulled up when frost is eminent and hung in a warm place for the remaining tomatoes to ripen.

After harvesting and washing, all these vegetables have appreciable shelf lives particularly if they are kept cool, although all can be kept at room temperature for significant periods of time, up to a week, without undue deteriorate. All these vegetables are eaten fresh out of the garden, boiled, and stir fried. They are commonly incorporated into salads and other cooked and uncooked dishes. In some cases they may also be used as garnishes and to add color to meals or dishes.

4.12.5   World Production of Other Vegetables

Countries producing the majority of the vegetables we have labeled as other are China, United States, Turkey, India, Egypt, Italy, Spain, Argentina, France, and Mexico. These countries produce 1.6 to 10 times as much of these vegetables as the other countries in the world, which represents 62 to 91 percent of the world's production. This is calculated based on world production of artichoke, cauliflower, eggplant, and tomato.

The major producer of this group of vegetables, on the basis of artichoke, cauliflower, and tomato, is China, which produces 30 percent of the world's production of these types of crops.[***,2] India produces 14 percent of the world's production, making it the second largest producer. Other large producers are Italy, Spain, Egypt, United States, and Turkey, which produce between 2 and 5 percent of the world's production.

4.13   TRANSPORT AND STORAGE OF VEGETABLES

Vegetables are very perishable and thus losses between harvest, sale, and consumption are significant. Losses can be minimized by three different methods. Careful handling during harvest, washing in clean water with soap, and providing protection during transport are all simple ways to prevent or retard spoilage. Farmers all over the world can use these methods. Controlling the environment of the picked vegetables, such as cooling, can also prevent degradation and loss. Treating vegetables with chemicals, natural or synthetic, is another way vegetables are protected between harvest and table.

[***]Calculated from FAOSTAT (http://faostat.fao.org) data for 2005 on the basis of hectares of asparagus.

In the first case vegetables must be handled carefully, washed, and transported in packing materials, which protect them from microorganisms and insects as well as physical damage. Care during harvesting prevents or minimizes bruising and cuts, which can lead to entry of microorganisms, which decompose the vegetable. In addition these types of imperfections make the vegetable less appealing to the buyer. Washing to remove soil, insecticides, and the like from vegetables will also remove deleterious organisms, and, if cool water is used, will cool the vegetable thus further extending its shelf life. Packing materials such as straw can protect vegetables from bruising during transport while keeping them cooler and protected from the sun.

It is common to control the environment of vegetables to prolong their shelf life. This can be as simple as keeping then moist or as complex as refrigeration and atmospheric control. When vegetables are harvested, they begin to loose water, and thus the quality begins to decline and make the vegetable less appealing to buyers and to sellers if the sale is on a fresh weight basis. On a fresh weight percentage basis the nutrient content increases. Applied moisture in hot, dry, and cold climates or while refrigerating can significantly prolong the shelf life of vegetables. However, in hot humid climates moisture, especially without refrigeration, can lead to increased degradation of vegetables. Refrigeration without freezing will increase shelf life of vegetables significantly; not only does it retard the loss of water it also retards the growth of microorganisms and inhibits insects.

The atmosphere surrounding the vegetables can be controlled by changing the ratios of oxygen, nitrogen, and carbon dioxide in it. This requires two things; first the vegetable must be in a container that will maintain the gas composition. Second, the gas composition must be tailored to the specific vegetable and the conditions under which it is harvested. This type of packaging and storage is used in developed countries but is not generally available in developing countries. Note that while the vegetables packaged this way are generally ready to eat they will cost significantly more.

Losses of vegetables can also be controlled by the use of various chemical means. Chemicals can be used to either increase or decrease the rate of ripening or color development. They can also be used to kill microorganisms that degrade the vegetables. One way this is done is by fumigation with various chemicals in a closed container.

4.14 CONCLUSIONS

Vegetables are considered an essential part of any diet, but they generally do not supply large amounts of nutrition. They are distributed widely throughout the world, although some do have more restricted environmental needs. Cultivation of vegetables is much the same as with other crop seeds planted in a prepared seedbed and weeded as necessary. Unlike other crops, vegetables can be harvested at varying times and even before maturity. Some vegetables have growth habits along the ground, or they can be supported by trellises or other means. Although there is significant national and international trade in vegetables, there are also important local and family production and use.

QUESTIONS

1. Which vegetables from the groups would you try to grow in Greenland? Why did you choose these vegetables?

2. Which of the vegetables in the groups discussed would you try to grow in Borneo? Why did you choose these vegetables?

3. Make a list of one vegetable from each group with the most calories.

4. Make a list of the one vegetable from each of the groups with the most carbohydrates.

5. Is there any discernable pattern between the amount of calories and the amount of carbohydrates a vegetable has?

6. Which vegetable in each of the groups has the most and which the least protein?

7. Is it possible that a vegetable not listed as being important in world agricultural trade would be widely grown and eaten? Explain how this might happen.

8. If you had the chance to introduce one vegetable to people not eating any of the vegetables in the groups, which would it be? Explain.

9. From a production standpoint why might trellises be advantageous?

10. Considering total food value, which one vegetable from each of the groups is best? Justify your answer.

REFERENCES

1. G. Lucier and A. Jerardo, *Vegetable and Melons Outlook*, Economic Research Service, U.S. Department of Agriculture, VC-S-312, Dec. 16, 2005.

2. Food and Agriculture Organization of the United Nations, FAOSTAT. Available at: http://faostat.fao.org/.

BIBLIOGRAPHY

A. R. S. Ballantyne, R. Stark and J. D. Selman, Modified Atmosphere Packaging of Shredded Lettuce, *Int. J. Food Sci. Tech.*, **23**, 267–274, 1988.

M. Fin, Safety of Modified Atmosphere Packaged Vegetables, *Hygiene Review*, 1997. Available at: http://www.sofht.co.uk/isfht/irish_97_atmosphere.htm.

R. J. Holmer, *Sustainable Vegetable Production for Small Farmers on Problem Soils in the Highlands of Bukidnon (Philippines) for Fresh Market and Processing*, Schriftenrelhe Agrarwissenschaftliche Forschungsergebnisse, Hamburg, Germany, 1998.

S. W. Huang and F. Gale, China's Rising Profile in the Global Market for Fruits and Vegetables. Available at: http://www.ers.usda.gov/AmberWaves/April06/DataFeature/.

Postharvest Losses of Fruit and Vegetables in Asia. Available at: http://www.fftc.agnet.org/library/ac/1993d.

J. M. Stephens, Soil Preparation and Liming. Available at: http://edis.ifas.ufl.edu/VH024.

M. Yamaguchi, *World Vegetables Principles, Production and Nutritive Values*, AVI Publishing, Westport, CT, 1983.

5

ROOT CROPS

5.1 ROOT CROPS ON THE THREE FARMS

On the three farms Donio grows two kinds of root crops but only on land around the edge of his farms, Aída and Octavio grow potato as their major crop, and Steve does not grow any root crop. Donio grows a kind of sweet potato, which is also called boniato and locally called *camote*. He also grows *gabe* (taro). Both are grown on land surrounding his house and along the edges of his fields; see Figure 5.1. In addition to harvesting the tubers, the leaves of *camote* are harvested, boiled, and eaten as a vegetable component of the main part of a meal. He uses cuttings from already established plants to produce new plants, but otherwise he does nothing to cultivate this crop. Root crops being sold in the market in Baybay, Philippines, near where Donio farms is shown in Figure 5.2.

On the Ecuador farm Aíde and Octavio grow two varieties of potato, Cecilia and Catalina (Fig. 5.3) as their major crop, on a 1-ha portion of their main field close to their house and produce two crops a year. The area planted to potatoes is rotated between maize, small grains, and legume crops, but this is the only crop for which outside inputs—seed, fertilizers, and chemicals—are purchased in the nearby small town. All planting, harvesting, and transport out of the field is done by hand owing to the

World Food: Production and Use. By Alfred R. Conklin, Jr. and Thomas Stilwell
Copyright © 2007 John Wiley & Sons, Inc.

Figure 5.1. Sweet potato (*camote*) growing along edge of Donio's rice field. (Courtesy of Henry Goltiano.)

steep slope of the land. Harvested potatoes are used both as a food source for the family and are sold in the local market.

Potatoes are a main and important crop, both in terms of income and food source throughout the Ecuadorian highlands. More than 60 verities of potatoes are gown and

Figure 5.2. Market in Baybay Philippines showing root crops being displayed for buyers. (Courtesy of Henry Goltiano.)

Figure 5.3. Some of the many different varieties of potatoes grown in Ecuador.

readily available in local markets. Nine of the more common potatoes are shown in Figure 5.3. Although it appears that they are simply different sizes of the same variety, this is not the case; each is a different variety. The different sizes, shapes, and colors of potatoes are identifying characteristics of different varieties. For example, the red color and white spots of gabriela and the black skin and yellowish interior of leona negra makes these varieties instantly recognizable as different. Two other interesting points are that leona negra is reputed to have medicinal properties and some varieties, for example, chola, are sold both washed (to remove adhering soil) and unwashed.

On the Ohio farm, Steve does not grow potato or any other root crop, including his son in his garden.

5.2 INTRODUCTION

Root crops can be divided into three general groups: those that provide a large amount of carbohydrate in the diet, those that are used as vegetables, and those that are used primarily for flavor or as a garnish. Potatoes, sweet potatoes, cassava,* yams, and taro supply large amounts of carbohydrates in the diets of much of the world's population. Beets, carrots, and turnips are used as vegetables in many parts of the world, and garlic, onions, and ginger add flavor to many dishes including soups. In this chapter we will only be considering those crops that provide large amounts of carbohydrates in the diet. Vegetable root crops are discussed in Chapter 4.

---

*Cassava is also called yuca (sometimes yucca) in North and South America [Axel Schmidt, CAIT (Centro Internacional de Agricultura Tropical) personal communication].

5.3 ROOT CROPS

Root crops include all crops where the main edible portion grows below ground.[†] In this chapter we will discuss tubers, which are modified roots, enlarged roots, and the below-ground modified enlarged stems called corms. Of the root crops, worldwide the potato and sweet potato, shown in Figure 5.4, which are tubers, are the best known and most widely grown. Examples of the major types of root crops discussed in this chapter are shown in Figure 5.4.

In addition to potatoes and sweet potatoes, there are many other types of root crops that are important and form a substantial portion of the diet in much of the world. Yam (a tuber), cassava (a root), and taro (a corm) are examples of three that are both common and important foods in many tropical countries, Pacific and Caribbean islands, and in countries in Africa. Yam is eaten like and often substituted for sweet potato in cooking; cassava is used to make flour used in baking breads, cakes, and pies. Taro, also called coco yam, is cooked by boiling, baking, frying, or roasting. It is also used to make poi in Hawaii and is considered a sacred plant to native Hawaiian islanders. In addition to the below-ground portion of the plant, boniato and taro leaves are commonly boiled and eaten. Cassava leaves are also eaten, however, this use is restricted primarily to areas of Africa.

While the potato is a well-understood name that applies to the same root crop around the world, the same cannot be said for the other root crops. The same name may be applied to two different crops or different names may be applied to the same crop. For instance, sweet potato may be called yam or boniato and cassava may also be called yucca. In addition each local language may have several names for the same crop, and different names may be used in different regions. It is not within the scope of this book to try and separate and identify all the different names and ways these names are used. The names that are commonly used (Fig. 5.4) will be the ones used throughout this chapter.

Figure 5.4. Examples of tuber crops, potatoes, sweet potatoes, boniato, yam, and taro.

[†]The peanut grows below ground but is generally not considered a root crop because it grows on a stem that grows into the ground from the above-ground portion of the plant.

TABLE 5.1. Botanical Description of Root Crops

Crop	Description	Figure
Potato	Initially upright growth; later growth along ground	5.9
Sweet potato, boniato, yams	Vines growing along the ground	5.10
Cassava	Woody stems with lobed leaves on top	5.16
Taro	Heart-shaped leaves on top of long stem	5.20

5.4 BOTANICAL DESCRIPTIONS

Root crops have very different physical appearances, which are summarized in Table 5.1. The potato, sweet potato, boniato, and yam all grow on vine or vinelike plants. Cassava has a woody stem with leaves at the top. Taro is very different in that it has a green stem, a heart-shaped leaf, and it can grow both with its roots submerged in water and in well-drained soil.

5.5 TUBERS

Tubers are enlarged, modified, and specialized structures occurring along and within roots of certain crops, most notably potato and sweet potato (here we will use the term tuber to mean only potato, sweet potato, and boniato). These structures serve as reproductive portions of these plants. Tuber contains "eyes," shown in Figure 5.5,

Figure 5.5. Potato eye from which a new potato plant can grow.

Figure 5.6. Potato with a sprouted eye.

each of which will sprout (Fig. 5.6) to produce a new potato plant. Tubers are the primary crop and food staple in many countries, such as those in South America and northern Europe. Tubers are high in carbohydrates, thus providing energy in the diet, but are generally low in protein, fat, fiber; however, some may contribute significantly to the vitamin and mineral content of the diet.

5.5.1 Climatic Adaptation

Tubers are grown on all continents and in all climates, although the potato and sweet potato are more universally grown than boniato. The only restriction is that the soil must be warm and moist enough for the plant to grow and the growing season long enough for tubers to form.

5.5.2 Importance to Human Nutrition

Tubers are eaten in all parts of the world and are the main carbohydrate source in many countries. In cases where they are not the main carbohydrate source, they are commonly eaten and contribute to the carbohydrate nutrition of the particular peoples. In addition to carbohydrate tubers can add significant amounts of both vitamins and minerals to the diet. A more detailed discussion of tuber contribution to nutrition is given below.

5.5.3 Cultivation

Soil is worked by plowing or other cultivation methods to loosen it before planting. All three tuber crops can be grown from the tuber either whole or from pieces, "seeds"

Figure 5.7. Potato planter pulled by a tractor. (Courtesy of Kitty O'Neil.)

containing what are referred to as eyes. The eyes are growing points that are visible on the surface of the tuber. A section of the tuber, of approximately 55 g, containing the eye shown in Figure 5.5 is treated to prevent fungus infections and then stored for a week to allow the cut side to form a hardened, protective layer.

A mechanical potato planter, shown in Figure 5.7, plants these seed pieces, which are placed in the hopper, the upper part of the planter. The lower part of the planter opens a furrow, allows the seed pieces to drop into the furrow, which is then covered with soil and packed. During planting, the seed pieces may also be sprayed to prevent insect damage.

Plants may also be propagated from the tops or vines, and this is a particularly prevalent method of reproduction of sweet potatoes, yam, and boniato in tropical countries. Reproduction using vines is done in two ways. A portion of a vine can be buried in the soil, and the buried portion will develop roots that subsequently produce tubers. Alternately cuttings of the vines may be taken and planted in a new seedbed.

Root crops are susceptible to infection by numerous disease organisms including viruses. Planting materials, both seed potatoes and vines, may harbor these organisms. Thus, a number of methods are used to control diseases in potatoes. The most important is to use disease-free planting materials, which can be obtained from commercial seed produces. Unfortunately, much of the world lacks the development of a seed industry, which means that farmers do not have access to disease-free planting materials. This in turn results in decreased yields worldwide.

Other methods of disease control are crop rotation, between potatoes and other crops, and soil fumigation. In addition control of pests and insects that transmit disease organisms is also important.

All root crops grow best in soil that farmers describe as being light. That is, it is easy to work and does not have high clay content. These soils would have a sandy or

Figure 5.8. Young potato plants growing in a raised bed.

loamy texture, be organic soils, or soils with high organic matter content (Chapter 9). Sandy soils, and organic soils will also produce well except that sandy soils hold little water and so irrigation may be needed. Organic soils are often associated with low-lying areas and must therefore be well drained before being used for root crops except for taro, which can be grown submerged in water as described below. Addition of organic matter to any soil, other than organic, will improve the soil for tuber production.

Areas to be planted to root crops are frequently mounded into either ridges (Fig. 5.8) or hills into which the cuttings, of either tubers or vines, are planted. Furrows between ridges can be used for irrigation and application of fertilizer. Potatoes can also be planted on flat ground with minimum cultivation. In many parts of the world

Figure 5.9. Potatoes growing where the plant tops enter the soil.

Figure 5.10. Sweet potato plant showing stem connecting potato to the plant tops and other roots.

root crops are planted on marginal land, hillsides, and other less productive areas and allowed to grow with little or no care. The tubers are harvested after some time and the land left fallow until another root crop is planted. In this way tubers are available to even the poorest of farmers or families.

As the plants grow, tubers form along and among the roots, as seen in Figures 5.9 and 5.10. The connection of tubers to roots is obscured in Figure 5.9 but is obvious in Figure 5.10. As shown, the tubers are located close to the place where tops emerge from the soil.

5.5.4 Fertilization

Root crops require a soil that has high levels of potassium, and thus potassium fertilization is essential for optimum production of quality tubers, which are essential for the production of food items such as potato chips. Phosphate is also essential and is used in fertilization of tuber crops. Although also essential, if used in excessive amounts nitrogen can lead to decreased tuber quality, and so nitrogen fertilization must be carried out carefully and monitored closely.

5.5.5 Diseases, Insect Pests, and Their Control

Microorganisms and insects attack the leaves and stems and the roots of tuber crops. Both types of attacks can lead to serious losses of produce. Numerous types of fungi commonly attack both leaves and tubers, causing losses both during production and after harvest. Destruction of leaves causes a decrease in photosynthesis and thus a decrease in yield. These types of diseases commonly cause wilting of the plant and thus are called wilts or sometimes rots. Microbial attack, by fungi or bacteria, of tubers causes discoloration and soft spots that can eventually destroy the whole tuber.

There are also a number of types of insects, with various common names, that attack the leaves of tuber crops. The most common of these are the various potato beetles. Tubers can also be attacked by nematodes, which are microscopic worms, wire worms, and grubs all of which can injure or completely destroy the tuber.

There are three relatively simple approaches to minimizing disease and insect pest in tuber crops. First, crops are grown in rotation with other crops such that there are four or more years between tuber crops on the same piece of land. Second, only disease-free planting materials, are used. Third, the pH of the soil can be kept low such as to inhibit the growth of various microorganisms. To protect tuber sections used as planting material, the open sides are often coated with sulfur. This accomplishes two things. The sulfur kills bacteria and fungi, and it also causes the soil around the cutting to become more acid as bacteria oxidize the sulfur to sulfuric acid. Both effects of sulfur improve the vigor of the tuber plant and increase the amount of disease-free harvest.

Various general-use insecticides and other pest control agents can be used to control disease and pest in potatoes. This includes dusting or spraying plants with insecticide to control potato beetle.

5.5.6 Harvesting

Harvesting tubers can be done manually by digging the individual plants or mechanically using a tractor (Fig. 5.11). Hand harvesting is common in many parts of the world, particularly where the crop is on a steep hillside. Figures 5.9 and 5.10 show that tubers are located just under the soil where the tops emerge. Thus it is possible to dig potatoes without damaging the tubers because their location is easily determined. Once the soil is loosened, tubers can easily be collected by hand and carried out of the field.

Figure 5.11. Tractor pulling a potato haravester. Potatoes are removed from the soil, soil separated from potatoes, and potatoes transferred to wagon for removal from field. (Courtesy of Spudnik Equipment Company LLC.)

Sometimes dug potatoes may be left in the field so that the soil can dry, making it easier to remove.

Simple plowing with animal or tractor power followed by manually collecting the tubers is also done in many localities. This method requires careful and relatively deep plowing to expose the tubers without causing damage to too much of the harvest. In Figure 5.8 the line at the bottom of the ridge, of the bed, is the depth to which soil is plowed or dug.

Many types of mechanical tuber harvesters are available. Some dig plants and tubers with a scooplike harvester, that lifts the plants and tubers out of and separates them from the soil on a moving screen conveyer and leaves them piled in a row, called a windrow, in the field. Complete mechanical harvesting is done with a similar scoop that cuts below the plants (Fig. 5.11) and lifts them out of the ground and onto a moving conveyer screen where soil and rocks fall through the screen and the tubers are collected in a wagon at the side of the conveyer (Fig. 5.11).

In most cases tubers harvested by mechanical means are ready for sale immediately after harvesting. However, complete removal of stones and rocks from soil is not always accomplished, and so some additional sorting may also be needed before further processing of the tubers.

Once harvested, potatoes are dried and stored in a cool dark storage facility. In the past this has been a dark, cool cellar of a house and was called a root cellar. Storage at temperatures below about $7°C$ is usually effective in inhibiting the growth of disease-causing bacteria and fungi. However, freezing can cause damage to tubers, and thus the temperature cannot be allowed to get too low.

5.5.7 Consumption

In tropical climates dried tubers can be stored for short periods of time, although they are more commonly eaten fresh. In most cases plants are left in the field and tubers harvested as needed for food or when they are to be sold in the local market. Additional information about consumption is given below.

5.5.8 World Production

Potatoes and sweet potatoes are native to South America while the yam is reported to be native to both South America and Africa. The origins of other root crops are largely unknown.

Potatoes (*Solanum tuberosum*) shown growing in Figure 5.9, sweet potatoes (*Ipomoea batatas*), in Figure 5.10, and boniato and yams (*Dioscorea cyenensis*) are all tubers of vines. Although sweet potatoes and yams are often confused and the names used interchangeably, they are, as seen by their scientific names, entirely different plants. Of these tubers boniato is an exception in that it is classified as *Ipomoea batatas* and so is another variety of sweet potato, although as seen in Figure 5.4 the sweet potato and boniato look quite different.

As shown in Figure 5.12, potatoes are grown on every continent and in almost every country in the world. On the other hand the production of other root crops, for

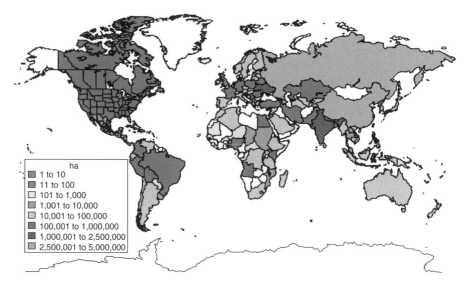

Figure 5.12. Map of potato-producing countries. (Data abstracted from http://faostat.fao.org/faostat.)

example, taro and yam (Figs. 5.13 and 5.14), is much more limited, as they are produced primarily in Africa, South America, and the Pacific islands.

In addition to the varieties of potatoes, there are many varieties of other tubers, for example, sweet potatoes and yams—large, small, red, white—but they are all cultivated,

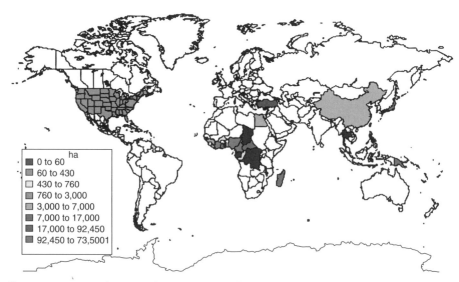

Figure 5.13. Map of taro-producing countries. (Data abstracted from http://faostat.fao.org/faostat.)

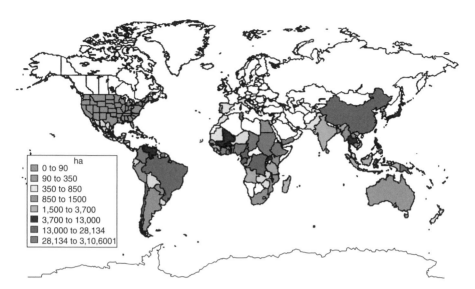

Figure 5.14. Map of yam-producing countries. (Data abstracted from http://faostat.fao.org/faostat.)

harvested, stored, and eaten similarly. This is also true for boniato, but internationally there are fewer varieties of this crop, although in local areas numerous varieties may be available. All are grown from tropical to temperate climates with good productivity.

In the United States the names sweet potato and yam are often used interchangeably, although as noted above they are completely different plants. Most yams are produced, sold, and used locally while sweet potatoes are processed and caned. In both these cases the sweet potato may be labeled as being yam and visa versa. Because of exports of various tuber products, the terms may also be used incorrectly in other parts of the world particularly outside of Africa.

China is the major producer of potato followed by the Russian Federation, Ukraine, and India. Poland, Belarus, and the United States are also major producers. Africa produces more than 95 percent of the world's yams with Nigeria being by far the largest producer in Africa followed by Ghana and Côte d'Ivoire.

China is also a major producer of sweet potatoes, producing 80 percent of the world's supply. However, sweet potato is both commonly grown and eaten in all parts of the world from tropical to temperate regions. Yams are grown extensively in Asia and the Pacific islands and are also important in Africa.

5.5.9 World Trade in Tuber Crops

As can be seen in Figure 5.12, most countries in the world produce potatoes. It is logical to assume that the major producers would be large countries and that major producers would be in South America where the potato originated. Figure 5.12, shows that neither of these is true in that the major producers are in Europe and Asia and that some smaller countries for example, Ukraine and Poland, are major producers.

TABLE 5.2. Major Exporters of Potatoes (metric tons)

Country	Exports (E)	Imports (I)	Difference (E − I)
Netherlands	1,696,616	1,633,569	63,047
France	1,434,335	363,728	1,070,607
Germany	1,320,537	538,017	782,520
Belgium	972,842	1,046,590	−73,748
Canada	428,064	163,650	264,414

Source: FAOSTAT data (archives), 2005, http://faostat.fao.org/.

One might expect that the major producers of potatoes might be the major exporters. However, as seen in Table 5.2, the major producers are not the major exporters of potatoes. Likewise, nonproducers were not major importers (Table 5.3), as can be seen by the fact that the Netherlands and Germany are both large exporters and importers of potatoes.

On the world map in Figure 5.15 sweet potato production is seen to occur in far fewer countries than potato. In Tables 5.4 and 5.5, it can be seen that of the 10 major importing and exporting countries, only 4 countries are both exporters and importers of sweet potatoes. Of these only the United States, Malaysia, and France are both major exporters and importers. China, Indonesia, and Israel are all major exporters of sweet potatoes while Canada, Japan, and Malaysia are major importers. Thus with respect to exports and imports, sweet potato is similar to potato in that the list of countries is not the same.

It is important to note that the quantities of sweet potatoes being exported and imported is dramatically lower than potatoes. Exports of sweet potatoes are from 55 to 58 times less than potatoes, and some countries that export do not import any sweet potatoes, for example, Israel and Dominican Republic. In addition, generally those countries exporting sweet potatoes only import relatively small quantities.

Importing countries import from 46 to 69 times less sweet potatoes than they do potatoes. All importing countries are also exporters of sweet potatoes, and the amounts are small except for Malaysia and France, which have significant amounts of exports compared to imports, that is, exports are 26 and 19 percent of imports for the two countries, respectively. As with potatoes countries may both export and import sweet potatoes.

TABLE 5.3. Major Importers of Potatoes (metric tons)

Country	Imports (I)	Exports (E)	Difference (I − E)
Netherlands	1,633,569	1,696,616	−63,047
Belgium	1,046,590	972,842	73,748
Spain	778,493	232,518	545,975
Italy	616,643	183,349	433,294
Germany	538,017	1,320,537	−782,520

Source: FAOSTAT data (archives), 2005, http://faostat.fao.org/.

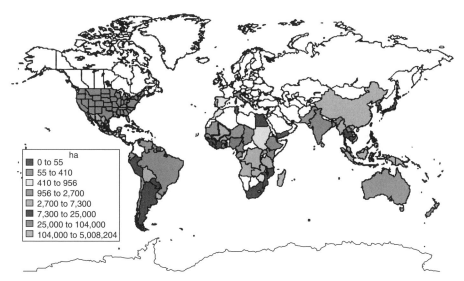

Figure 5.15. Map of sweet-potato-producing countries. (Data abstracted from http://faostat.fao.org/faostat.)

TABLE 5.4. Major Exporters of Sweet Potatoes (metric tons)

Country	Exports (E)	Imports (I)	Difference (E − I)
United States	29,155	5,141	24,014
China	27,218	26	27,192
Israel	13,368	0	13,368
Indonesia	11,822	3	11,819
Dominican Republic	7,576	0	7,576

Source: FAOSTAT data (archives), 2005, http://faostat.fao.org/.

TABLE 5.5. Major Importers of Sweet Potatoes (metric tons)

Country	Imports (I)	Exports (E)	Difference (I − E)
Canada	35,373	62	35,11
United Kingdom	15,426	134	15,292
Malaysia	9,113	2,382	6,731
Japan	9,002	139	8,863
France	7,704	1,497	6,207

Source: FAOSTAT data (archives), 2005, http://faostat.fao.org/.

One other significant difference between sweet potatoes and potatoes is that there are no countries that export more than they import nor import more than they export of sweet potatoes, but Belgium imports more potatoes than it exports (Table 5.2) and the reverse is true of Germany (Table 5.3).

As seen in Figures 5.14 and 5.15, there is a similarity in the countries producing yams and sweet potatoes. But there are fewer countries exporting and importing yams. However, because of the confusion in naming (between sweet potato and yam), it is hard to obtain good data on yam trade. The same thing holds true for boniato and so no data for this crop is presented here.

The question then becomes how might these differences in import and export come about? And why might a country both export and import the same food item?

In some cases a single country might record both exports and imports of the same crop for many reasons. Two of these would be geographic location and variety differences. Some areas of a country may be closer to larger markets or markets offering higher prices to producers in neighboring countries than in the producing country. Or it may just be simply a closer market. In other cases one variety may be favored in one country and another in another country and so are traded across borders. Potatoes come in various sizes and colored skins. One country, country A, may prefer small white potatoes over red-skinned potatoes. In this case country B may produce red potatoes and ship them to country A while importing white potatoes from this same country.

5.6 CASSAVA: MODIFIED ROOTS

Cassava, shown in Figure 5.16, is different from most other root crops in that the harvested portion is not a specialized root extension but is a swollen part of the root itself (Figs. 5.17 and 5.18). It is not a vine. It grows as a woody brushlike plant. All

Figure 5.16. Cassava plants growing in a field. (Courtesy of the Philippines Root Crop Research and Training Center.)

Figure 5.17. Cassava roots attached to stem. (Courtesy of the Philippines Root Crop Research and Training Center.)

varieties contain glucosides, which have a cyano or nitrile group that is released as cyanide, CN^-, which is toxic. There are two generally recognized varieties of cassava, sweet and bitter, the latter has more and former less cyano-containing glucosides.

5.6.1 Climatic Adaptation

Cassava is adapted and largely restricted to production in the humid tropics. It is commonly found in low-lying areas receiving high rainfall. Although it responds to improved methods of cultivation and fertility, it is often grown on poor soil.

5.6.2 Importance in Human Nutrition

For many people in the humid tropics cassava is a staple in their diet. It can be cooked and eaten in many of the same ways as potatoes, with the precautions noted below. It is made into a flour that is used in baking much as wheat flour is used. Additional information about the nutritional value of cassava is given below.

5.6.3 Propagation and Cultivation

Cassava is different from other tubers in that it is almost exclusively grown from stem cuttings. When harvested, the tops of cassava are cut and piled for burning. During this process a small portion of the stalk, containing at least one node, is cut off the stalk and is planted to produce the next crop. Soil is prepared by plowing or hoeing before planting. Initially, weeding is carried out by mechanical cultivation or by hand, but as plants mature they shade the ground so that weeding is minimized.

As with all root crops a light soil, that is, sandy to loamy, is best for cassava. In the case of sandy soils irrigation may be needed during dry periods, although cassava is highly resistant to drought. Good production of cassava requires fertile soil, which is obtained by using a field that has been fallow, use of manures, or commercial fertilizer. Much of the world production of cassava is on small areas by farmers who do not have access to inputs, and thus fertilization of cassava using commercial fertilizers is not common.

5.6.4   Diseases, Insects, and Their Control

Cassava is subject to attack by viruses, fungi, and bacteria, including root rots just as are tuber crops. Mites, mealybug, and grasshoppers attack plant leaves and can cause decreases in yield. As with other tuber crops, crop rotation and disease-free planting material can decrease the incidence of disease.

Diseases, insects, and poor cultural practices can lead to large reductions in harvested portions of the crop. Most planting materials are obtained locally and from previous crops because there is no commercial, disease-free, source of these materials. Thus, since cassava is propagated by stem cuttings, the careful selections of stems to be used for subsequent crops is an effective method of limiting these types of losses.

5.6.5   Harvesting and Storage

Cassava will continue to grow even after the roots have reached their maximum size. Thus harvesting is not a question of allowing the plants to mature but of when the maximum root production has occurred. This will be different in different localities, but often plants are allowed to grow from 8 to 16 months. If grown for longer periods of time, roots may become fibrous and thus not desirable for use as a food source. Harvesting involves cutting the tops off and pulling the roots; once harvested, roots do not store well, especially in the tropical conditions where it is most frequently grown. Thus roots are dried and cooked within a few, 2 to 3, days of harvest and used as food or dried to produce cassava flour.

Figure 5.18. Cassava showing white interior.

To increase storage times cassava roots can be washed and treated to kill adhering microorganisms. Subsequently, they can be wrapped in plastic, coated with wax, and refrigerated.

5.6.6 Preparation

Before this crop is safe to eat, the root must be cooked. Commonly, it is pealed, ground, washed, boiled, and dried. During this processing, usually during the washing and drying steps, toxic components, glucosides, are decomposed by enzymes in cassava itself and by microorganisms associated with the roots, and cyanide is washed out during the washing process. These procedures, if carried out carefully, are usually sufficient to remove all or enough cyanide precursors and cyanide such that the resulting meal (ground tuber) from the interior (Fig. 5.18) once dried is safe to eat.

5.6.7 World Production

Nigeria is by far the largest producer of cassava, followed by Brazil, the Democratic Republic of the Congo, and Indonesia (See Fig. 5.19). In the Congo not only are the roots used as a source of carbohydrates, but the leaves are also eaten. However, leaves of cassava are not commonly eaten in most cassava-producing countries.

5.6.8 World Trade

For 2004 there was virtually no trade in cassava. Only one country, the Congo, reported exports of only 1 metric ton of cassava while no other countries reported any import or export of this crop. It should be kept in mind that this does not mean that there was no movement of cassava across country borders just that it was either very small or was outside normal commercial channels.

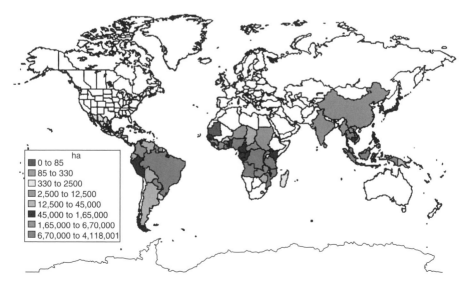

Figure 5.19. Map of cassava-producing countries. (Data abstracted from http://faostat.fao.org/faostat.)

Figure 5.20. Taro plant.

5.7   TARO: CORM

Taro (Figs. 5.20 and 5.21) is thought to have originated in Southeast Asia where it remains an important crop today, particularly in Southeast Asia and the Pacific islands. Although it is grown and consumed in parts of Africa, its principal

Figure 5.21. Taro corms.

consumption is in Southeast Asia and the Pacific islands. In many places it is an important crop of subsistence farmers.

5.7.1 Climatic Adaptation

Taro is hardy and will survive freezing when other tropical root crops will not. Despite this it is generally grown in tropical countries with high rainfall and that do not have freezing conditions. In part this is due to the fact that it can be grown in both water saturated and dry land soil.

5.7.2 Importance in Human Nutrition

Nutritional value of taro is much like other root crops and is described in detail below. It is the staple of many tropical island peoples of the Pacific Ocean including Hawaii in the United States.

5.7.3 Propagation and Cultivation

Taro is propagated from cuttings from side suckers on the main plant, small corms (shown in Fig. 5.21) obtained during harvest, corm pieces, and from huli. Corm pieces are obtained by cutting a corm into pieces that are each then used as a source of a new plant. This is somewhat similar to obtaining potato planting material from tubers. Corm pieces may then be started in a nursery before transplanting to the field. The use of huli, which is the top 1 to 2 cm of a corm plus 15 to 20 cm of the stems attached to it, is unique to taro production. The use of huli is a preferred propagation method in many places because little of the eatable portion of the taro is used.

In addition to the use of huli in propagation, taro is unique in that it can be grown as a lowland, flooded, or upland, dry-land crop. In this regard it is similar to rice, although the plants are vastly different as can be seen from Figure 5.20. For flooded production of taro the field is prepared much the same as for rice and water levels increased or decreased as needed during the growing season for various operations such as weeding and fertilization. The field is allowed to drain before harvest. Under these conditions taro is grown as a sole crop. Yield is significantly higher under flooded conditions, although time to maturity is increased.

Upland production of taro is also common. This type of production depends on a long 6- to 9-month rainy season or adequate irrigation water for this period of time. Land is prepared by plowing and harrowing as with most crops, and subsequently the field may be ridged and the taro "seed" stock planted on the top or in the furrow between ridges. The field may also be mulched to conserve water.

Weeding and fertilization are required during the growing season of taro. Weeds must be controlled for maximum production; however, with taro weeds may be controlled by either traditional weeding methods or by flooding if sufficient water is available. Although taro responds to fertilization, most is grown by subsistence farmers who do not have either the money or access to fertilizer.

5.7.4   Diseases, Insect Pests, and Their Control

The fungus *Pythium* is the major microbial disease of taro root and can cause significant losses in yield. The fungus *Phytophthora* is the major disease of taro leaves. Tools used in harvest are frequently disinfected with bleach, alcohol, or hydrogen peroxide, especially when huli to be used for a future crop are harvested. This prevents the spread of disease organisms. Only disease-free huli are then used for subsequent crop production.

Both root aphid and leafhoppers are major insect pest of taro, although there are also a number of other common pests such as mealybugs, snails, and slugs, which can also cause yield losses, and nematodes, which will attack corms causing yield loss and decrease in corms.

As with all crops, rotation and leaving fields without crops for a period of time are effective in decreasing the incidence of disease and insects. However, there are also some recommended practices such as isolation of fields, planting in small fields, intercropping, and maintaining the soil pH close to neutral, which are practices different from those commonly applied to potatoes, sweet potatoes, cassava, and the other tuber crops.

5.7.5   Harvesting and Postharvest Practices

Dry-land production takes from 5 to 12 months, while flooded production takes from 12 to 15 months. A general yellowing of the leaves indicates that the crop is ready for harvest, which is commonly done by hand. Dry-land harvesting is easer than flooded, as the roots in flooded conditions must be cut or broken to remove the corms. Once harvested, the corms are dried and can be stored for short periods of time in this condition.

Generally, taro is eaten shortly after harvest and thus is not generally stored. Sometimes processed into flakes or flour, taro can be stored for some time in this condition. However, this is not a common practice, and taro in these or other processed forms is not commonly nor regularly found.

5.7.6   World Production

The largest producers of taro are Nigeria, Ghana, and Côte d'Ivoire, all in Africa. This, however, should not lead one to the conclusion that taro is not important in the Pacific islands; it is. As can be seen in Figure 5.13, taro production is mostly limited to the tropical countries of the Pacific Ocean, Africa, and South America, although some is grown in the United States and China. Also total hectareages planted is quiet small when compared to other root crops. Nevertheless some peoples in these areas depend on taro as their major source of carbohydrates in the diet.

5.7.7   World Trade

On a worldwide bases there is little trade in taro, as can be seen in Table 5.6. China is the largest exporter followed by Fiji Islands, which exports less than one tenth the

TABLE 5.6. World Trade in Taro (metric tons)

Country	Exports	Imports
American Samoa	0	3,000
China	130,581	397[a]
Dominica	680	NR[b]
Fiji Islands	10,020	0
Ghana	5	NG
Japan	NR	51,320
Niue	190	NR
Saudi Arabia	NR	2
Samoa	457	NR
Tonga	4	NR
Trinidad and Tobago	1	655
United States	1,700	43,199

[a]Includes Macao SAR.
[b]NR = not reported.
Source: FAOSTAT data (archives), 2005, http://faostat.fao.org/.

amount exported by China, and the United States exports less than one tenth the amount exported by Fiji Islands. There are only six other countries reporting exports of taro. In terms of imports Japan is the largest importer followed by the United States, both of which import about the same quantities. American Samoa imports less than one tenth the amounts imported by Japan and the United States.

5.8 OTHER COMMON ROOT CROPS

Arrowroot, yam-bean, and tannia are three root crops that are less well known but are common in many tropical countries. Each has some characteristic that makes it specifically interesting and valuable as a root crop. Arrowroot has particularly fine carbohydrate and is thus easy to digest, yam-bean is a nitrogen-fixing plant that alleviates the need for nitrogen fertilization, and tannia is used by millions of subsistence people around the world but is not well known because little produce gets to market.

5.8.1 Arrowroot

Arrowroot is like yam and sweet potato in that there are a number of different plants, for example, *Calathea allouia* and *Maranta arundinacea*, which are called arrowroot. Also, like potato, arrowroot comes in a number of different colors mostly red and white, which are *Maranta arundinacea* varieties, and purple, which are *Calathea allouia* varieties. In both cases plants produce tubers or tuberous roots that are high in starch.

The arrowroot starch is of high quality, is easily digested, and its amino acid content is good except for a lack of the amino acid cystine (also Chapter 2). In addition to being a source of starch in the diet, it is also described as an herb, a health food, a herbal medicine, and its leaves used as clothing for infants.

Arrowroot is propagated from rhizomes or from suckers and after planting is generally given little care. It is often grown in association with other crops, and little preparation of the planted area either before or after planting is carried out. It is generally grown for 10 to 12 months before harvesting. Once harvested it can be cooked, eaten raw, or cooked and processed into flour. The name arrowroot is applied to all parts of the plant including flour produced from the tubers.

5.8.2 Yam-Bean

The yam-bean, also frequently called jicama (Fig. 5.22), is both unique and interesting because it is a legume and produces both a tuberous root and seeds in beanlike pods. As a legume, it is an efficient nitrogen fixer and is thus grown on poorer soil and in rotation with other crops. It is also grown intercropped, in rotation, and commercially as a single crop. Cultivation practices vary widely around the world. When intercropped with beans and maize (corn), it is sold to generate income while the maize and beans are used to feed the family.

The plants may be grown from seed or from cuttings. Plants can be pruned to encourage tuber, production, shown in Figure 5.22. Tubers are harvested after about 5 months and once harvested can be stored for 20 to 30 days in well-ventilated buildings or in the sun. During this time carbohydrates in the tubers are converted to sugar, making them sweeter. Tubers are sensitive to cold temperatures and thus cannot be stored at less than about 12°C without damage.

In most cases harvested tubers are sold at local markets, the exceptions are Mexico and Thailand where significant commercial production is practiced. Tubers can be cleaned and trimmed to remove roots and stems, washed, and sterilized using highly concentrated chlorine solutions. At this point they are ready for further storage and shipment.

Figure 5.22. Yam-bean or jicama.

All parts of the yam-bean plant, except the tubers, contain rotenone and various other rotenoids. These toxic compounds are usually at low concentrations in young plants and plant parts, and thus young or immature plant parts are used for food. However, toxic levels in mature leaves, stems, pods, and seeds prevent their use by either humans or animals. Because of trimming and harvesting when the plant parts are young and because local populations are aware of the toxicity problems, cases of toxicity are rare. Rotenone can be isolated from seeds, making them valuable as a source of this chemical.

5.8.3 Tannia

Tannia is related to cocoyam, also called taro, being from the same Araceae family. Because of this, it is sometimes called the new cocoyam, although this name can cause confusion. Tannia originated in northern South America but has spread to other parts of the world, notably Africa. Where grown, it is a staple of subsistence farmers. The tuber contains both calcium oxalate (calcium oxalate is a common form of kidney stone) and saponins, which can be both toxic and digestive irritants. For these reasons it is roasted or otherwise cooked to remove some or all these components before eating.

Tannia is grown from corms planted in well-prepared soils, that is, plowed and harrowed. Weeding is important for good yields. After between 10 and 12 months the plants die back and the corms are dug and harvested. Corms can be washed, disinfected, and subsequently stored in cold rooms.

5.9 GENETICALLY MODIFIED ROOT CROPS

As with most crops, those grown as major crops in developed countries have gotten the most investigation, in terms of genetic engineering, to produce genetically modified crops. With root crops the potato has had by far the most number of genetically modified varieties developed. As with most genetically modified crops, the first concern has been development of crops resistant to pest, primarily insects and microorganisms. This is particularly true for potatoes. However, potatoes genetically engineered to be herbicide resistant and higher in protein have also been developed.

Interestingly enough, although genetically engineered potatoes have been developed and released to the farming community, they have subsequently been withdrawn because of concerns over their safety, particularly in relationship to processed potato products. There has been some considerable controversy over the safety of potatoes genetically engineered to have more protein. Another interesting development is genetically engineered potatoes designed to be a source of starch for paper making.

The genetic modification of sweet potatoes has made little progress, and even less genetic engineering has been done on yams and taro. In the case of taro, in addition to consumption concerns, there are religious concerns since taro is a religious symbol in some Pacific island nations and the state of Hawaii.

Additional information about genetically modified crops is given in Chapter 11.

5.10 NUTRITIONAL VALUES OF CARBOHYDRATE-RICH ROOT CROPS

Starchy root crops—cassava, yam, potato, and sweet potato—are a major source of calories for most of the world's population. In some cases they constitute the major food intake on a daily basis.

As seen in Figure 5.23, cassava has the highest carbohydrate and consequently a relatively lower content of other nutrients. For this reason cassava is considered to be a particularly poor source of amino acids and protein. However, all these crops have low contents of protein, fat, and fiber, although comparatively speaking sweet potato is the most balanced. Fat is lowest in jicama and potato, being 0.09 and 0.08 percent, respectively, and highest in taro, 0.74 percent. There is a surprisingly high content of fiber in sweet potato, considering its other characteristics.

Because of its high carbohydrate content cassava is highest in energy, as seen in Figure 5.24. The energy content of jicama is lowest, mostly because of its high water content. Energy content of the other root crops gradually increases from potato to yam.

In terms of total vitamins root crops are similar except for sweet potato and taro, which are lower with sweet potato being lowest. Inspite of this sweet potato is significantly higher, up to 100 timers higher, in vitamin A than most root crops. Likewise taro is significantly higher, up to 100 times, in vitamin E than most other root crops.

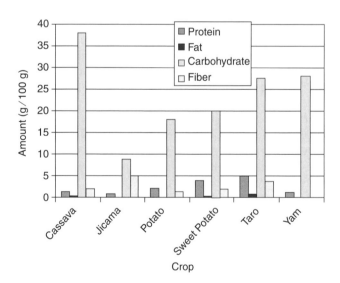

Figure 5.23. Nutrient value of root crops. (Data for graph from the Agricultural Research Service Nutrient Data Laboratory USDA National Nutrient Database for Standard Reference, Release 17, http://www.nal.usda.gov/fnic/foodcomp/Data/SR17/reports/sr17page.htm.)

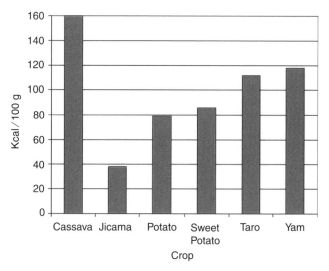

Figure 5.24. Energy content of root crops. (Data for graph from the Agricultural Research Service Nutrient Data Laboratory USDA National Nutrient Database for Standard Reference, Release 17, http://www.nal.usda.gov/fnic/foodcomp/Data/SR17/reports/sr17page.htm.)

5.11 PREPARATION AND CONSUMPTION

Table 5.7 gives examples of how these root crops are prepared and eaten. In all cases they constitute the major calorie intake and may also be the major constituent in snack foods. In tropical countries storage is a problem, and so root crops are harvested and eaten shortly thereafter. Root crops grow and can be harvested on a year-round bases so there is no great need for storage.

TABLE 5.7. Methods of Preparation and Consumption of Carbohydrate-Rich-Root Crops

Root	Preparation[a]	Consumption
Cassava	Crushed, boiled, and water removed[b]	Processed into flour used for breads and cakes
Yam	Peal and boil until tender	Eaten whole or mashed
Potato	Washed and cooked peeled or unpeeled by boiling or frying	Eaten whole, mashed or fried
Sweet potato	Cooked peeled or unpeeled by boiling	Eaten whole or mashed

[a]There are many other methods of preparation of most root crops; see, particularly, E. Schneider, *Uncommon Fruits and Vegetables: A Commonsense Guide*, HarperCollins, New York, 1986.
[b]Different cassava varieties contain different amounts of cyanide, which is removed by boiling and washing. Extreme caution in preparing cassava must always be exercised.

5.12 CONCLUSIONS

Root crops are grown around the world and are an important and sometimes an essential part of many diets. They are relatively easy to cultivate and often are grown on marginal land. Harvesting in many countries is a labor-intensive activity as it involves exposing the roots of plants by digging. Except for cassava, all types of root crops can be stored for longer periods of time than most fruits and vegetables, but shorter than grains. Diseases and insect damage can be limited by simple cultural practices such as crop rotation and selection of disease-free planting materials.

While root crops are a major source of carbohydrates in the diet, they can also serve as a significant source of some vitamins and minerals. This is especially true in situations where they are eaten in large quantities.

QUESTIONS

1. Explain the difference between a tuber, a root, and a corm.
2. What characteristics do taro and rice have in common both in terms of production and as a food?
3. While attending a birthday party in Papua New Guinea a person suffers sever abdominal pain after eating a piece of cake that she assumes is made from wheat flour. What other flour might this cake contain?
4. Describe in some detail the different ways in which root crops may be propagated using examples from this chapter.
5. Explain why some root crops, grown by subsistence farmers, might not be found in local markets. You may want to refer to Chapter 1 in answering this question.
6. Explain how a country might be both an exporter and an importer of the same crop. Give at least three ways in which this might happen.
7. Root crops can have importance above and beyond their use as food. Give an example of a root crop that has this characteristic.
8. Only two common crops can be grown under flooded conditions; rice is one; name the other. Why might this be advantageous to a farmer in the high rainfall tropics?
9. Minor root crops may be an important crop in a localize area. Explain two reasons why this might be so.
10. Hypothesize how it might be that cassava can be reported to be exported by a country but no countries report any imports.

BIBLIOGRAPHY

G. Belanger, J. R. Walsh, J. E. Richards, P. H. Milburn, and N. Ziadi, Nitrogen Fertilization and Irrigation Affects Tuber Characteristics of Two Potato Cultivars, *Am. J. Potato Res.* **79**, 269–279, 2002.

M. Bokanga, Cassava: Post-harvest Operations. Available at: http://www.fao.org/inpho/content/compend/text/ch12.htm.

Cassava. Available at: http://www.iita.org/.

M. P. Cereda and M. C. Y. Mattos, Lanamarn: The Toxic Compound of Cassava. *J. Venom. Anim. Toxins*, **2**, 6–12, 1996.

W. W. Collins, Root Vegetables: New Uses for Old Crops, in J. Janick and J. E. Simon, eds., *New Crops*, Wiley, New York, 1993, pp. 533–537.

D. C. Giacometti, J. León, and J. Costa, Tannia, Yautia, Neglected Crops: 1492 from a Different Perspective, in J. E. H. Bermejo and J. León, eds., *Plant Production and Protection Series*, No. 26, FAO, Rome, 1994, pp. 253–258.

R. P. Griffin, Sweet Potato and Irish Potato Insect Pests, Clemson Extension HGIC 2215. Available at: http://hgic.clemson.edu/factsheets/HGIC2215.htm.

M. Kluepfel, J. H. Blake, and A. P. Keinath, Irish and Sweet Potato Diseases, Clemson Extension HGIC 2214. Available at: http://hgic.clemson.edu/factsheets/HGIC2214.htm.

T. Kucharek and P. Roberts, Florida Plant Disease Management Guide: Sweet Potato. Available at: http://edis.ifas.ufl.edu/PG058.

M. Lamberts and S. M. Olson, *Tropical Root Crop Production in Florida*, Document HS965, IFAS, Florida Cooperative Extension Service, Gainesville, FL, 2006, pp. 433–438.

L. A. Manrique, *Manrique Tropical Root Crops*, Manrique International Press, Honolulu, 1996.

M. Meyhuay, Potato Post-harvest Operations. Available at: http://www.fao.org/inpho/content/compend/text/ch17_01.htm.

L. M. Moor and J. H. Lawrence, Cassava Manihot esculenta Crantz, *NRCS Plant Guide*. Available at: http://plants.usda.gov/.

C. R. Noda, C. R. Bueno, and D. F. S. Filho, Guinea Arrowroot. Neglected Crops: 1492 from a Different Perspective, in J. E. H. Bermejo and J. León, eds., *Plant Production and Protection Series*, No. 26, FAO, Rome, 1994, pp. 239–244.

I. Onwueme, Taro Cultivation in Asia and the Pacific, RAP Publication No. 1999/16 (FAO) Bangkok, Thailand, 1999.

J. Ooka and B. M. Brennan, Crop Profile for Taro in Hawaii. Available at: http://www.ipmcenters.org/cropprofiles/docs/hitaro.html.

L. Opara, Yams Post Harvest Operations. Available at: http://www/fao.org/inpho/content/compend/test/ch24_04.htm.

E. Schneider, *Uncommon Fruits and Vegetables: A Commonsense Guide*, HarperCollins, New York, 1986.

E. H. Simonne, D. M. Maynard, G. J. Hochmuth, M. L. Lamberts, C. S. Vavrina, W. M. Stall, T. A. Kucharek, and S. E. Webb, Sweetpotato Production in Florida, Document HS738UF/IFAS, Florida Cooperative Extension Service, Gainesville, FL, 2004, pp. 293–299.

M. Sørensen, Supercrop: The Yam Bean, a Tuber Undaunted by Drought, Poor Soil, or Insects, Produces Astonishing Yields, The Crop is the Focus of a Worldwide Effort to Unlock Its Potential, *Natural History*, **113**, 38–43, 2004.

M. Sørensen, *Yam Bean (Pachyrhizus DC). Promoting the Conservation and Use or Underutilized and Negelected Crops*, Vol. 2, Institute of Plant Genetics and Crop Plant Research, Gatersleben/International Plant Genetic Resources Institute, Rome, 1996.

J. M. Stephens, Arrowroot—*Maranta arundinacea* L, Document HS542, Cooperative Extension Service, Gainsville, FL, 2003.

J. G. Vaughan and C. A. Geissler, *The New Oxford Book of Food Plants*, Oxford University Press, Oxford, UK, 1999.

H. Wagner, Researchers Get to the Root of Cassava's Cyanide Producing Abilities, Ohio State Research. Available at: www.osu.edu/units/research.

R. Wright, G. Hein, W. W. Hoback, and A. Pavlista, Biology and Management of Potato Insects, Nebraska Cooperative Extension EC02-1565.

6

FRUITS, BERRIES, AND NUTS

6.1 FRUIT, BERRY, AND NUT PRODUCTION ON THE THREE FARMS

Of the three farmers only Aída and Octavio grow large amounts of fruits and berries, and none of the three grows nuts for either home use or sale.

Donio does not grow any berries or nuts; he does, however, have some bananas growing along the edges of his fields, as seen in Figure 6.1. He will often use these as either a snack or as a source of carbohydrate in his family's diet. Donio does not practice any kind of management other than to plant the banana plants and harvest the bananas. Suckers* are taken from an existing banana plant and transplanted to produce a new banana plant, or the suckers are allowed to grow in place and produce what bananas they will. There is no weeding nor any fertilization of bananas.

Steve does not grow any fruits, berries, or nuts, although there are wild berries and nut trees growing on his farm and the lands he rents. He does not harvest these and does not use them as a source of food.

Aída and Octavio grow seven different fruits that are only eaten by the family: *Persea* spp. (avocado/*aguacate*†), *Carica* spp. (papaya/Babaco†), *Fragaria* spp.

---

*A banana plant will produce only one stalk of bananas. Young plants, suckers, will grow from the roots of the old banana plant, and these will produce more bananas. Suckers may be left to grow around the old plant or be transplanted.

†First name in parentheses is the common English name, after the slash is the common local name. Also see Glossary for alternate names.

World Food: Production and Use. By Alfred R. Conklin, Jr. and Thomas Stilwell
Copyright © 2007 John Wiley & Sons, Inc.

Figure 6.1. Bananas and coconuts growing along the edges of one of Donio's fields. (Courtesy of Henry Goltiano.)

(strawberry/*Fresa*[†]), *Rubus* spp. (blackberry/*Mora*[†]), *Passiflora* spp. (passion fruit/ *taxo*[†]), *Cyphomandra* spp. (tree tomato/*tomate de l'arbol*[†]), and *Physalis* spp. (husk tomato/*uvilla*[†]). They are grown in the garden and so are also irrigated using the drip irrigation system used for vegetables and bear at different times throughout the year. All are eaten as snacks or as part of a meal.

Both tree tomato (*tomate de l'arbol*; Fig. 6.2) and blackberry (*mora*; Fig. 6.3) are widely grown and eaten in Ecuador. The tree tomato looks, inside and outside, much like and is used in much the same ways as a tomato. The blackberry is eaten fresh, sweetened or unsweetened and is used to make desserts, such as

Figure 6.2. Tree tomato fruit.

Figure 6.3. Blackberry growing in Aída and Zorro's garden.

sorbet. Both are squeezed to extract their juices, which are used as a beverage at meals or for snacks.

No nuts are grown on any of these farms. All three families, however, do eat peanuts as snacks.

6.2 INTRODUCTION

In this chapter the terms fruits, berries, and nuts are applied as commonly used even if scientifically they might be classified otherwise. For example, some fruit might not be classified as fruit, some berries might be fruits, and some nuts could be, in some cases, considered fruits. Also, to make it easier to follow the development of this chapter, edible portions of plants that are commonly called fruits or tree fruits will be discussed under these terms even if, as in the case of papaya, they do not technically grow on a tree. In a similar fashion, berries come from both low-growing bushes, brambles, and other types of plants. Coconut, which is neither a tree nor nut, and peanuts, which are also not nuts and grow on annual plants, will be discussed separately in Sections 6.9 and 6.11.

Fruits, berries, and nuts grow in isolated patches, or individual trees, as wild plants, and in orchards, farms, or plantations. This is true for most of the world and is unique to these sources of food as opposed to tubers, vegetables, and grains, which are, with few exceptions, planted and harvested in definite gardens or fields. Blueberries, shown in Figure 6.4, and walnuts would be good examples of wild or semiwild growth of common berries and nuts.

Because these sources of food can be found wild, untended or unclaimed, they can make a substantial addition to the diets of subsistence farmers and others living at a subsistence level. In addition many of these food sources are relatively inexpensive especially those fruits and berries that do not ship well.

Figure 6.4. Blueberries, inset shows ripe berries.

6.3 TREE FRUITS

6.3.1 Climatic Adaptation

Tree fruits are divided into tropical, subtropical, and temperate. Thus, as the name implies, they are more or less restricted to these respective climatic zones. Tropical trees are restricted to tropical areas with sufficient rainfall to support tree growth. They are generally sensitive to both low temperatures and freezing and may be injured or killed by either or both. Subtropical trees may be found in some tropical and some temperate areas, although their major and best production is in the subtropical areas. Temperate fruit trees will often grow in subtropical and tropical areas but will not bear fruit or are not very productive. For example, apple trees will not flower unless they experience a period of freezing!

6.3.2 Importance in Human Nutrition

Fruits from trees are a common part of the diet of people around the world. Different climatic zones produce fruit very different in color, consistency, use, and taste. Fruit trees are found in large orchards covering many hectares and containing hundreds of trees or may be a single tree on a small lot. The former are commercial orchards; the latter typically supply fruit for a single family. Plantings of small numbers of trees are also common, and the fruits are gathered, used by the household, and then the excess is sold in local markets. Additional nutritional information is given below.

6.3.3 Cultivation

One reason for small plantings, as opposed to single trees, primarily intended for single household use, is that many fruit trees require a pollenizer.[§] That is, they require

_____

[§]A pollenizer is the pollen-donating plant and the pollinator is the pollen-transferring agent, for example, the honeybee.

another tree of the same or different species to the produce pollen needed for fruit production. The same flower cannot pollinate itself, nor can another flower on the same tree pollinate it, and thus a pollenizer tree is necessary. For this reason a single tree may not be able to produce fruit if no pollenizer is close by. To avoid this problem a pollenizer is included in the planting. The pollenizer will, in many cases, also produce fruit or nuts, and so the plant is productive in terms of providing food.

As with other food items only commercial production is counted when reporting a county's or an area's production, although total production may be significantly higher if single trees or small plantings of several trees are included.

Many fruit trees come in both a regular and a dwarf size, size being determined primarily by height with the dwarf being, typically, less than half as tall as the regular size tree. Dwarf size trees are the same variety as the regular size, producing the same size, taste, nutrition, and the like as the regular size trees. Two advantages of dwarf trees are that they tend to start producing earlier and fruit is easier to harvest. For some varieties there are also semidwarf varieties, which are in between in height. Unfortunately, dwarf varieties usually do not live as long as standard height trees.

6.3.4 Plantings

Fruit trees are often produced by grafting or budding a particular, usually a named tree type, onto a specific rootstock, chosen because it is from a disease-resistant tree of the same type or the roots impart some other desirable characteristic such as dwarfing. These grafted plants will be grown for several years in a nursery before transplanting into an orchard. This process is particularly important if one wishes to be assured of a high-quality crop with known characteristics. Planting seed will often produce plants with varying characteristics and will bear fruits with different characteristics. To have an orchard that will produce the desired fruit of the specific type needed or wanted, grafted plants are necessary. This applies primarily to tree fruits and nuts but not to other types of fruit-producing crops.

However, tropical fruit trees are commonly grown from seed. Black walnuts are also commonly grown from seed. This approach may be used when the nuts to be produced are not intended to be the major crop or the characteristic of the nut is not important. In the case of black walnut the wood of a mature tree is highly valuable and so the planting may have the primary purpose of producing wood rather than nuts.

6.3.5 Required Area Characteristics

As with any tree fruit orchard, as shown in Figure 6.5, plantation or plantings are long-term commitments because the trees live and produce for a long time. There is typically a 5- to 10-year period before trees bear fruit in sufficient quantities to be economical. Thus, care must be taken in selecting a suitable site. Soil, rainfall, climate, water availability, and drainage all must be suitable for the trees that are to be planted (additional information about the soil requirements for tree plants is given in Chapter 9). When a suitable site is found, land is cleared and suitable tree stock is planted.

Trees are planted equal distances from each other, the distance depending primarily on the nature size of the trees. The larger the tree the farther apart they are planted

Figure 6.5. Peach orchard in Georgia, United States.

and vice versa. The very large mango trees may be planted 12 m apart while smaller peach trees and dwarf trees may be 2 m apart. In some cases, trees are planted in rows where the trees are closer together in rows than the distance between rows. For instance, trees may be 6 m apart in rows while the rows are 8 m apart. Variations on the distance will depend on soil and other characteristics, as noted above, and on the equipment to be used, tractors, sprayers, and the like. Today there is a trend to plant trees closer together in rows with more space between rows.

During this initial growth of trees, they must be cultivated and protected. Weeding and fertilization must be done as well as spraying or other activities to protect the young trees from damage. In addition some trees will be "trained," often by pruning, to induced growth in a certain desired way, for example, to make them able to bear without limbs breaking or to follow a certain form that will facilitate harvesting.

6.3.6 Long-Term Care

Individual trees or orchards must have enough water and be fertilized if there is to be a continuing supply of fruit from the trees. If there is not sufficient rainfall to support the trees throughout the year, irrigation will be needed. For those areas where rainfall is sufficient for only part of the year or where there is insufficient rainfall, irrigation water in quantities sufficient to meet tree needs must be available, and a method of supplying it to trees at appropriate times must be in place. As with all irrigation, care must be taken to prevent salt buildup in the soil. This is accomplished by adding enough irrigation water to leach salts out of the soil profile (Chapter 9).

Many people think of trees as just growing wild without the need for human interference. When growing fruit trees, collecting the produce results in removal of nutrients from the soil, and these nutrients must be replaced for continued long-term, sustainable, and satisfactory production. In addition nutrients removed by weeding and pruning are replaced by fertilization that can be manure but is more often chemical fertilizer. As with other crops the nutrients of greatest concern are nitrogen, phosphorous, and potassium. The amount of fertilization needed will depend on the crop being

Figure 6.6. Protecting young mangoes with paper coverings; inset shows closeup of covered fruit. (Courtesy of Petra Conklin.)

grown and the innate fertility characteristics of the soil. Also the time of fertilization will be determined by the type of tree, climate of the area, and the fruiting pattern. More information about fertilizers is available in Chapter 10. An interesting example of the relationship between fruit production and fertilization is found in the mango. Mango trees are induced to flower by applying nitrate fertilizer.[1]

6.3.7 Protection of Fruit Trees

There are three approaches to protect the fruit trees. Roots from varieties resistant to various pests can be used as the graft stock for the desired varieties. This then protects the tree from soil-borne diseases. Physical barriers such as wrapping the trunk of trees can protect them from damage by animals, particularly deer and rodents. The trees themselves can be sprayed with pesticides and the area between the trees can be sprayed with herbicides. It is also possible to wrap individual fruit, as shown in Figure 6.6, to protect it from attack by insects.

6.3.8 Harvesting and Postharvest

Some fruit can be harvested before it is mature and will mature or ripen during storage and transport to market. This is common with bananas and avocados. Other fruit is picked when it is ripe, such as is the case with apples. In both cases care must be taken not to damage the fruit during harvest as this will lead to bruising or entry of decay organisms into the fruit, making their shelf life short. Commercially, fruit is often hand picked and placed in boxes in the field. Boxes are then transported to storage facilities where fruit is washed, sorted, and reboxed. If the market is close, storage times will be minimal and storage problems will be minimized.

Any time fruit is handled it may become bruised and thus must be handled carefully during packing, unpacking, and transport. This includes even short-distance transport from the orchard to the home. Packing materials such as straw and sawdust, rice hulls and wrapping with paper are all common ways of protecting fruit during transport. After picking and before packing, fruit must be washed. This helps remove soil and other contamination, which carries organisms that cause damage. Both of these activities are common to most fruits and berries. Cleaning is also applicable to nuts, although because of their shells, they do not need a lot of protection during shipment unless they are removed from their shells.

Most tropical fruits are picked and eaten fresh off the tree with little storage. The time between picking and eating is often measured in hours or at most days. Because of the fragility of tropical fruit and the general tropical conditions of high temperature and in many countries high humidity, fruits are not stored and methods of storage have not received a great deal of attention. In many areas of the world the lack of refrigeration facilities means that cooling and storing at reduced temperature is not a option. However, even if refrigeration is available, many tropical fruits are injured by exposure to low temperatures. In addition ripening may not continue and undesirable changes in the fruit may occur. For these reasons other methods of preparing for storage and methods of storage of tropical fruits are used.

In some cases dipping tropical fruits in hot water is effective in killing disease organisms and prolonging storage life. Recommended methods vary in time and temperature, between 5 minutes at 55°C to 30 minutes at 42°C and 20 minutes at 49°C, depending on the fruit. In addition various standard methods of spraying with pesticides, fumigation with various fumigants, and controlled atmospheres of various types are used (Chapter 10 for more about these methods of prolonging storage life).

Subtropical fruits particularly the citrus, can be stored at room temperature for some time without deterioration. As with all fruits, longer storage is best done at lower temperatures after cleaning. Oranges are unique in that they may be green when ripe and are often exposed to ethylene gas to cause them to turn orange. This process is usually carried out during storage.

Temperate fruit storage life can be extended by storing them at low temperatures that prevent the growth of disease organisms and slow or prevent the continued ripening of fruit. This storage method has been used for a long time and started even before the development of refrigeration with the use of cellars and icehouses. Controlling humidity is also important in prolonging the storage of temperate fruits.

6.3.9 History

As with most agricultural production the first uses of fruit trees as a source of food is lost in the past. Tropical fruit trees are thought to have originated in either tropical central or South America, Asia, or Africa. Mangos developed on the Indian subcontinent, kiwifruit in China, and carambola in Sri Lanka and the Moluccas. Pineapple, papaya, and guava developed in south, central, and tropical America, respectively.

Temperate tree fruits also have a variety of points of origin. Apples, pears, cherries, and plums developed in the Caucasus, southeastern Europe, Caspian–Black Sea, and Caucasus–Black Sea areas, respectively. Peaches and apricots on the other hand developed in China and western China, respectively.

Figure 6.7. Mature coconut.

Subtropical fruits developed in the Orient with the exception of the grapefruit, which developed from the shaddock also known as the pummelo (many variations on the spelling are given in the Glossary). However, the exact development of the grapefruit and the origin of the shaddock are both unknown.

6.3.10   Other Fruit-Producing Plants

In addition to trees, fruits are produced on several other types of plants. In the tropics banana and coconut are grown on treelike plants. Both banana and coconut are shown in Figure 6.1. Figures 6.7 and 6.8 show the outside and inside of a coconut, respectively.

Figure 6.8. Inside of mature coconut showing husk on the outside, shell and meat on the interior.

Figure 6.9. Papaya plant with unripe papaya fruit.

Figure 6.10. Pineapple nearing maturity. (Courtesy of Petra Conklin.)

Papaya fruit on a papaya plant is shown in Figure 6.9. Certain fruits are produced on bushes or bushlike plants, while pineapple (Fig. 6.10) is its own unique type of plant.

Some berries are produced on brambles while others are produced on low-growing vine-type plants. Each of these, and other unique fruit, berry, or nut production characteristics will be discussed in the appropriate section below.

Cultivation, fertilization, protection, harvesting, transport, and storage characteristics of other fruits and berries are basically the same as for tree fruits as discussed in above sections.

6.4 TROPICAL FRUITS

Tropical-fruit-producing countries are shown in Figure 6.11, which was produced by summing FAO data on world production of avocado, banana, kiwi, mango, papaya, and pineapple. Tropical fruits are produced on trees and on a number of other different types of plants, which in some cases look like trees although they are botanically very different from a true tree.

6.4.1 Tropical Tree Fruits

Tropical fruits, three common examples of which are shown in Figure 6.12, are similar in size and stature to other fruits and grow on trees similar to other fruit trees. There is one striking difference in that some of these fruit trees bear fruit on stems on the trunk of the tree rather than on stems along or on the end of branches. One common occurrence of this type of fruit production is the jackfruit, as pictured in Figure 6.13. Another difference is that the fruit comes in forms that are quite different from one another.

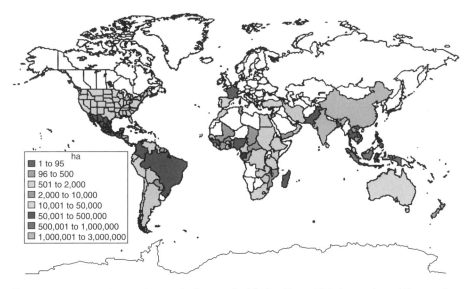

Figure 6.11. Map of countries producing tropical fruits. (From FAO data at http://faostat.fao. org/default.aspx.)

Figure 6.12. Three common tropical fruits: mango, guava, and carambola. Seeds in each type of fruit is indicated.

Avocado, mango, and many others have skin that surrounds flesh that is eaten and have a large seed in the middle. In other cases the fruit may have a thick rind surrounding a collection of flesh that is eaten but that also surrounds a large seed. An example of this arrangement is found in the jackfruit; see Figure 6.13. In other cases the seed may be contained in the edible portion of the fruit and the fruit and seed are both eaten, for

Figure 6.13. Jackfruit with thick stem originating from trunk of tree and fruit shown in inset (connection unseen).

TABLE 6.1. Selected Fruits of Tropical Trees: Characteristics and How Eaten

Common and Scientific Names	Characteristics	How Eaten
Avocado, *Persea americana*	Green/purple skin with flesh inside and roundish seed in middle	Cut in half, remove seed, and eat flesh with spoon. Sweetener may be added
Carambola,[a] *Averrhoa carambola*	Green-yellow cross section star shaped	Whole—skin, flesh, and seed
Guavas, *Psidium* spp.	Skin surrounding flesh with imbedded seeds	Whole—skin, flesh, and seed
Jackfruit, *Artocarpus heterophyllus*	Thick bumpy peal, fleshy fruit around large seed[b]	Peal removed, fleshy fruit removed from seed and eaten
Mango, *Mangifera* spp.	Thin skin yellow when ripe	Pealed, flesh around seed eaten
Pummelo, *Citrus maxima*	Very similar to grapefruit	Sections removed, peeled, and the inner flesh eaten

[a]Also called star fruit.
[b]Seed can also be eaten.

example, guavas (Fig. 6.12). Table 6.1 gives a summary of selected tropical fruits, their characteristics, and how they are commonly eaten.

Tropical fruit trees can be very large, often as large or larger than any other common tropical tree. This makes protecting and harvesting fruit difficult. In the case of mango the upper most reaches of the tree are not readily accessible by normal spraying equipment. Fruit can, however, be protected by covering it with paper bags (Fig. 6.6). Harvesting is done by placing ladders in the trees (not only resting on the ground extending into the tree but also the whole ladder between branches of the tree; Fig. 6.6) and climbing throughout the tree on these ladders. Fruit can also be protected by smoke. A smoky fire is burned under the tree and the resulting smoke drives insects away from the fruit.

All tropical tree fruits are low in protein; this means that they cannot be a significant source of protein in the diet (Fig. 6.14). Avocado has the highest amount of fat, which is indeed the largest component of this fruit. Durian also has some significant fat content while the others, excepting pummelo[¶] which has almost no fat, have minimal amounts, and jackfruit are relatively high in carbohydrate. Breadfruit, durian, and jackfruit have significant carbohydrate contents while guavas, kiwi, and mangoes have moderate levels, and the remainder have low carbohydrate contents comparatively speaking. All the fruits have small amounts of fiber with avocado having the largest amount.

Fruits are noted for contributing vitamins to the diet. Of the tropical fruits under discussion, guavas have the highest amount of vitamin C and the second highest values for both A and B vitamins. Papaya contains the highest amount of vitamin A

[¶]Pummelo has many different spellings that are given in the Glossary.

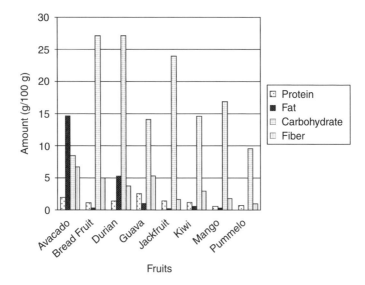

Figure 6.14. Major nutrient content of fruits of tropical trees. (Data in USDA National Nutrient Database for Standard Reference, Release 17, http://www.nal.usda.gov/fnic/foodcomp/Data/SR17/reports/sr17page.htm.)

while mango is highest in B vitamins. All these fruits except pummelo have significant amounts of all the B vitamins.[||] In terms of minerals, avocado is highest in iron and potassium while papaya is high in calcium. These fruits thus contribute significantly to the vitamin and mineral needs of peoples living in tropical climates.

It is very common to see people in tropical climates eating fruits as snacks several times a day. In addition, fruit will be used as part of a meal both by itself, mixed with other components of the diet, and cooked with dishes to add flavor, color, or texture. Because of the regular consumption of multiple servings of fruit a day, people in these climates obtain significant amounts of vitamins and minerals from this source, at least in the seasons when the fruits are available.

Green Mangoes
In all mango-growing areas, Asia, Africa, and Central America, one will commonly see people eating green mangoes. They are cut into sections, commonly dipped in salt or a salty mixture of small fish and eaten as a snack. It is the crunchy, sweet/sour taste that makes green mangoes a favored snack.

6.4.2 Other Important Tropical Fruits

Countries producing other tropical fruits are shown in Figure 6.15, a map produced on the basis of FAO data on world production of banana, papaya, and pineapple. The

---

[||]On the basis of a summation of all the B vitamins present.

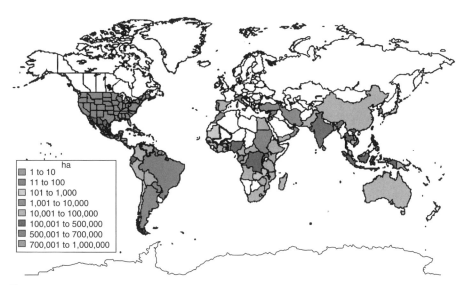

Figure 6.15. Map of countries producing banana, papaya, and pineapple. (Mapped from FAO data at FAOSTAT at http://faostat.fao.org/default.aspx.)

largest amounts are produced in Brazil and India with other tropical countries in South America, southern Africa, and Asia producing smaller amounts. Table 6.2 describes the characteristics of the fruits and how they are eaten. The banana plant is basically all stem and leaf while the papaya is classified as an herb; see Figures 6.1 and 6.9. Pineapple grows as a unique short bushy plant, as seen in Figure 6.10. All three are extremely important fruit-bearing plants in the tropics.

As with all tree fruits both banana and papaya can be found in plantations (bananas), orchards, or other large plantings and as individual or a small number of plants around the house.

TABLE 6.2. Selected Other Tropical Fruits: Characteristics and How Eaten

Common and Scientific Names	Characteristics	How Eaten
Banana, *Musa* spp.	Skin around flesh; no seeds[a]	Peel and eat
Papayas, *Carica papaya*	Much like a melon with outside skin and interior flesh and seeds (seeds are not imbedded in flesh)	Seeds scraped out and inner part eaten ripe; whole fruit can be eaten when not ripe
Pineapple, *Ananas comosus*	Rough out side with inner flesh inside and a central fibrous core	Outside carved off, "eyes" removed, and flesh to core eaten (core of some pineapple can be eaten)

[a]Occasionally seeds are found, but they are typically not viable.

Banana is very different from other fruits in a number of aspects. First, each stalk or trunk bears only one bunch of bananas in its lifetime. Second, new plants come from young plants growing from the base and roots of an old plant. These are called suckers and can be removed and transplanted to a new location to produce new productive plants. Banana fruit almost never has seeds and the seeds that do occur are not useful for producing new plants. In addition to the fruit, the flower at the end of the banana stalk can be used to make salads.

The papaya has many different varieties, and is unique in its cultivation; see Figure 6.9. The common papaya has both male and female plants and requires at least one of each for production of fruit on the female plant. In practice, however, a single male plant will be used to pollinate several female plants. There are also types of papaya that are said to be a hermaphrodite, that is, they are self-fertile, can self-pollinate, and thus do not need a separate male plant. Both these and the common papaya are usually propagated from seed. There is also a seedless variety that must be propagated vegetatively. However, all can be propagated by tissue culture techniques**, including genetically engineered species.

In terms of plant growth, fruit size, and other characteristics, there is a great variety of papayas. There are both papaya plants that can grow to a height of 7 m or more (23 ft), thus making harvesting difficult, and dwarf papayas that are easier to harvest. Fruits are usually green, turning yellow as they mature. They are often picked before they are completely ripe because ripe fruits are often attacked by birds and fruit bats. The fruit itself comes in several sizes from small, called solo or individual, since one is enough for one person, to the large, which may be as large as a watermelon although of a different shape. The color of the ripe fruit can range from light yellow to dark yellowish red.

Although most people eat papaya after it has ripened, many prefer it unripe when it is still crunchy to eat. In addition both the ripe and unripe papaya can be used in various dishes including salads. Grated papaya can be pickled and used as garnish on various dishes.

Pineapple is also unique in its propagation; see Figure 6.10. The small growth at the top of the pineapple fruit, called a crown, can be cut off and planted to produce a new pineapple. Suckers, young plants growing from the roots of an older plant (similar to banana), and slips, which are similar to crowns but grow from the stem or on the sides of the fruit itself, are other ways of producing new pineapple plants. The latter two methods of production are faster growing and thus preferred. All these are commonly called vegetative reproduction.

As is seen in Figure 6.16 all these fruits are highest in carbohydrates and relatively low in other major nutrients. Banana is highest in fiber followed by papaya and pineapple. Protein is slightly more than 1 percent in banana, but the other fruits are less than 1 percent and all have less than 1 percent fat.

Papaya has the highest vitamin, calcium, and potassium content while banana has the highest iron, potassium, and fiber content. Pineapple is the lowest in all the

---

**Tissue culture involves taking some tissue from a plant, placing it on special sterile growth media, and producing a whole new plant from these few cells.

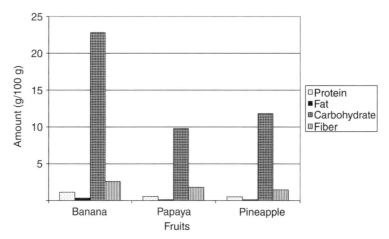

Figure 6.16. Major nutrient content of nontree tropical fruits. (From data in USDA National Nutrient Database for Standard Reference, Release 17. http://www.nal.usda.gov/fnic/foodcomp/Data/SR17/reports/sr17page.htm.)

categories of vitamins and minerals. Of these tropical fruits papaya is the highest in B vitamins due to its relatively high levels of niacin, folic acid, and pantothenic acid. However, banana has a higher content of niacin, B_6, and pantothenic acid than does papaya.

6.5   SUBTROPICAL FRUITS

World production of subtropical fruits, from FAO data on the basis of dates, figs, grapefruit, lemons, limes, olives, and oranges, is shown in Figure 6.17, which also shows that most production is in the Americas and Asia with the highest production in countries surrounding the Mediterranean. Table 6.3 gives the characteristics of some selected subtropical fruit and how it is most commonly eaten.

Subtropical fruits are favored around the world by all peoples. Production is typically not possible in temperate zones and is not common in humid tropical zones except at high-altitude zones in tropical regions where they can be successfully produced. Both oranges and passion fruit are grown commercially in Zimbabwe at the higher altitudes. Passion fruit is commonly shipped from Zimbabwe to Europe.

The most widely grown subtropical fruit is citrus, which grows on trees. Many citrus trees are small in stature when compared to tropical and temperate fruit trees particularly the mango and apple trees. Because of their small stature, they are relatively easy to spray to control insects or other pests. In addition to citrus there are many other important subtropical fruits such as figs and olives, which grow on trees; see Figure 6.18. While these are preeminent fruits of the Middle East and Mediterranean, their production in all other parts of the world is minimal or nonexistent. These fruits

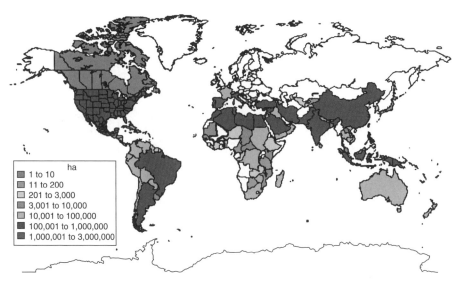

Figure 6.17. Map of countries producing subtropical fruits. (Mapped from FAO data at FAO-STAT at http://faostat.fao.org/default.aspx.)

are grown on trees and are propagated and cultivated as described above for tropical fruit trees.

Two common subtropical fruits that do not grow on trees are dates and passion fruit. Dates grow on date palms, which, like coconut, look like trees but are not. The second is passion fruit, which grows on vines rather than trees. There are a wide variety of passion fruits produced in both Africa and South America.

TABLE 6.3. Selected Fruits of Subtropics: Characteristics and How Eaten

Common and Scientific Names	Characteristics	How Eaten
Dates, medjool, *Phoenix dactylifera*	Fruit around central seed	Flesh around seed, often dried with sugar added
Figs, *Ficus* spp.	Thin skin surrounding edible flesh and many seeds	Flesh and skin eaten raw or cooked
Grapefruit, *Citrus paradisi*	Skin surrounding sectioned flesh with seeds imbedded	Skin removed and sections eaten (seeds spit out)
Kiwi, *Actinidia deliciosa*	Skin surrounding seeds imbedded in flesh	Whole fruit eaten
Olives, *Olea europaea*	Skin surrounding flesh around seed	Seed often removed and skin and flesh eaten
Orange, *Citrus sinensis*	Skin surrounding sectioned flesh with seeds imbedded	Skin removed and sections eaten (seeds spit out)
Passion fruit, *Passiflora edulis*	Outer skin, wrinkled when ripe	Outside of fruit removed and interior eaten

Figure 6.18. Olive tree with fruit shown in inset. (Courtesy of Ali Shehadeh.)

Subtropical fruits are low in protein and fat; see Figure 6.19. The protein and fat content are slightly over 2 percent or less, with most being less than 1 percent. The exception would be olives, which have a relatively high oil content. Of these fruits lemons have relatively high vitamin C content while grapefruit and olives have relatively high B vitamins. Dates have by far the highest amount of B vitamins while lemons have about the same amount, although somewhat lower than dates. In terms of minerals dates have the highest levels of iron and potassium while lemons have the highest amount of calcium.

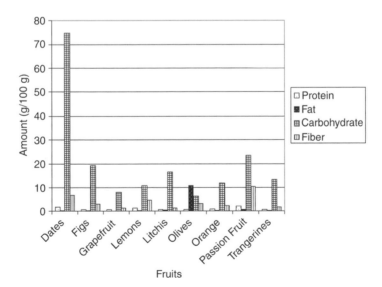

Figure 6.19. Major nutrient content of subtropical fruits. (From data in USDA National Nutrient Database for Standard Reference, Release 17. http://www.nal.usda.gov/fnic/foodcomp/Data/SR17/reports/sr17page.htm.)

6.6 TEMPERATE FRUITS

Temperate-fruit-producing countries are shown in Figure 6.20 on the basis of FAO data on the world production of apples, apricots, cherries, nectarines, and pears. Most of these countries occur in the Northern Hemisphere and to a lesser extent in South America. However, Australia also produces large amounts of temperate fruits. Other countries producing large amounts of temperate fruits occur in northern and South Africa and in a limited number of countries in Europe and Asia. Characteristics of temperate fruits and how they are commonly eaten are given in Table 6.4.

Temperate fruit trees are of moderate height compared to tropical fruit trees. They are grown in large orchards and in small plantings of two or more trees. It is uncommon to find single trees without close neighbors. In part this is done because temperate fruit trees commonly require a pollenizer to produce fruit. This may be another tree of another variety of the same fruit, which is very common, and this tree is chosen as the pollinator because it produces large amounts of pollen. Apples are a good example in that pollen from flowers on the same tree will not pollinate each other, even on different parts of the same tree; and pollen of the same variety will not pollinate another tree of the same variety effectively.

Also because of the importance of pollination, honey bees are frequently kept in hives in orchards. The honey produced is sold to provide additional income to the fruit farmer.

Pruning is commonly practiced on temperate fruit trees and has several purposes. In younger trees it is used to produce a stronger tree less likely to break under heavy fruit production. It also tends to make the trees smaller, thus making spraying and

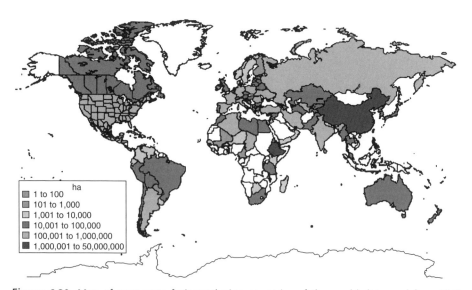

Figure 6.20. Map of temperate fruit-producing countries of the world. (Mapped from FAO data at FAOSTAT at http://faostat.fao.org/default.aspx.)

TABLE 6.4. Selected Temperate Zone Fruits: Characteristics and How Eaten

Common and Scientific Names[a]	Characteristics	How Eaten
Apple, *Malus domestica*	Thin skin with flesh surrounding core, which contains seed, many different colors	Skin and flesh eaten; core discarded
Apricots, *Prunus armeniaca*	Skin surrounding flesh, which surrounds seed in center	Skin and flesh around seed eaten
Cherry, *Prunus apetala*	Thin skin with flesh surrounding seed in center, may also be seedless, may be red or yellow	Skin and flesh usually eaten
Nectarines, *Prunus persica*[b]	Skin around flesh, which surrounds seed	Skin and flesh are eaten
Peach, *Prunus persica*[b]	Thin skin with flesh surrounding large seed	Skin and flesh eaten; sometimes skin is removed before eating
Pear, *Pyrus communis*	Thin skin with flesh surrounding core, which contains seed, may be green, yellow or brown	Skin and flesh eaten; core discarded
Plum, *Prunus maritime*	Thin skin with flesh surrounding large seed	Skin and flesh usually eaten

[a]Only one scientific name is given and many fruits have many species and subspecies.
[b]Peaches and nectarines are the same genus and species but different cultivars (cultivars are distinct variations of the same genus species).

harvesting easier (cherries growing in Fig. 6.21). In older trees it may be used to revitalize tree growth. Pruning may decrease total yield and yet result in an increase in marketable fruit.

Apples, apricots, nectarines, and pears are all relatively high in carbohydrate, while apples, apricots, and pears have the highest levels of fiber; see Figure 6.22. All these fruits are low in protein and fat. Apricots and plums are relatively high in vitamin C while apricots are high in vitamin A and all the others are relatively low. Apricots, nectarines, and peaches are relatively high in B vitamins while plums have moderate levels and the others low levels. Apricots are highest in iron and potassium while nectarines[††] and peaches have moderate levels of these two minerals. Apricots and pears have high levels of calcium and fiber while the other fruits have moderate levels.

[††]Nectarines are actually peaches that lack the genetics for fuzziness.

Figure 6.21. Cherries on tree and ripe cherries in inset.

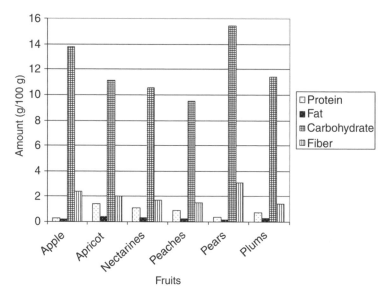

Figure 6.22. Major nutrient content of temperate fruits. (From data in USDA National Nutrient Database for Standard Reference, Release 17. http://www.nal.usda.gov/fnic/foodcomp/Data/SR17/reports/sr17page.htm.)

6.7 BERRIES

6.7.1 Climatic Adaptation

Berries are most often grown in temperate regions, although they are also grown to a lesser extent in subtropical areas. In high-altitude tropical areas they may also be grown successfully.

6.7.2 Importance in Human Nutrition

Berries do not constitute a major portion of most people's diets, although significant quantities may be eaten when they are in season. In these cases berries may provide significant amounts of carbohydrates, vitamins, and minerals depending on the type of berry. The nutritional characteristics of berries is given below.

6.7.3 Plantings, Cultivation, and Long-Term Care

Berries may bring to mind fruit growing on thorny stalks or low-growing bushy plants. However, there are a diversity of plant types that produce what are commonly called berries. Blackberries and raspberries (Fig. 6.3) do grow on thorny stalks. However, of the berries in Table 6.5, only mulberries grow on trees, while blueberries (Fig. 6.4) grow on bushlike plants. Strawberries (Fig. 6.23) and cranberries are produced by low-growing bushy plants, cranberries being grown in bogs or in marshy areas. Grapes are most often trained to grow on trellises (see Chapter 4 for more on trellises).

Tree fruit such as mulberries and bush fruit such as blueberries are long-lived plants that produce for many years. They are often planted, protected, and harvested much like any tree fruit, and spacing between plants is dependent on plant size and the equipment used in berry production. Blackberries and raspberries grow on thorny stalks. The first year's growth is vegetative and one-year-old stalks bear fruit. After fruiting the old stalks are removed. However, a planted area is kept in the same berry production for many years.

Both cranberries and strawberries are vinelike plants, but this is where the similarity ends. Cranberries are evergreen plants that grow in acid peat bogs while strawberries require slightly acid well-drained soils. In addition the fruits are quite

TABLE 6.5. Selected Common Berries: Characteristics and How Eaten

Common and Scientific Names	Characteristics	How Eaten
Blackberries, *Rubus* spp.	Rounded with sections containing seeds	Stems removed; berries eaten whole
Blueberries, *Vaccinium corymbosum*	Round fruit usually without seeds	Stems removed; berries eaten whole
Cranberries, *Vaccinium oxycoccus*	Small round berries with thin skin and few seeds	Eaten whole and jellies
Grapes, *Vitis vinifera*	Skin surrounding flesh containing seeds	Skin and flesh eaten
Mulberries, *Broussonetia papyrifera*	Round with sections containing seeds	Stems removed; berries eaten whole
Raspberries, *Rubus* spp.	Round with sections containing seeds	Stems removed; berries eaten whole
Strawberries, *Fragaria virginiana*	Conical shaped with flesh containing seeds	Stems removed and whole berry eaten

Figure 6.23. Ripe strawberries, raspberries, and blueberries as presented in the market in the United States. (Courtesy of A. Conklin, Jr.)

different in their size, shape, and morphology. Cranberry vines are used to plant production areas and take several years to start production. Once producing, berry production may continue for 100 years or more.

In temperate regions strawberry plantings are made from young plants, which produce runners that produce new plants, and the planting produces berries for about 5 years before needing to be replanted, which does not occur in the same field until the land has "rested" for a year to help control diseases and weeds. In the hot subtropics, for example, Florida, planting occurs in November and berries are harvested the following December to May after which the plants are plowed under.

6.7.4   Harvest and Postharvest Care

Typically, berries change color when ripe and are then harvested. In many cases the berries are hand harvested and transported from the field in boxes to a processing area where they are washed and packaged for sale. Washing and cooling leads to increased storage or shelf life. All berries are highly perishable and must be handled with great care to keep them in good condition until sold.

6.7.5   Consumption

Berries are most typically eaten fresh and without cooking. Although they may be eaten as part of the dish of a main meal, they are more often added to main dishes as garnishes. They are also eaten as desserts either fresh or in combination with other dessert items such as ice cream.

6.7.6   World Production

World berry production is shown in Figure 6.24 on the basis of FAO data on the world production of berries. The largest berry production is in the United States and the Russian Federation with lesser, but still large, production in Canada, Germany, several eastern European countries, and some countries in the Middle East. The characteristics of selected berries and how they are commonly eaten is given in Table 6.5. In addition to being eaten as listed in Table 6.5, berries are also eaten as a snack. They are also used in salads, as toppings for other foods, particularly breakfast cereals in the

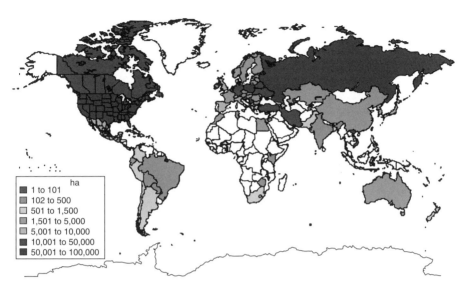

Figure 6.24. Map of countries producing berries. (Mapped from FAO data at FAOSTAT at http://faostat.fao.org/default.aspx.)

United States, and in the preparation of both pies and other pastries, such as tarts and donuts.

6.7.7 History

Berries are usually restricted by their soil and climate requirements. Thus, areas where they are commonly grown in large quantities are considered to be their place of origin. This would be particularly true of blueberries and cranberries but is less true for strawberries and blackberries, which are grown under a wider range of environmental conditions. Strawberries are grown on all continents and are a high value crop.

> **Strawberries**
> Although wild strawberries are harvested in some places, the modern cultivated strawberry is a hybrid of northeastern North America and the Chilean Andes strawberries, which were first crossed in a greenhouse in France. (Personal communication, Malcolm Manners, Horticulture Department Florida Southern College.)

6.7.8 Nutritional Value

Blueberries, cranberries, grapes, and raspberries have a relatively high level of carbohydrates while the remaining berries have moderate levels (Fig. 6.25). Both blackberries and cranberries have relatively high levels of fiber while the remainder of the berries have lower levels. All the berries have low protein and fat contents. Mulberries have the highest level of vitamin C while blackberries and raspberries

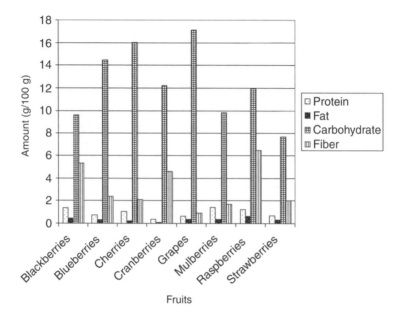

Figure 6.25. Major nutrient content of berries. (From data in USDA National Nutrient Database for Standard Reference, Release 17. http://www.nal.usda.gov/fnic/foodcomp/Data/SR17/reports/sr17page.htm.)

also have high levels, and grapes have the lowest level. Raspberries have the highest level of vitamin A while blackberries have the second highest level, grapes the third highest, and all the others have relatively low levels. Raspberries and blackberries have the highest levels of B vitamins with all the other berries having moderate levels.

Mulberries have the highest level of iron and a high level of potassium. Blackberries and raspberries also have higher levels of iron and potassium. Grapes have a high level of potassium while the other berries have a moderate level. Mulberries also have the highest levels of calcium while raspberries and blackberries have a high level and the other berries have from moderate to low levels.

6.8　NUTS

6.8.1　Climatic Adaptation

Nut trees are grown in all climatic conditions, however, each type of nut is typically restricted to one climate. Walnuts and chestnuts are typically grown in temperate regions while cashews are grown in tropical areas.

6.8.2　Importance in Human Nutrition

Most nuts are eaten as snacks and thus do not add greatly to the dietary intake of most people. However, when eaten, nuts can provide needed dietary inputs as seen in the section on the nutritional characteristics of nuts given below.

6.8.3 Plantings

As with fruit trees, nut trees are often produced by grafting a particular, often named, tree type bud onto a specific rootstock. Often the rootstock is chosen from a disease-resistant tree of the same type. These grafted plants will be grown for a year or more in a nursery before transplanting into an orchard. This process is particularly important if one wishes to be assured of a high-quality crop with known characteristics. Planting from seed will produce plants with varying characteristics and will bear fruits with different characteristics. Thus, to have an orchard that will produce the desired nut of the specific type, grafted plants are necessary; see Figure 6.26.

In a limited number of cases, for instance, black walnuts, trees are grown from seed. This approach may be used when the nuts to be produced are not intended to be the major crop or the characteristic of the nut is not that important. In the case of black walnut the wood of a mature tree is highly valuable, and so the planting may have the primary purpose of producing wood rather than nuts.

Some nut species require a pollinizer tree for nut production. This is a tree, sometimes considered a male tree, that produces pollen for the crop. In the case of almonds, for instance, cross pollination with another cultivar[§§] of almond is necessary. Because different cultivars produce pollen at different times, it may be necessary to have two different cultivars in addition to the main planting. Typically, only a limited number of trees are needed for cross pollination.

6.8.4 Cultivation

As with any tree, nut plantations or plantings are long-term commitments. Because of this, care must be taken in selecting a suitable site. Soil, rainfall, climate, water availability, and drainage all must be suitable for the trees that are to be planted (additional information about the soil requirements for tree plants is given in Chapter 9). When a

Figure 6.26. Pecan nut orchard. (Courtesy of Georgia Farm Bureau: Shot by Jennifer Whittaker.)

[§§]A cultivar is a distinct type of the same species of plant that has unique characteristics, that are predictably displayed.

suitable site is found, land is cleared, and suitable tree stock is planted. The trees are weeded and protected from insect and animal damage. This is often a 5- to 10-year commitment before trees bear nuts in sufficient quantities to be economical.

6.8.5 Long-Term Care

Nut trees may produce nuts for long periods of time, up to 100 years (up 250 years for black walnuts!). For high productivity, the plantings must be cared for by keeping the area under the trees clean and free of debris; see Figures 6.27 and 6.28. Some nut trees may benefit from pruning, although some may not. It may also be necessary to spray the trees from time to time to protect them from insects.

6.8.6 Protection

There are three approaches to protection of nut trees, although in general they need less protection than fruit trees. Roots from varieties resistant to various pests can be used as the graft stock for the desired varieties. This then protects the tree from soil-borne diseases. Physical barriers such as wrapping the trunk of trees can protect them from damage by animals such as deer. Trees can be sprayed with pesticides to control insects and herbicides between the trees to control weeds.

6.8.7 Harvest and Postharvest

When ripe, many nuts fall naturally from the tree and can simply be picked up from the ground. This may be a satisfactory procedure for situations where the nuts are intended for family or local use. For commercial production, trees may be mechanically shaken (Fig. 6.27) to remove nuts that then can be swept up (Fig. 6.28) and collected for processing. Care must be taken not to damage the tree while carrying out this operation.

In some cases the nuts must be harvested by hand because they do not fall off the tree naturally or they are highly perishable and must be handled carefully. This would be the case with chestnuts.

Each type of nut is unique, and several, such as the cashew and chestnut, have very unusual characteristics. The cashew nut forms on the end of a fruit called an "apple." In some places the apple is used as food; in others it is discarded. The nut is encased in a shell that contains a liquid that can produce skin blisters and swelling. This liquid must be carefully removed before the nut can be used as food. The chestnut has a spiky covering that breaks open to release the nut when ripe. In many cases the nut itself is surrounded by a fleshy outer covering, as with black walnuts, that is removed and discarded before the nut is cleaned for in-the-shell sale. The shell is removed when the nut is eaten or processed for use in food production.

Handling after harvest will depend on the nut and its particular characteristics. In some cases the soft outer vegetative covering may need to be removed as in the case of cashews and black walnuts (the outer vegetative portion of cashews is edible and the outer portion of black walnuts can be eaten by domesticated and wild animals). In

Figure 6.27. Tree shaker used to shake ripe nuts from tree so that they can be collected as shown in Figure 6.28. (Courtesy of Georgia Farm Bureau: Shot by Jennifer Whittaker.)

Figure 6.28. Sweeping up fallen pecans after shaking trees; see Figure 6.27. (Courtesy of Georgia Farm Bureau: Shot by Jennifer Whittaker.)

some cases the nuts will need to be dried before transport or storage as is the case with chestnuts and macadamia nuts.

Nuts, having a shell, are handled in two different ways. The shell may be left on and cleaned before selling. Figure 6.29 shows nuts in the shell, shell opened, and nut

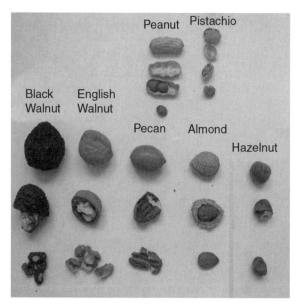

Figure 6.29. Nuts.

out of the shell. In this way the meat inside is protected during transport and storage. Otherwise, the shell may be removed and the meat harvested and packaged for sale. In this latter case the meat must be protected from spoilage.

6.8.8 History

Many nuts have limited areas of production compared to fruits, berries, and other foods. Commonly, less than 20 countries (Fig. 6.26) are reported to be major producers of most nuts. This makes identifying the origins of many nuts well established with the great exception of the coconut. Cashews are from the Amazon basin of Brazil, pistachios from south-central Asia, chestnuts northern North America and Europe, the walnut from the Mediterranean basin, and macadamia nuts from Australia.

On the other hand, the coconut is found in all tropical countries of the world, 90 countries in all, making identification of its origin impossible, although it is commonly believed that the Malaysia/Indonesia area is the center of origin. Because of its widespread occurrence and extreme usefulness, the coconut is produced in far greater quantities than other nuts (see below).

6.8.9 World Production

World nut production is highest in Asia, followed by South America and Africa. Within these regions, as shown in Figure 6.30, India, Brazil, Turkey, Indonesia, and Nigeria, in descending order, are the top producing countries. In 1979 Woodroof reported almonds as being produced in the largest quantities worldwide, followed by walnuts,

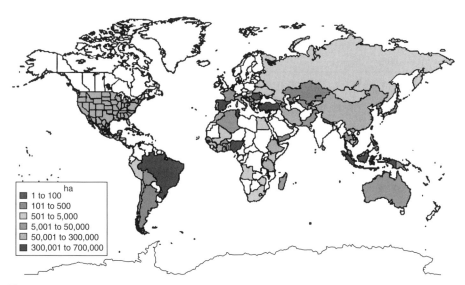

Figure 6.30. Map of counties producing nuts. Data is summed for almonds, cashew, chestnuts, pistachio, and others (does not include peanuts or coconuts). (Mapped from FAO data at FAO-STAT at http://faostat.fao.org/default.aspx.)

cashews, and chestnuts in descending order. In 2004 the FAO reported cashews being produced in the largest quantities, followed by almonds, walnuts and chestnuts.[2,3] Although the relative position of the nuts in terms of world production has changed, these four nuts are by far the most important. However, both coconut and peanut (which are not really nuts) individually are produced in far larger amounts than all other nuts combined; see sections below.

Table 6.6 gives the characteristics of selected nuts and how they are commonly eaten. Nuts, are often cooked or roasted in various ways before eating, often significantly changing the nut's taste. They may be cooked in hot oil or dry roasted and after cooking may be seasoned in various ways, commonly by adding salt but also including other spices.

Coconut and peanut (groundnut) have nut in their names and are thought of as being nuts, although neither is a nut and neither grows on a tree. Because of their unique characteristics, both of these important foods will be discussed separately in sections below.

6.8.10 Nutritional Value

Except for chestnuts, all nuts in Figure 6.31 are highest in fat, which in this case might better be called oil. These oils are commonly used as cooking oils, and in some cases, such as coconut, as fuel for cooking. Chestnuts are high in carbohydrates with cashews being second highest followed by pistachio and the other nuts being relative low. Almonds, cashews, hazelnuts, and pistachio are also relatively high in protein.

TABLE 6.6. Selected Nuts: Characteristics and How Eaten

Common and Scientific Names	Characteristics	How Eaten
Almonds, *Prunus dulcis*	Thin shell surrounding flesh	Shell cracked open and flesh halves eaten
Hazelnuts (filbert), *Corylus* spp.	Thin shell surrounding nut	Shell broken open and rounded nut eaten
Cashew, *Anacardium occidentale*	See explanation on page 221	Eaten after washing
Chestnuts, *Castanea* spp.	Thin relatively soft shell inner flesh eaten usually after cooking	Shell removed and flesh eaten
Coconut, *Cocos nuifera*	Outer husk surrounding shell with flesh adhering to inside of shell. Center filled with liquid	End cut off; liquid removed for drinking. Shell cut open and flesh scraped off and eaten
Macadamia, *Macadamia intgrifolia*	Rough or smooth shell around nut	Shell removed and nut eaten
Peanut, *Arachis hypogoea*	Outer shell removed; nuts inside removed	Eaten raw but more commonly boiled or roasted in the shell.

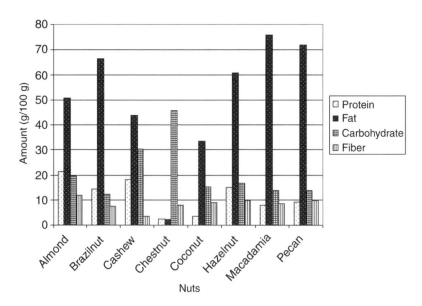

Figure 6.31. Major nutrient contents of nuts. (From data in USDA National Nutrient Database for Standard Reference, Release 17. http://www.nal.usda.gov/fnic/foodcomp/Data/SR17/reports/sr17page.htm.)

Nuts are relatively low in vitamins and have highly variable amount of minerals, particularly calcium. The one exception is chestnut, which has the highest level of vitamin C. However, all the other nuts are much lower in both vitamins C and A than fruits. They are about the same level or higher as fruits in terms of B vitamins and iron. Nuts on the whole have higher levels of potassium and calcium, with pistachio having a particularly high level of potassium, than many fruits.

6.9 COCONUT

Although coconut has nut in its name, it is not a nut and it does not grow on a true tree; see Figures 6.1, 6.7, and 6.8.

6.9.1 Climatic Adaptation

The coconut palm is ubiquitous and flourishes throughout the low-elevation humid tropics and is grown extensively in large plantations in South East Asia and the Pacific. In these areas it is recognized as a source of all that people need. It is not adapted to any other climate. Although it may grow in other areas, it will either not produce or produce very small yields.

6.9.2 Importance in Human Nutrition

For people living in coconut-producing areas it supplies an important part of their dietary needs. Although the coconut "water" and "meat" are often eaten only as a snack or dessert, coconut oil is very important in cooking and is essential in food preparation. It also provides calories in the diet. In addition to providing food in various forms, young coconut, mature coconut, coconut water and milk, and alcoholic beverages, the plant provides materials needed for shelter, fuel, and even clothing.

Coconut "Tree"
The coconut plant is more closely related to the grasses than trees. The coconut trunk is composed of a hard bark outside and a center containing a bundle of ropelike fibers. The trunk contains no true wood, although it can be cut into planks that are used as beams in building houses and other structures in tropical countries. However, during their productive life coconut trees are too valuable to be used as lumber.

6.9.3 Plantings, Cultivation, and Long-Term Care

As with fruit and nut trees, coconut has several varieties including both tall and dwarf. The dwarf varieties start bearing earlier and do not live as long. Because coconuts grow to as much as 25 m tall (82 ft), dwarf varieties are desirable because the fruits are closer to the ground and thus easier to harvest.

Coconuts germinate and grow just like any other seed. Often coconuts are allowed to germinate in a nursery before planting. They are placed on the ground, kept moist, and partially covered, allowing them to germinate. Once growing the young coconut plant is transplanted to plantations that may cover 100 or even thousands of hectares. In addition coconuts are planted along roads and on the edges of rice paddies and fields.

Areas around coconut trees are kept free of weeds, fronds, and damaged immature coconuts that fall from the trees. Coconuts may be grown without the addition of fertilizer, and little is done to protect trees or nuts from insects. However, rats are a particular problem as they can climb the coconut trunk and chew holes in nuts destroying them. Pieces of thin metal 20 to 30 cm wide are often attached to the trunks of coconuts to prevent rats climbing to the top. Young coconut trees may take up to 15 years to reach full production and produce for 50 years. Thus, areas planted to coconut must be suitable for such long-term use.

6.9.4 Consumption

In tropical areas coconuts are usually eaten or used as food in their young or immature stage; see also Table 6.6. In all other regions the mature coconut meat is used as a food. The coconut has liquid in the center, which is called coconut milk in some places, but is called coconut water in coconut-growing areas. Coconut meat can be ground and mixed with water to produce coconut "milk," which is also called coconut cream. The milk or cream is separated from the remaining meat by filtering.

The young coconut can easily be cut using a large knife. A hole is made in the husk and shell, and the water is removed and drunk or used for cooking. The coconut is then cut open and the meat, which is soft, scraped out and eaten (Fig. 6.32) or mixed with other ingredients to make various foods. Both the water and young coconut meat are favorite snacks in coconut-producing areas.

In other regions the mature fresh or dried coconut meat is used in preparation of food, mostly as a garnish or to add coconut flavoring.

6.9.5 Nutritional Characteristics

Raw coconut meat is high in fat and carbohydrate and relatively low in protein and has significant fiber content; Fig. 6.31. It has low levels of both vitamin C and the B vitamins. In mineral content it is highest in potassium and phosphorus and relatively low in all the others. The coconut water is highest in carbohydrate, although still low, and relatively low in all other constituents including vitamins and minerals.

6.9.6 History

Because of the ubiquitous nature of coconut in Asia, the Caribbean, and South and Central America, the origins of the coconut is not certain. It is assumed that mature coconuts dropped in the ocean and were transported (floated) to neighboring land masses where, after washing ashore, they sprouted and grew. Because of its ability to survive floating in ocean water, this is a probable method of dispersal of coconut.

Figure 6.32. Inside young coconut being removed by scraper, made form the husk of the coconut husk. Young coconut being eaten.

6.9.7 World Production

In terms of world production coconut is produced in far larger amounts than any other food with nut in its name, including peanut, which is also not a nut (Section 6.10). Indonesia and the Philippines are the world's largest producers of coconut, followed by India, Brazil, and Sri Lanka. All other countries produce one tenth or less the amounts produced by Indonesia and the Philippines.

6.10 OTHER USES OF FRUIT AND NUT TREES

Both fruit and nut trees have uses other than the production of fruit and nuts. In most cases they also produce high-quality, high-value lumber. Thus, when they are at the end of their productive lives, they can be harvested for lumber, providing the farmer with additional income. Planted on a staggered or rotational schedule, some trees can be harvested for lumber every year and be replaced with young trees. With this type of plan a fruit or nut orchard can be a continuing source of both produce and lumber.

In situations where there is a need to protect sloping or mountainous land, feed people, and produce lumber for housing and other needs, both fruit and nut trees, including coconut, can be the answer. Coconut is not a tree, but its long trunk is woodlike and is commonly used as an inexpensive building material. In addition such plantations can act as watersheds and thus provide a continuing source of water for the local population.

6.11 PEANUTS

6.11.1 Climatic Adaptation

Peanuts can be grown under any environmental conditions providing a growing season long enough and warm enough to produce nuts. After planting, it takes 30 to

40 days for flowering to begin and 70+ for mature nuts to be produced. Thus, the growing season must be longer than 110 days to produce significant numbers of peanuts. In addition the temperature needs to be 25 to 30°C for good production. Peanuts are grown in temperate, subtropical, and tropical regions.

6.11.2   Importance in Human Nutrition

Peanuts are highly nutritious, as shown below, but are most often eaten as snacks rather than a main part of a meal. However, some cultures do make significant use of peanuts in their main meals, such as in parts of India and Africa. In these cases peanut can provide significant calories and protein to the diet. They may also contribute significant amounts of vitamins and minerals.

6.11.3   Cultivation

Peanuts are grown much like any annual crop; see Chapters 3, 4, and 5. The field is plowed and harrowed, and ridges on which the peanuts will be planted may be produced. During these operations liming and fertilization is done as needed. The peanut seed may be planted shelled or unshelled, although the latter are slower to produce. As the plant develops it flowers and once pollinated (flowers are self-pollinating), a stem develops and grows into the ground as shown in Figure 6.33, where the nut is produced. Because the nut is produced underground, sandy or other light, low-bulk density, soils are preferred for peanut production. In spite of this preference peanuts are regularly and successfully produced on soils with significant clay content.

Peanut is grown as either a rainfed crop or under irrigation. Lack of water can lead to large reductions in yield, and thus in areas of uncertain rainfall irrigation is essential for assured yields.

During the growing season, plants must be protected from insect damage and weeded to remove competition. Losses from weed competition can be between 18

Figure 6.33. Peanut plant showing nut formation below ground.

and 96 percent, depending on the type of peanut and the growing conditions. At maturity the plants are dug or plowed out of the ground and the nuts collected and dried. Equipment for digging, removing soil, and inverting the peanut plants with nuts attached is available for large-scale production. Turning plants after digging allows for some drying of nuts in the field, which is essential for storage.

6.11.4 Postharvest Handling

Within limits there is an inverse relationship between peanut moisture content and storage. The dryer the peanut the longer it can be stored without damage. Peanuts dried to 6 to 8 percent moisture and left in the shell can be stored for a year or more without cooling if they are protected from moisture, insects, and rodents. Shelling can be done by hand or by threshers of various types; however, losses will occur if the nut is too dry when shelled.

6.11.5 Nutritional Characteristics

Peanuts are eaten fresh, boiled, or roasted either in the shell or shelled; see also Table 6.6. They may also be ground to produce peanut butter and to extract the oil, which is a high-value product. They may also be part of main dishes at meals or eaten as snacks. Use as a snack is particularly common in all areas where peanuts are grown.

Peanuts are an excellent food as they contain both a high amount of protein and provide a high energy content, in the form of oil with around 25 percent protein and 49 percent oil. They also contain a relatively low carbohydrate content, 16 percent, and add significant fiber, 8 percent, (these figures are on a raw basis). They are relatively high in minerals and B vitamins but low in other vitamins.

6.11.6 History

Peanuts, which originated in South America, are also called groundnuts. The name peanut is used almost exclusively in the United States while internationally the term groundnut is most commonly used. They are not nuts and are very different from true nuts and coconuts in that they are produced by an annual, low-growing, legume (Fig. 6.32). The edible seed (the "nut") inside a fibrous shell is shown in Figure 6.29. There are many other types of similar groundnut-bearing plants in Africa, but none is as widespread as is the peanut, which is produced on all continents and from tropical to temperate climates.

6.11.7 World Production

World peanut production is greatest in tropical, subtropical, and the warmer temperate regions of the world. By far the largest producers are China and India followed by Nigeria and the United States (the U.S. Department of Agriculture places the United States third rather than fourth). The major exporters are China, Argentina, and the United States. However, only about 5 percent of world peanut production is traded.

6.12   CONCLUSIONS

Fruits, berries, and nuts can be an important and nutritious component, providing essential proteins, carbohydrates, fats, fiber, vitamins, and minerals to people's diets. They can be found as individual plants or trees or as large commercial orchards. Tree fruits, nuts, berries, and other fruits are long-term crops because plants produce over long periods of time. This minimizes the amount of land preparation that must be done each year. All these types of produce must be protected from insects and disease organisms during growth and from damage during picking, transport, and storage. They can be eaten fresh off the tree or washed and transported to local markets where they provide income to local farmers. All can be treated in various ways, most commonly by cooling to extend storage life.

Coconut and peanuts are very different from tree fruit and nuts and other fruits and berries. Coconuts grow on tall treelike plants while peanuts grow as an annual, relatively short-seasoned crop. Both are important crops; however, coconut is primarily eaten as a snack while peanut in its various prepared forms is eaten as both a snack and as part of a main meal. Peanuts are very nutritious and add significantly to the protein and energy content of any diet.

QUESTIONS

1. Considering only carbohydrate, which type of temperate fruit would be best to grow for augmentation of this component in the diet? If vitamin content were your primary concern, would you recommend growing the same or a different fruit?

2. Which tropical tree fruits would you grow to provide iron and potassium in a person's diet?

3. Of the nuts (as the term is used in this book), which would you chose to grow if your primary purpose was to produce cooking oil?

4. Explain why both harvested fruit and pruned branches must be considered in determining how much fertilizer to add to a fruit orchard each year.

5. Name two tropical fruits that do not grow on trees but on treelike plants.

6. Describe other uses of fruit and nut trees besides the production of food.

7. Explain the purpose or need of planting several trees of the same or different species in a fruit or nut tree orchard or individual house planting.

8. Which of the tropical fruits appears to have both higher protein and fat content?

9. Explain how soil may affect both the need for water and fertilizer application in an orchard. (You may need to refer to Chapters 9 and 10 to answer this question.)

10. Why is fruit and nut production a long-term commitment?

BIBLIOGRAPHY

Z. M. Ali and H. Lazan, Papaya, in *Tropical and Subtropical Fruits*, P. E. Shaw, H. T. Chan, Jr., and S. Nagy, eds., Agscience, Auburndale, FL, 1998.

H. Lazan and Z. M. Ali, Guava, in *Tropical and Subtropical Fruits*, P. E. Shaw, H. T. Chan, Jr., and S. Nagy, eds., Agscience, Auburndale, FL, 1998.

N. Narain, P. S. Bora, R. Narain, and P. E. Shaw, Mango, in *Tropical and Subtropical Fruits*, P. E. Shaw, H. T. Chan, Jr., and S. Nagy, eds., Agscience, Auburndale, FL, 1998.

C. O. Perea, H. Young, and D. J. Beever, Kiwifruit, in *Tropical and Subtropical Fruits*, P. E. Shaw, H. T. Chan, Jr., and S. Nagy, eds., Agscience, Auburndale, FL, 1998.

A. Rahnman, A. Shuklor, A. A. Faridah, H. Absullah, and Y. K. Chan, Pineapple, in *Tropical and Subtropical Fruits*, P. E. Shaw, H. T. Chan, Jr., and S. Nagy, eds., Agscience, Auburndale, FL, 1998.

D. K. Salunkle and S. S. Kadam, eds., *Handbook of Fruit Science and Technology Production, Composition, Storage and Processing*, Marcel Dekker, New York, 1995.

P. E. Shaw and C. W. Wilson, III, Carambola and Bilimbi, in *Tropical and Subtropical Fruits*, P. E. Shaw, H. T. Chan, Jr., and S. Nagy, eds., Agscience, Auburndale, FL, 1998.

J. S. Shoemaker and J. E. Teskey, *Tree Fruit Production*, Wiley, New York, 1959.

REFERENCES

1. T. Yeshitela, P. J. Robbertse, and P. J. Stassen, Effects of Various Inductive Periods and Chemicals on Flowering and Vegetative Growth of "Tommy Atkins" and "Keitt" mango (*Mangifera indica*) cultivars. *New Zealand J. Crop Horticult. Sci.*, **32**, 209–215, 2004.

2. J. G. Woodroof, *Tree Nuts: Production, Processing Products*, 2nd ed, AVI Publishing, Westport, CT, 1979.

3. Available at: http://www.fao.org/es/ess/top/commodity.html.

FARM ANIMALS AND FISH

7.1 ANIMALS ON THE THREE FARMS

Subsistence Farmer. Donio and his wife, Rosita, only own a few chickens such as the one shown in Figure 7.1. They are "free-range" chickens that roam the neighborhood. Occasionally, Donio will feed the chickens with surplus grain or rice bran. The hens provide eggs and, on special occasions, meat. The family does not normally sell a chicken, unless there is urgent need for money. Sita cares for a neighbor's buffalo, and Donio cares for an 8-month-old buffalo owned in partnership with a neighbor. Donio provides the feed, cares for the animal, and trains it for field work.

Family Farmer. Aída and Octavio have recently added commercial swine (shown in Fig. 7.2) production to their farm. They have constructed a series of pens of concrete block for the young pigs. They also have rabbits and chickens for sale and for family use. The pigs are primarily for sale in the local market, but sometimes the family will use one for a fiesta. They are too valuable for everyday food. They have guinea pigs that are primarily for family use. These provide small amounts of meat that the family can use one to two times each week.

World Food: Production and Use. By Alfred R. Conklin, Jr. and Thomas Stilwell
Copyright © 2007 John Wiley & Sons, Inc.

Figure 7.1. Chicken nesting in the kitchen in the Philippines.

Commercial Farmer. Steve Murphy and his brother have beef cows and a contract hog operation. There are about 40 cows of a mixed breed. They are fed on pasture all year with some mineral supplement. Their calves are raised on grass and, before being sold, are fed some ground maize from the farm with some added vitamins and minerals. The contract swine operation will feed about 1200 weaned pigs starting when they are 6 weeks old. They are raised in a special house, shown in Figure 7.3, with slatted floors (to permit manure to drop through). All feed is supplied by the contracting company for 15 weeks. After 15 weeks, the pigs are taken back by the contracting company and the house is cleaned and prepared for the next batch of pigs. With two batches of pigs each year, they will provide around 2400 pigs yearly.

Figure 7.2. Swine on the Tipán farm in Ecuador.

Figure 7.3. Swine on the Murphy farm in United States.

7.2 IMPORTANCE IN WORLD AGRICULTURE

Animals and their products contribute around 29 percent of the total value of agriculture, forestry, and fisheries in the developing world. This varies from 19 percent in sub-Saharan Africa to over 38 percent in the Caribbean. This does not include a substantial input to grain farming as draft animals. An estimated 250 to 300 million animals are used as work animals in the world.[1] Over 90 percent of the work performed is in developing countries. Most animals in the developing world are on small farms. Given the increasing costs of petroleum-based energy sources, it is likely that the use of draft animals will increase. In contrast, animal producers in Europe and North America tend to be large and specialized. There is very little use of draft animals for farming. Most trade in animals is on the local scale, involving fresh meat, milk, and eggs. International trade is primarily in processed animal products

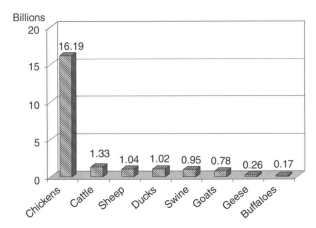

Figure 7.4. Numbers of farm animals in the world. (From *Global Livestock and Health Atlas*. FAO.)

such as leather, frozen meat, and dried milk. Animals kept for commercial and family use are varied, but the most common are chickens, as shown in Figure 7.4.

Nearly 40 percent of the world fish production enters international trade. The use of fish in the human diet ranges from 10 percent in Latin America to over 30 percent in Asia.[2] A significant portion of fish consumed by humans is produced by aquaculture: "farming of organisms that live in water, such as fish, shellfish, and algae."[3] The largest share of this production is in China. Many developing countries are increasing their production of fish, and for some it is an important source of national revenue. The primary products traded on the local level are fresh fish, sometimes with little processing. Products traded on the international market include fresh and processed fish and fish meal for livestock feed.[4]

7.3 ANIMAL PRODUCTION SYSTEMS

Many types of livestock are raised around the world, but most fit within the classification system developed by FAO shown in Figure 7.5.[5] The first major group is defined as only livestock production systems. In this system the farm income comes almost entirely from livestock. Within this system there are landless and grassland-based systems. Landless systems found in Europe, North America, Asia, and other developed countries account for 36 percent of the meat produced in the world. In the United States, the large feedlot is typical of the landless production system. The major species raised in this system are cattle, chickens, and pigs. This system is capital intensive and larger operations benefit from economies of scale. The grassland-based livestock system produces only 9 percent of the meat in the world and is found mostly in Central-South America, North America, Europe, and other developed countries. Examples of this system are the steppes of Mongolia, New Zealand's sheep systems, and the llama/sheep grazing systems of the Altiplano in Peru and Bolivia.

Mixed farming systems produce livestock in association with other crop enterprises. In these systems, livestock produce less than 90 percent of the total value of farm income. A significant amount of their feed comes from crops raised on the same farm. Over 54 percent of the meat produced in the world comes from this system in countries of Asia, eastern Europe and Commonwealth of Independent States (CIS), Europe, and other

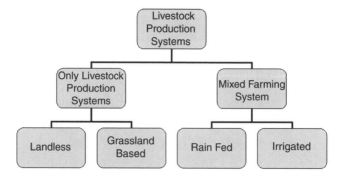

Figure 7.5. Livestock production systems.

developed countries. Over 90 percent of world milk production takes place within mixed farming systems. The mixed farming system is most common in the United States, Europe, and northeastern Asia.

As population increases, some systems are coming under pressure to modify. Intensive landless systems are on the decline in Europe, while production in grassland and mixed farming systems is increasing in most of the world.

7.4 CHICKEN, DUCK, AND GOOSE PRODUCTION

Chicken

> *Latin Name*: *Gallus gallus* or *Gallus domesticus* (Fig. 7.6)
>
> *Other English Names*: Rooster, hen, cock, cockerels, pullets, chick

Duck

> *Latin Name*: *Anas platyrhynchos* (Fig. 7.7)
>
> *Other English Names*: drake, duckling, hen

Goose

> *Latin Name*: *Anser anser* and *Anser cygnoides* (Fig. 7.10)
>
> *Other English Names*: gander, gosling

Chickens are found in all countries of the world, with the greatest concentrations in China, the United States, Indonesia, and Brazil (Fig. 7.8). Since they are kept in close proximity to humans, chickens are found in nearly every climate where humans live.

Figure 7.6. Farmyard chickens in Ecuador.

Figure 7.7. Penkin ducks in the United States.

Ducks are less widely distributed than chickens, being mostly raised in North and South America, Eurasia, and Australia. As shown in Figure 7.9, the largest populations are found in China, Hungary, Indonesia, and Vietnam. They are almost absent in sub-Saharan Africa. Because ducks are aquatic birds, they are most commonly raised near ponds, though open water is not necessary for their production.

Geese (shown in Fig. 7.10) are the largest bird commonly raised by humans. They tend to be raised in the cooler climates, though they can survive tropical conditions. Figure 7.11 shows the leading producer countries are China, Egypt, Romania, and Madagascar.

Modern chickens are mostly flightless, due to their long history of domestication. Chickens are grouped by classes related to their geographic area of development. They are raised for both eggs and meat. Often chickens are kept for egg production, but when this declines, they are sacrificed for meat.

Ducks are closely related to their wild relatives, sometimes having colorations similar to wild species. Ducks are less prolific egg layers than chickens, so they are generally raised for meat purposes in the United States. In China and the Philippines, eggs are the most valuable product. Ducks are generally considered to be less susceptible to diseases than chickens. In some Asian rice-producing countries they are raised as part of an integrated rice–carp–duck production system.

Geese are multipurpose farm birds. Their large size and rapid growth make them a preferred meat source in some countries. They are guard animals, being territorial and quite noisy when strangers enter and are useful for weeding certain crops. They are also a source of feathers for insulated jackets and bedcovers as well as meat and eggs.

7.4.1 Biology

Major characteristics that distinguish chickens from other birds are the presence of a comb on the top of the head and wattles under the chin. These secondary sex

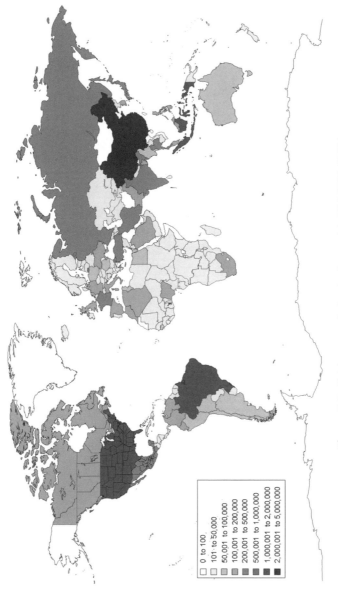

Figure 7.8. Chicken population in the world. (FAOSTAT, 2005.)

0 to 100
101 to 50,000
50,001 to 100,000
100,001 to 200,000
200,001 to 500,000
500,001 to 1,000,000
1,000,001 to 2,000,000
2,000,001 to 5,000,000

239

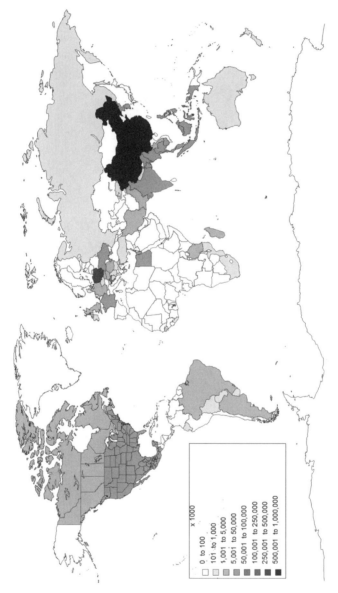

Figure 7.9. Duck population in the world. (FAOSTAT, 2005.)

x 1000

0 to 100
101 to 1,000
1,001 to 5,000
5,001 to 50,000
50,001 to 100,000
100,001 to 250,000
250,001 to 500,000
500,001 to 1,000,000

Figure 7.10. Geese in the United States.

characteristics are more prominent in the male but also are present in the female. The genus name *Gallus* is derived from the Latin name for comb. They are covered by feathers with some hairs on the body. The hairs are normally singed off during processing. The two legs are covered with scales. Chickens do not have teeth. Grinding is done in the gizzard by muscles and small pebbles they pick up from the ground. Adult chickens weigh from 450 g (Dutch) to 4.5 kg (Jersey Giant).[6] Egg size varies with the weight of the hen and come in colors from white to brown and even blue. A caged chicken raised under ideal conditions will produce 250 to 300 eggs each year. Under less than ideal conditions only 30 to 60 eggs per year can be expected.

7.4.2 Chickens

Chickens are believed to be descended from the red jungle fowl (*Gallus bankiva*) of Southeast Asia. There are indications that chickens were domesticated as early as 3200 BC in India, while records show chickens in China and Egypt starting in 1400 BC. These showed chickens being used for cockfighting rather than for food, but it is unlikely the hens and their eggs were wasted. Starting in 1849, poultry exhibitions became popular in the United States. This resulted in many breeds and varieties being preserved and new ones being developed. There are about 60 breeds of chickens based on body shape, number of toes, and color. The variety of chicken within a breed is further distinguished by minor variations on coloration and comb. A still more detailed distinction can be made by strain. Commercial strains exist that have been developed by companies for specific desirable meat or egg characteristics.

Where are your chickens?
There are approximately two chickens for every human being in the world.

7.4.3 Ducks

Domestic ducks are descended from wild ducks, and some breeds still bear the markings of their wild relatives. Ducks are more adapted to free-range production than chickens. They feed on snails and insects in rice paddies during the day and return

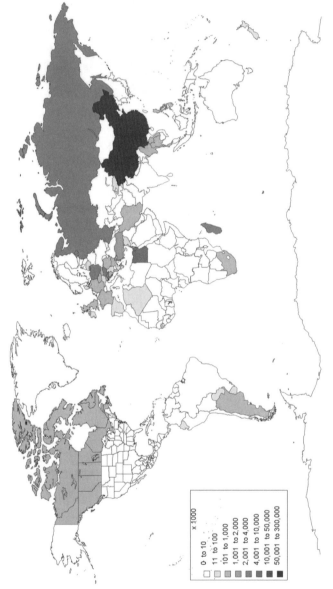

Figure 7.11. Goose population in the world. (FAOSTAT, 2005.)

to their enclosure at night. Their wide beak contributes to feed wastage when fed concentrated feed rations. Their anatomy is similar to the chicken, except that the beak is flattened, there is no comb or wattle, and there is webbing between the toes. The body weight ranges from 1.4 kg (Mallard drake) up to 4.5 kg (Muscovy drake).[7] Though classed as a duck, the Muscovy breed is genetically related to geese. It prefers to feed on grasses and roosts in trees in its native Amazon basin.

In most of the world ducks are raised for their meat and eggs. Ducks are the fastest growing poultry animal, weighing 3 kg in only 6 to 7 weeks.[8] They are generally less prolific egg layers than chickens, producing up to 300 eggs per year under ideal conditions. Under more common, free-ranging conditions, each duck will produce 40 to 60 eggs per year. An advantage of raising ducks for eggs is that they will lay their eggs within 3 hours after sunrise and then forage for food. Chickens will lay eggs for up to five hours after sunrise and only then start to forage for food. The most common breed of duck is the Pekin. There are an unknown number of breeds, mostly in Asia. They are classed as being for meat, eggs, herding ability, and other characteristics important in a particular production system. Crosses between ducks and the Muscovy are usually sterile (mule ducks) and are created specifically for meat purposes. These crosses typically have a faster growth rate than chickens.

7.4.4 Geese

The two major domestic species of goose are descended from the wild Greylag goose in Europe and the wild Swan goose of Asia.[9] Both species interbreed and some commercial breeds have resulted. A total of 96 breeds are recognized by FAO, though there are certainly more. The *Anser cygnoides* breeds are distinguished by a knob at the upper base of the beak. Most domesticated species are white. Brown pinfeathers tend to discolor the skin when harvested, producing a dark speckled appearance. Males (ganders) weigh from 2.8 kg (Philippine Domestic) to 12 kg (Shitou) when mature. Eggs of most species weigh from 120 to 220 g each. Geese have gizzards similar to chickens and ducks. The heavy muscles of the gizzard can grind more fibrous materials than ducks or chickens and promotes digestion in the cecum. The microbial fauna in the lower stomach also help to extract 15 to 30 percent of the nutrition in many feeds.

The goose is able to utilize different food materials than chickens and ducks because of its digestive system. Geese prefer grasses and grass seeds for feed. This has led to their use as weeders in specific vegetable crops such as strawberries, grapes, and onions. Geese feed on grasses and grass seeds, leaving most broadleaf plants untouched.

7.4.5 Feeding Requirements

The care and feeding of most types of poultry fall within the mixed farming system of animal production (Fig. 7.5). Within this system, poultry production can be classified as free-range extensive, backyard extensive, or semiintensive. In Africa, Asia, and Latin America, over 80 percent of the farms produce poultry using free-range or backyard extensive systems. Free-range extensive production provides little or no shelter, and the birds scavenge for food on all parts of the farm. Backyard extensive production poultry range free during the day but are confined during the night in some type of shelter. They are usually fed some small amount of grain in the morning. Semiintensive production of poultry confines the birds to some space with access to an enclosed outside area. Feed is available inside the housing unit. Intensive systems are found in both mixed and landless systems in all countries of the world. All landless systems use some type of intensive production system characterized by battery cages (for egg production), slatted floors, or deep litter. The advantage of intensive systems is a high productivity and efficiency. For example, a free-range extensive chicken will produce 20 to 30 eggs per year while a chicken raised in a battery cage will produce 180 to 220 eggs per year.[7] Geese and ducks are rarely raised under intensive systems.

Chickens require a mixed diet of relatively concentrated feeds. Nearly half of a typical diet is composed of coarsely ground grains of maize, sorghum, barley, oats, wheat, or rice. Around 10 percent of wheat or rice bran (a milling by-product) helps to add fiber and bulk to the diet. Much of the protein in the diet is provided by the meal of soybean, peanut, cottonseed, sunflower, safflower, or sesame remaining after oils have been removed. Dried and ground alfalfa should be fed if the chickens cannot feed on fresh pasture. Salt must be provided in the feed. About 2 percent of the diet should be bone meal to provide calcium for bone growth. Ground limestone, marble chips, or ground oyster shells should be fed to help with grinding in the gizzard.[10] The nutrition requirements given above are general and change according to the age, purpose, and confinement of the chickens.

Ducks kept in confinement have similar feed requirements as chickens. A recommended commercial feed ration consists of 46 percent crushed grain, 18 percent bran, 18 percent pollard, 10 percent meat meal, 2 percent soybean meal, 3 percent alfalfa meal, and 2 percent milk powder for calcium.[11] Smaller flocks of ducks raised under free-range conditions have a much more varied diet. In fact, in integrated rice–duck systems, ducks are a major factor in control of harmful insects and snails, as shown in Figure 7.12. In Japan, the Aigamo breed is used for rice production. Nearly 400 seven-day-old ducklings per hectare are released into a rice field about 10 days after transplanting. The floating, nitrogen-fixing fern, *Azolla*, serves as food for the ducklings. As they grow, they also feed on snails and grassy weeds, reducing the total weed biomass 52 to 58 percent and reducing the population of the golden apple snail (a rice-eating snail) by 74 to 84 percent.[12]

Geese have different dietary requirements than chickens and ducks. Under intensive production systems young goslings are started on a prepared ration of ground grains and other ingredients similar to that of chickens. Green plants are mixed in with the feed and by 5 to 6 weeks of age the birds should be feeding primarily from pasture grasses. They prefer more succulent clovers, bluegrass, timothy, and bromegrass. A good-quality pasture will support 50 to 100 geese on 1 ha. Under less intensive, free-range conditions, geese may be fed a limited amount of small grains after hatching but then live exclusively on grasses in the neighborhood. For the production of goose liver (foie gras) in Europe, adult geese are force fed wet maize to enlarge the liver before killing.

Figure 7.12. Ducks searching for snails in a rice paddy.

7.4.6 Pests and Diseases

Animals are affected by numerous pests and diseases. Some are endemic to particular regions of the world, while others are present in nearly all countries. All animal diseases cause economic losses, and some are transmissible to humans, causing illness or death. The OIE (World Organization for Animal Health) classifies animal diseases as list A, list B, or list C. List A diseases have the potential for rapid spread, having serious public health consequences and are of major importance in international trade. List B diseases are of public health importance and are significant in international trade. List C diseases are all other diseases not included in List A or B.[13] Several diseases affect all classes of poultry but only the ones in list A will be mentioned here.

Avian Influenza. Avian influenza is a Type A influenza virus. Within the Type A virus there are 15 subtypes classified as H or N. Several of the H types can infect humans and other animal species. Avian Influenza has been reported in Eurasia, Africa, and North and South America. Symptoms include nasal discharge, coughing and sneezing, diarrhea, and sudden death. The Type A virus can be low pathogenic or high pathogenic. Low-pathogenic forms display few symptoms. Low-pathogenic forms sometimes mutate to high-pathogenic forms, causing extensive losses. An epidemic in Pennsylvania poultry in 1983–1984 caused by the H5N2 virus started as a low-pathogenic infection but within a few months mutated to a high-pathogenic form causing up to 90 percent mortality in flocks. Control of the outbreak resulted in the destruction of 17 million birds at a cost of around $65 million.[14] The H5N1 strain is a high-pathogenic form causing up to 70 percent mortality in humans.

Newcastle Disease. The second most important viral disease affecting chickens around the world is Newcastle disease, known as Ranikhet disease in Asia. It is an airborne virus that spreads quickly and has a high mortality among young birds. Wild, migrating birds can carry this disease and hot, dry periods favor the airborne spread to other birds. While ducks and geese do not seem to be susceptible to Newcastle disease, they are carriers and can infect other species. Although an effective vaccine exists for Newcastle disease, most viral diseases are controlled by destroying infected flocks.

7.4.7   Marketing

The life span of a chicken can be as much as 10 to 15 years. A female chicken (hen) will start laying eggs at about 6 months of age. The presence of a male in the flock has no effect on number of eggs produced. Unfertilized eggs have a longer shelf life than fertilized eggs, but some markets prefer fertilized eggs. In commercial flocks, layers are kept for only 12 to 14 months of production. After this, egg production drops, and the layers are sold for human consumption as stewing chickens. Chickens produced specifically for meat (broilers or fryers) are usually sold when their weight reaches 1 to 1.8 kg. Under modern production conditions this weight is reached as early as 7 weeks after hatching.

Sale of chickens, ducks, and geese for meat depends on the presence of refrigerated transport and storage. In developing countries where refrigeration is not widely available, birds are sold alive. They are harvested shortly before consumption. Frequently middlemen purchase birds from individual farmers and market them at larger, wholesale markets in cities. In some cases birds may pass through five to six merchants before arriving at the final consumer. The farmer typically receives 60 to 65 percent of the final market price. In developed countries with reliable refrigeration, poultry is often harvested on, or near, the farm and only processed meat is marketed. This requires a well-organized system of processing, grading, and packaging.

Eggs are a product widely consumed in many cultures. Eggs also require cooling to maintain quality from the producer to the final consumer. This cooling may be by electrical or mechanical means. Marketing of eggs in developing countries is often by weight since grading and standardized sizes have not been adopted. There may also be checks for quality by candling or flotation. Candling involves holding an egg in front of a candle, or other light source, to judge the size of the air bubble in the large end of the egg. An egg with a large air bubble will float in a bucket of water. In either case, a large air bubble indicates an old egg and is usually rejected. Adoption of size and weight standards for eggs usually accompanies sale by quantity. This is common in most developed countries where eggs are sold by the graded dozen.

Feathers of both ducks and geese are an added source of income for farmers. Geese are the source of down, used in jackets and bedding. Down feathers are the immature breast feathers. These may be plucked from geese harvested for meat or plucked from live geese. Growing geese undergo a process of molting (shedding) of feathers every 6 weeks. Careful timing permits plucking down feathers during molting to harvest around 100 g of down from each goose.

In some European countries, fatty liver (foie gras) is an important product of geese. To increase the size of the liver before harvest, the geese are force fed with wet maize five to six times a day over a 2- to 3-week period. This causes the liver to nearly triple in weight.

7.5 CATTLE AND BUFFALO PRODUCTION

Cattle

Latin Name: *Bos taurus*

Other English Names: Cow, bull, heifer, calf, steer, ox, bullock, beef

Buffalo

Latin Names: Bubalus arnee and *Bubalus carabanesis*

Other English Names: Swamp buffalo, river buffalo, carabao, water buffalo

The Latin name *Bos taurus* is commonly accepted as the official name of cattle shown in Figure 7.13. However, some taxonomists consider domesticated cattle to be direct descendents of aurochs and use their Latin name, *Bos primigenius*, as the name for domesticated cattle. *Bos indicus* is another Latin name sometimes used for the Brahma breed of cattle. Since all cattle are considered to have a common ancestor, *Bos taurus* is the only Latin name in common use for all cattle.

The countries with the most cattle are Brazil, India, China, United States, and Argentina, as shown in Figure 7.14. The production systems in Brazil, United States, and Argentina are varied, while those of India and China are predominantly mixed farming systems. Because cattle utilize grasses and other vegetation unsuitable for human consumption, they are often raised in areas not suited for grain crop production. Dry prairies and hilly topography are not useful for maize or wheat production but are well suited for grazing cattle. Cattle are adapted to climates where other animals do not adapt well. Scottish Highlander cattle have coats of long hair for protection from cold during the frigid winters. In contrast, the Ovambo breed of Namibia is well adapted to heat and poor-quality forage.

Figure 7.13. Beef cow and calf.

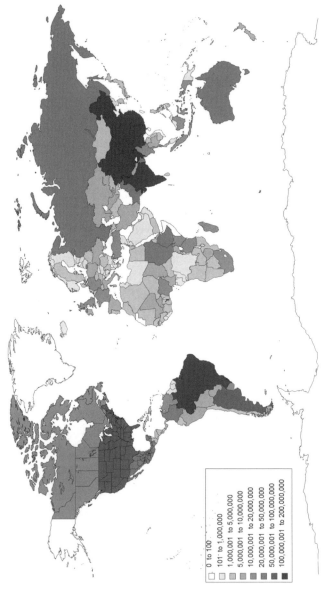

Figure 7.14. Cattle population in the world. (FAOSTAT, 2005.)

0 to 100
101 to 1,000,000
1,000,001 to 5,000,000
5,000,001 to 10,000,000
10,000,001 to 20,000,000
20,000,001 to 50,000,000
50,000,001 to 100,000,000
100,000,001 to 200,000,000

Figure 7.15. Water buffalo in the Philippines.

Buffalo are quite different from the American bison. The misnaming of bison as buffalo is attributed to early French explorers calling them "le boeufs," which was later mispronounced to form "the buffs" and "the buffels."[15] Similar to cattle, buffaloes are ruminants with many of the same nutritional needs and management practices. In contrast to cattle, buffaloes are adapted to wet environments, as evidenced by the two main types: water buffalo (shown in Fig. 7.15) and swamp buffalo. Most breeds have a sparse hair coat and do not tolerate cold well. The exception is the river buffalo, which has some cold tolerance. Buffaloes are confined mostly to Asia (96 to 97 percent). The leading countries for buffalo production are India, Pakistan, and China, as shown in Figure 7.16. Egypt and the Philippines also maintain significant populations. Buffaloes are found in the United States but are not as popular as cattle or bison.

There is some disagreement on the Latin names for buffaloes as well. Some experts give the Latin name of *Bubalus arnee* to all types of buffalo with two subspecies of *Bubalus arnee bubalis* for the river buffalo and *Bubalus arnee carabanesis* as the swamp buffalo. Others consider them to be separate species since they have different chromosome counts. Swamp buffaloes are from Southeast Asia and tend to be used primarily for draft purposes. River buffaloes are from India and Pakistan and are used for milk and meat production in addition to draft purposes.

7.5.1 Biology

Cattle and buffaloes are ruminants. The stomach of ruminants has four chambers. The first two are the rumen and reticulum, where the fibrous foods are stored and broken down by bacteria and protozoa. The broken-down food then passes through to the omasum, where water is removed. The last chamber, the abomasum, is similar to the human stomach and breaks down proteins and simple carbohydrates. This

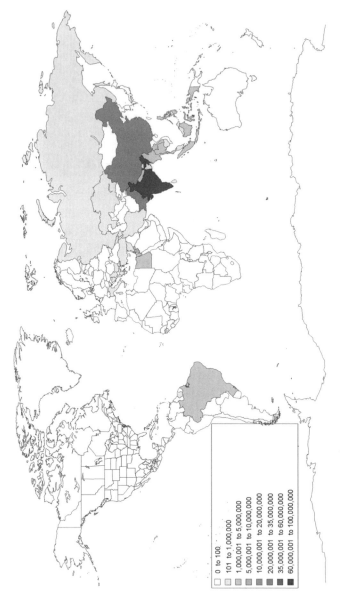

Figure 7.16. Buffalo population in the world. (FAOSTAT, 2005.)

| 0 to 100 |
| 101 to 1,000,000 |
| 1,000,001 to 5,000,000 |
| 5,000,001 to 10,000,000 |
| 10,000,001 to 20,000,000 |
| 20,000,001 to 35,000,000 |
| 35,000,001 to 60,000,000 |
| 60,000,001 to 100,000,000 |

enables ruminants to use fibrous plant parts for food. Humans have only a limited ability to absorb nutrients from very tender plant leaves such as lettuce. Ruminants are able to use coarse stems and leaves of plants, which are completely indigestible by humans.

Cattle were domesticated by 6500 BC or earlier in Turkey and the Near East. There are some indications that domestication occurred much earlier. Modern cattle evolved from aurochs, a wild ox. The last known surviving aurochs was killed by a poacher in 1627 in a Polish game park.

Cattle occupy different roles in many societies. In some African cultures, cattle are synonymous with wealth. They are kept not as a business enterprise but as a type of savings account. Calves are the "interest" and when cash is needed, one is sold. In India, cows are considered sacred by Hindus and killing a cow is a mortal sin. It should be noted that buffalo are not treated the same as cows and can be used as food. In Spain and many Latin American countries, bulls are considered to be a symbol of strength. Thus, a man who defeats a bull in the ring (bullfighting) is considered a hero. For most farmers in the world, cattle serve several tasks: providing milk, meat, and as a draft animal. Special breeds of cattle have been developed for each task. The Holstein breed has been bred to produce large quantities of milk. The Black Angus was developed specifically for beef production. The Brahma breed has a distinct hump behind the head that makes it useful for pulling a plow or cart.

Some breeds have developed tolerance, or even resistance, to ticks that enable them to survive in specific ecosystems. The Brahma is a breed well known for its tolerance to heat, ticks, and ability to utilize coarse grasses. They have been widely crossed with temperate climate beef breeds to obtain more productive meat breeds for tropical areas.

In contrast to cattle, buffaloes have received relatively little systematic effort for breed improvement. As a result there are large variations between animals for meat and milk production. In some countries there are only a few recognized breeds. The most distinct breeds are found in India and Pakistan. India and Pakistan have at least five generalized groups of buffalo breeds named as Murrah, Gujarat, Uttar Pradesh, Central Indian, and South Indian. Within each group are several distinct breeds.[16] Although there are genetic differences between swamp buffalo and river buffalo, they will interbreed and their offspring are fertile. This is one of the arguments for considering them as one specie.

The major role of buffaloes is as draft animals in rice-producing countries. Buffalo males are often neutered, or castrated, to enhance their performance as draft animals. The result is that there is a selection of faster growing, large animals for work, and these are removed from the gene pool. Mostly smaller bulls are left for breeding. Over the years there has been a tendency for many breeds to grow slower and mature smaller. There are very few large herds of buffalo. Most are kept on small, mixed farms as a work animal and to supply incidental milk and meat.

Comparison with cattle shows that buffalo have greater resistance to internal and external parasites than cattle. Buffalo calves are born at a lighter weight than cattle, but they tend to grow faster due to the high butterfat content of the milk. The mature animals are able to subsist on a diet of relatively poor-quality forage compared to cattle. Buffaloes are typically very docile animals, often being led around by small children in villages.

7.5.2 Feeding Requirements

Ruminants occupy a complementary place in the human food production system. They produce food for human consumption, utilizing many materials that are not usable as food by humans. In contrast, chickens and pigs require many grains that could be eaten by humans. Most grasses and forage plants grow quite well on land that is too hilly for mechanized farming. The only alternative use would be forestry. Some areas of Africa are too dry to produce crops, but the small amount of green plants that grow there are sufficient to nourish hardy breeds of cattle.

The best diet for cattle depends on the specific breed, the production system, and the end product desired. Cattle for milk production are fed a ration of forage and concentrates rather than letting them graze pastures. Generally, cattle trample as much as 30 percent of the lush forage in alfalfa or clover pastures while searching for green leaves and stems to eat. They also use valuable energy walking over hectares of pasture searching for food. The most efficient dairy herds keep cows in small fields or lots and bring the feed to them. This may be in the form of dry hay, moist silage, compressed pellets of alfalfa, or freshly cut maize plants. A part of the feed ration is also composed of ground grains, such as maize, barley, or rice.

A major objective of high-production diary herds is to provide high-quality forage so that each cow will produce 1 kg of milk for each 1.4 kg of dry matter fed.[17] A lactating cow will consume 16 to 22 kg of dry matter each day depending on the stage of lactation. Under proper management, a single cow will produce about 25 kg of milk each day.[18] This varies widely for different breeds and management systems. Intensively managed Holstein dairy cows will produce milk up to 300 days after giving birth to a calf. Their gestation period is about 9 months. Before breeding the cow again, it goes through a 60 day "dry," or rest, period. Intensive management systems try to have cows produce one calf each year.

Dairy cows in other parts of the world often produce less milk but on less expensive feed rations. Tropical areas account for only 36 percent of global milk production. In India, cows will produce 8 to 10 kg milk per day. In Brazil, cows in different regions of the country produce 12 to 22 kg milk per day. Regional differences can be seen in Figure 7.17, where the average cow in Europe produces 5100 kg of milk each year, and in sub-Saharan Africa where an average cow produces only 340 kg milk each year. Asia is slightly better, producing 900 kg milk per year, and Central and South America are a bit higher with each cow producing 1100 kg of milk each year.[19] These figures are for tropical lowland areas. In temperate, upland areas milk production is higher. Attempts have been made to introduce the Holstein breed into these countries, but production often declines because they are not adapted to the adverse temperatures, diseases, insects, and poor quality of forage.

Buffalo are significant milk producers for the small farmer in Asia. The fact that a buffalo cow can be used for light field work and give milk is a crucial factor for a small

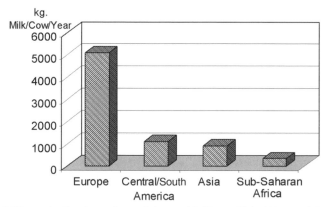

Figure 7.17. Milk production in regions of the world. (From *Global Livestock Production and Health Atlas*, 2006, FAO.)

family farmer. In addition, buffalo cows can do this while eating the poorest quality fodder. Most small farmers will feed buffaloes crop residue or pasture them on roadsides. Most dietary requirements of buffalo are similar to a cow except that a buffalo can utilize poor-quality forage more efficiently. Experiments in Thailand showed that buffalo cows will produce 2 kg of milk per day when fed rice straw.[20] A full-grown Holstein cow would slowly starve on the same diet.

Buffalo milk is higher in fat than cow milk. Because of this, it is preferred for making mozarella cheese. Buffalo cows will produce milk for 250 to 290 days after calving, nearly as long as Holstein cows. Some breeds will produce as much as 7 kg milk per day.

Cattle fed for meat purposes receive significantly different feed than milk cows. The young calf is weaned from the mother at 6 to 8 months of age when it weighs 225 to 275 kg. Male calves are typically castrated to increase weight gain. These animals are called steers. The calves (or steers) are then fed on pasture until they weigh about 350 kg. The time period needed to reach this weight depends on the quality of forage consumed. When this weight is reached, they may be switched to a more nutrient-rich diet containing grains such as maize, wheat, sorghum, soybeans, and barley. A wide variety of other feed sources are used, such as distiller's grains, soybean hulls, citrus pulp, and other waste products of the food-processing industry. A steer fed on concentrates will typically gain about 1 to 1.5 kg per day under feedlot conditions. When the animals reach a weight of 500 to 600 kg, they are ready for market.

Landless intensive beef systems will purchase animals at the 350-kg weight (called stocker calves) and feed them until ready for market. Usually these operations are the stereotypical feedlots producing thousands of animals each month. An alternate production system is extensive pasture-fed cattle as in Argentina, southern Brazil, and northeast Mexico. Relatively little beef is produced in mixed farming systems in developing countries. This is due to the traditional separation between pastoralists and settled farmers.[21] In addition, the relatively low demand for beef on local markets has not encouraged the development of large mixed farming herds.

Kobe beef costs over $300 per pound! It comes from Wagyu cattle raised in Japan. These steers are massaged daily, fed freshly cut grass with grains, and fed beer during the hot summer months.

A comparison of extensive and intensive production systems was done by P. Auriol at FAO. Some of the results are shown in Table 7.1.[21] Obviously, the least efficient production system is natural pasture without feed supplements. Five times more feed units are needed to produce 1 kg of carcass weight than the intensive landless system. The main advantage of the natural pasture system is that it requires relatively little cash input for seminomadic cattle herders.

Beef cows are bred every 12 months. Large landless intensive operations will typically use artificial insemination methods to guarantee that cows will deliver calves within a certain time period. This ensures a supply of calves when the demand is highest for feedlot finishing. The productive life of a beef cow is normally 7 to 9 years, at which time the cow also becomes beef. The time from birth to harvest ranges from less than 1 year to nearly 5 years for calves on poor-quality pasture. The longer time spent on pasture may involve less cash inputs for the farmer, but it also involves greater risk of loss and more expense on health care.

Similar to dairy cows, specialized breeds have been developed for beef purposes. The most widely known breed in the United States is the Black Angus. Some other beef breeds for temperate climates are Hereford and Shorthorn, which were developed in England, and Maine Anjou developed in France. Some breeds more adapted to tropical conditions are Brahma and Santa Gertrudis.

Buffaloes are used for meat purposes, but only after having served a useful life for milk and/or draft purposes. The countries in which buffalo are most numerous (India,

TABLE 7.1. Efficiency Comparison of Beef Production Systems

System	Age at Harvest	Feed Needed for Body Maintenance (%)	Feed Conversion Rate
Natural pasture without feed supplements	6–7 years	85	39 feed units/kg carcass
Pasture with feedlot for 150 days	35 months	66	16 feed units/kg carcass
Pasture with feedlot for 150 days and improved breeding	30 months	63	13 feed units/kg carcass
No pasture, high-energy ration, specialized breeds	13–14 months	54	7 feed units/kg carcass

Pakistan, and China) have a limited market for fresh meat due to the lack of a refrigerated distribution system. This limits the meat to local markets. A second limiting factor in India is the strong vegetarian tradition among Hindus, though buffaloes are not considered sacred, as are cows.

A third significant use of both cattle and buffaloes is for draft power. In most parts of the developing world oxen, or bullocks, are an essential source of farm tillage power. As can be seen in Figure 7.18, the hump behind the head is necessary to hold the yoke in place for pulling loads. This develops in male animals after about 1 year of age. They are then castrated, or neutered, to maintain a docile temperament.

In parts of Africa the presence of draft animals is limited by the presence of trypanosomiasis. This parasite, carried by the tsetse fly, is fatal to cattle. Several important diseases are also carried by cattle ticks. These problems have limited the spread of draft animals in sub-Saharan Africa. As a result, much of the tillage of agricultural lands is done by hand labor.

Animals for draft purposes have the advantage of being a renewable energy source and provide manure as fertilizer, milk, beef, and hides. When born on the farm, their initial cost is minimal compared to a tractor. A problem is that work animals are not needed for field work during most months of the year. Unlike a tractor, they must be fed and cared for even though they do no useful work. As a result, the farmer tends to ignore the animals during periods when feed is lacking and work is not needed.

7.5.3 Pests and Diseases

The OIE list A diseases affecting cattle and buffaloes are described here. The OIE method of classifying diseases is covered in Section 7.4.6.

Foot-and-Mouth Disease. Foot-and-mouth disease (FMD) is a highly contagious virus disease of cloven-hoofed animals. Those most affected are cattle, buffaloes,

Figure 7.18. Bullocks in India.

pigs, sheep, and goats. Human infection is rare. It is not very lethal to adult animals but causes severe loss in weight and milk production. Symptoms include high fever, blisters inside the mouth, and blisters on the feet. It is found in most countries of the world. Some countries have been able to eliminate the disease within their borders and maintain strict monitoring to prevent infection from other countries. When infected animals are found, the only remedy is to slaughter them and all animals that have had contact with them. In severe cases, even animals on adjacent farms are slaughtered, burned, and buried as a precaution. A vaccine is available but is not always effective since there are seven strains (serotypes) of the virus. Most vaccines only protect against three to four serotypes for periods from a few months to several years.

Vesicular Stomatitis. Vesicular stomatitis is a viral disease of cattle. To date it has only been reported in North and South America. Symptoms are similar to foot-and-mouth disease. It can infect humans handling infected animals. Most animals recover in 2 to 3 weeks. Though widespread, its most serious threat is confusion with food-and-mouth disease.

Rinderpest. Rinderpest is also known as cattle plague. It is caused by a virus affecting both cattle and buffaloes. Though previously widespread in Eurasia and Africa, it is now confined to a few areas of Africa and central Asia. The Global Rinderpest Eradication Programme of FAO aims to eliminate the disease by 2010. There is only one serotype but several strains have differing degrees of virulence. Depending on the strain, animals may recover or most will die. It does not affect humans.

Contagious Bovine Pleuropneumonia. Contagious bovine pleuropneumonia is endemic to Africa and Eurasia. It was eradicated from the United States in 1892. It is caused by the bacteria *Mycoplasma mycoides*, which are spread by droplets of moisture exhaled by infected cattle and buffaloes. It does not affect other animals or humans. Mortality ranges from 10 to 70 percent depending on environmental conditions. Cattle cannot be imported into the United States from countries where this disease is present.

Lumpy Skin Disease. Lumpy Skin Disease is a highly contagious viral disease affecting cattle in Africa and Eurasia. Symptoms are fever, lameness, and the development of lumps or nodules in the skin. It is spread by biting insects and through direct contact with infected animals. While it is not normally fatal, there is damage to the hides and cattle from infected areas cannot be exported. It does not affect humans or other animal species.

Rift Valley Fever. Rift Valley fever is caused by a virus transmitted by mosquitoes infecting primarily cattle. It is highly contagious to humans and may be fatal. Among cattle, it is most serious among young animals. In addition to high mortality, it causes abortions in pregnant cows. It is distributed throughout Africa and parts of the Middle East. Symptoms in cattle are fever, weakness, and death of young animals within 36 hours. Humans in contact with infected animals may be at risk of infection through airborne virus particles. A vaccine is available for humans.

7.5.4 Marketing

The principal products of cattle and buffaloes are milk, meat, and leather. Milk and meat are highly perishable products requiring special treatment to prevent spoilage.

Milk. Milk is taken from cows and normally sold within 1 day in most tropical villages. Typically it is the farmer or an intermediary who sells the milk to an established list of clients. An Indian milk vendor is shown in Figure 7.19. Such rural providers are often plagued by poor quality and irregular quantities. In most countries an established dairy industry exists to provide a uniform quality of milk to urban consumers. The quantity and quality of available milk varies with the animal. Dairy cattle bred for milk production give large quantities, but local breeds in many parts of the world have limited milk production (Fig. 7.17). In the case of buffalo, milk production is often a side product of an animal used primarily for field work.

Milk processing in North America, Europe, and other industrialized countries is highly mechanized and controlled. The procedure involves specialized trucks to collect milk from farms and its transport to a central processing facility. Processing involves separation of some fat from the whole milk and standardization to a desired fat content. The separated fat is turned into ice cream, butter, or buttermilk. The remaining liquid is designated as "beverage milk." Some treatment is given to reduce bacteria counts to acceptable levels. This may involve pasteurization, a short heating to 72°C for 15 seconds. Pasteurization is designed to reduce the content of pathogenic bacteria and to extend the keeping quality of the milk. Pasteurized milk can be kept at a temperature below 10°C for several days. Sterilization is a more severe treatment than pasteurization involving heating the milk to 110°C for 20 to 30 minutes after the bottle is filled and sealed. This gives the milk a longer shelf life but also gives it a cooked flavor and a

Figure 7.19. Rural milkman in India.

brownish color. Sterilized milk is not really sterile but can be kept for several weeks without refrigeration. Ultra-high-temperature (UHT) treatment involves heating the raw milk fluid to 135 to 150°C for only 2 seconds before packaging in sterile containers. UHT milk can be kept for several months without refrigeration. The final product of all treatment processes may be sold in metal cans, glass bottles, plastic bottles, plastic bags, or paper boxes.[22]

Because of the highly perishable nature of liquid milk, most cultures have developed products made from fermented forms of milk. These usually involve conversion of lactose to lactic acid, which halts further fermentation and acts to protect the milk from spoilage. These milk products may be classified as fermented milk, butter and ghee, and cheese.

Typical fermented milk products include yogurt, sour buttermilk, and curds (cottage cheese). Each culture has local variants on this process with unique names and flavors. In most cultures butter is made by churning fermented milk, as opposed to churning whole milk in European cultures. Ghee is a cooking product made in Asia and parts of Africa by heating unsalted butter from 110 to 120°C to remove moisture. After cooling, the remaining mixture is filtered through muslin cloth and bottled for sale.

The main purpose of cheese making is to preserve the milk over long periods of time without spoilage. Cheese is made by precipitating casein from whole milk by the addition of specific enzymes. Cheese is produced by nearly all cultures in the world, with the possible exception of some countries in southern and eastern Africa. The variety of cheese types is extensive.

Meat. It is interesting to note that with some animals the meat has a different English name than the animal. For example, chicken meat is called "chicken" while meat of a cow is called "beef." Similarly, calf meat is called "veal" and pig meat is called "pork." One theory holds that the names of the animals came from Anglo-Saxon farmers who used English names but that the people who cooked and served the meat used Norman French names. Therefore beef is related to the French *boeuf* and veal is derived from *veau*.[23]

Meat and meat products are more susceptible to bacterial contamination and degradation than milk products. This is often due to the unsanitary conditions of harvest and less than ideal methods of transport and handling. In many northern climates handling of meat is done under refrigerated conditions in the summer months and under naturally cold conditions in the winter. Tropical areas have no cold season, and rural areas often lack refrigeration for storage of meat products. One of the reasons for this lack of facilities is the low per capita meat consumption in many developing countries. Typically, a family in Africa or parts of Asia will consume less than 20 percent of the amount of meat products as a North American. With less demand, it is not surprising to see less effort in the marketing of meat products.

While meat products in North America are prepared and packaged under carefully controlled conditions, many rural butchers harvest animals and sell the meat in a local market the same day. These meat products are also consumed the same or the next day. Even where homes have refrigerators, the level of contamination is high enough to prevent storage for more than 1 to 2 days. In contrast, meat products sold in industrialized countries is packaged to prevent contamination and damage so that it can be kept in refrigerated conditions for up to 10 days without spoilage. For marketing purposes, this means that unsold inventory can be kept for the next day. If the butcher does not have refrigeration, unsold inventory spoils and is lost. Prices must be raised to compensate for spoilage and losses.

Methods of preservation of meat products have been developed and are commonly used in areas without easy access to refrigeration. These involve either drying or cooking.

Drying normally consists of cutting meat into thin strips to permit rapid dehydration. The strips may be treated with salt or lightly smoked to enhance the flavor and keeping qualities. In regions with high temperatures and low humidity, drying in the sun is often practiced. Where this is not possible, drying in heated boxes or rooms is required. After the meat strips are dried, they must be stored in dry, insect- and rat-proof conditions. When used for food, the strips are cut into small pieces to eat as snacks or rehydrated before cooking.

Cooking is practiced in more commercial settings where the meat products are to be stored in glass jars or metal cans. The basic process consists of heating the products to temperatures over 100°C or higher by combining hot water and steam under pressure. The goal is to nearly eliminate all microorganisms without severely altering the taste or texture of the meat. To destroy spores of *Clostridium botulinum*, it is necessary to heat the product to 121°C for 2.45 minutes. In practice, meat products cooked at 121 to 140°C and sealed properly can be stored as long as 4 years in tropical conditions.[24]

Leather. Leather is the treated skin of any animal. The majority of cattle and buffalo leather goes into the manufacture of shoes and footwear. Some is used as jackets and other garments in industrialized countries. The leather from buffaloes is especially useful the for manufacture of helmets.

7.6 SHEEP AND GOAT PRODUCTION

Sheep

> *Latin Name*: *Ovis aries*
> *Other English Names*: Ewe, ram, lamb, wether

Goat

> *Latin Name*: *Capra hircus*
> *Other English Names*: Doe, nanny, buck, kid, billy, wether

Domestic sheep are found in most countries of the world shown in Figure 7.20 with the top five countries being China, Australia, India, Iran, and Sudan. Most sheep production is in grassland-based production systems such as Mongolia's steppe areas, New Zealand's sheep enterprises, and the sheep grazing systems of the altiplano of Peru and Bolivia. Landless production systems are only found in parts of western

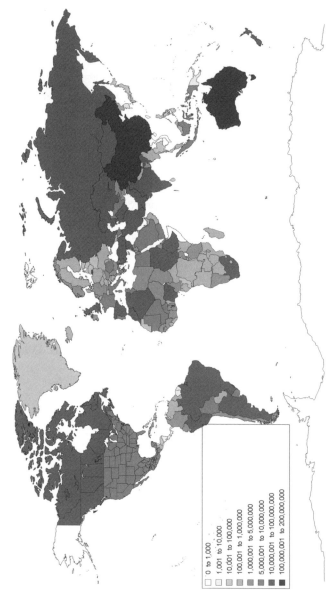

Figure 7.20. Sheep population in the world. (FAOSTAT, 2005.)

Legend:
0 to 1,000
1,001 to 10,000
10,001 to 100,000
100,001 to 1,000,000
1,000,001 to 5,000,000
5,000,001 to 10,000,000
10,000,001 to 100,000,000
100,000,001 to 200,000,000

Asia and northern Africa. Sheep are typically associated with marginal areas in the grassland-based systems and as such are less productive than the humid grassland-based systems for cattle.

Goats are also mainly found in more marginal grassland-based systems. As shown in Figure 7.21, China and India are the two leading countries with goat populations well over 100 million. The next five countries with large populations of goats are Pakistan, Sudan, Bangladesh, Nigeria, and Iran. Similar to sheep, goats are produced in grassland-based systems, often on marginal land. It is not unusual to see herds of goats in villages being pastured along roadsides or on poor fallow fields. Goats utilize different herbage than sheep often preferring plants rejected by other animals.

7.6.1 Biology

Sheep and goats are classified as small ruminants. Their digestive tracts are similar to cattle and buffaloes, consisting of four chambers. Goats differ from sheep and cattle in their method of eating and in the type of plant material they can digest. Cattle and sheep are grazers; they eat by cutting off tufts of grass and forage plants and chewing the stems and leaves. Goats are browsers, eating broad-leaved plants and stems of shrubs above the ground. When sheep and goats are turned into a pasture, it is common to see the sheep graze on grasses near the middle of the field, while the goats will roam along the fence, browsing on weeds and shrubs. Contrary to urban legends, goats will not eat tin cans. However, their curiosity will lead them to explore new items with their prehensile upper lip and tongue.

Sheep are descended from the mouflon (*Ovis musimon*). There are two remaining populations of mouflon in existence: the Asiatic mouflon in the mountains of Asia Minor and southern Iran, and the European mouflon in Sardinia and Corsica.[25] The first sheep were probably domesticated over 6000 years ago. Depictions of sheep appeared in the art of Mesopotamia and Babylon around 3000 BC. Over the centuries, human caretakers have selected their offspring for desirable wool, docile nature, and flocking instinct, making them easy to care for and keep in captivity. Over 200 breeds of sheep are recognized today. A typical wool and mutton breed is shown in Figure 7.22.

Sheep are generally grouped as either hair type or merino types. Hair sheep comprise only about 10 percent of the world sheep population. They are most adapted to tropical conditions because they shed hair in hot weather. They also have somewhat lower body temperatures than merino-type sheep and are primarily raised for meat purposes. Merino sheep have wool instead of hair and must be sheared each year. Wool is a fine, curly type of hair covering the sheep. It consists of fibers of keratin covered by very small overlapping scales. It does not conduct heat well and the crimping of the fibers helps entrap air, making it useful for clothing in cold climates. It absorbs moisture easily.

While not normally thought of as milk animals, sheep are milked in many countries. A typical daily yield ranges from 0.5 to 0.9 liters of milk per day during an 85- to 110-day lactation period in Hungary.[26] Specific breeds of sheep have better milk production than others, but most sheep milk production is from multipurpose breeds. The Awassi breed of sheep has higher milk production than many other breeds.

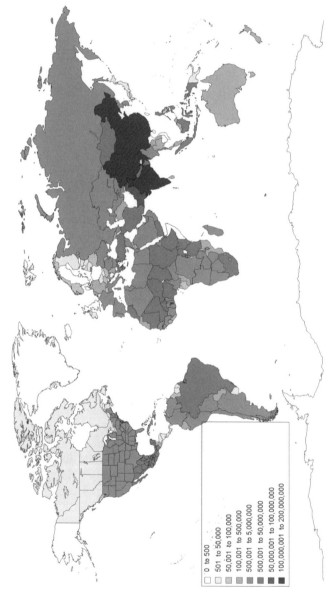

Figure 7.21. Goat population in the world. (FAOSTAT, 2005.)

Legend:
- 0 to 500
- 501 to 50,000
- 50,001 to 100,000
- 100,001 to 500,000
- 500,001 to 5,000,000
- 500,001 to 50,000,000
- 50,000,001 to 100,000,000
- 100,000,001 to 200,000,000

Figure 7.22. Suffolk sheep ewe.

Goats were probably domesticated about 10,000 years ago in the area of present-day Iran. They were descended from the wild bezoar (*Capra aegagrus*). Genetic studies indicate that the development of goats differed from that of cattle and sheep. Cattle and sheep have localized variations, but goats seem to be remarkably uniform in their genetic composition around the world.[27] This means that there has been continuous and extensive interbreeding among goat populations from different parts of the world since they were domesticated. Goats have, therefore, been a part of human trade and migration for thousands of years while sheep probably have not.

Goats are multipurpose animals, providing meat, milk, and hides. A typical meat-type goat is shown in Figure 7.23. Until the invention of the printing press in Europe, goatskin (parchment) was the most common writing surface in Europe. Goat breeds are divided into five categories: dairy, fiber, meat, skin, and pet. Fiber breeds of goats produce fine wool prized by the garment industry. Most goats have fine, soft fibers beneath surface guard hairs. Cashmere goats have been selected for more soft (cashmere) fibers and fewer guard hairs. Angora goats produce mohair, an equally fine and valuable type of wool used in clothing.

> Fainting Goats, Myotonic, or wooden leg, goats will fall over (faint) and lie still when frightened. This is caused by a genetic mutation that reduces muscle chloride conductance. When frightened, the goat cannot move and appears to faint.

Goat milk is considered easier to digest than milk of cows or buffaloes. It is frequently recommended for infants and adults with lactose intolerance. The milk lacks the protein agglutinin so it does not require homogenization (breaking up of fat globules). Goats bred specifically for milk production (such as Nubians) under good management will produce 3 to 4 liters per day for 200 to 250 days of lactation. Most goats in developing countries do not produce nearly as much milk as specialized breeds under

Figure 7.23. Boer meat-type goat.

good management. In Bangladesh, the black Bengal goat will produce an average of 294 mL milk per day for 72 to 83 days under favorable management. When kept under scavenging conditions, the same breed will produce only 121 mL per day for 98 to 105 days.[28]

7.6.2   Feeding Requirements

Sheep are grazers, preferring grasses and a few broadleaf species. To maintain growth and productivity, sheep need a continuous supply of nutritious forage supplemented with high-protein feeds at certain phases of their growth cycle. In industrialized countries ewes are kept on pasture until they are bred when supplements containing soybean meal, cottonseed meal, or fish meal are added to their diet. After lambing the protein level of their diet can be reduced to a maintenance level. Lambs for meat and wool are the primary market products of sheep. Lambs are fed on high-quality forage after weaning, frequently with additions of grain and protein supplements.

In developing countries, sheep are often kept on pasture year-round. After being bred, they are fed on high-quality grasses and forage legumes to produce healthy lambs. As soon as the lambs are weaned, the ewes are placed on a maintenance diet. Because breeding is seasonal (in response to day length), lambing often occurs in the spring when lush grasses and weeds are available. Weaning then occurs after the flush of summer growth ends and lower quality vegetation is available for forage. In parts of North Africa, sheep are traditionally fed dry barley straw during the summer and fall months. This is also a dry season when little or no green forage is available. The result is that ewes lose weight during this season.

Goats are browsers and have a reputation for eating anything they come across. In fact, goats are very picky eaters. They prefer the tips of shrubs and young trees as well as broad-leaved plants. These are the parts of plants that are most tender and have the highest digestible protein contents. When kept in pastures, goats must be managed similar to sheep and cattle, rotating fields to prevent overgrazing. They should have a variety of plant species for grazing. Feeding of goats in industrialized countries is similar to that of sheep. Nannies are fed some high protein supplements after breeding. Supplements are continued after giving birth and the kids are fed supplements after weaning. In developing countries goats give birth in the spring when rains have initiated heavy growth of grasses and shrubs. This also provides food for the new crop of kids as well as fattening them for sale.

A secondary use of goats is to control some weed shrubs. In parts of the United States goats are used to control problems of multiflora rose (*Rosa multiflora*). This is a thorny, woody, perennial shrub that was introduced as "living fences" and as wildlife cover in the 1930s but later found to be invasive, taking over productive pastures. Goats readily browse the growing tips of this shrub, eventually eliminating it from pastures.

7.6.3 Pests and Diseases

Goats are generally considered to be rustic: able to survive with a minimum of attention. Sheep have a reputation of being easily affected by sickness and are especially vulnerable to predators. This section covers the diseases on the OIE list A (having the potential for rapid spread, having serious public health consequences, and are of major importance in international trade).

Foot-and-Mouth Disease. This is the same as described for cattle in Section 7.5.3.

Peste des Petits Ruminants (PPR). PPR is a highly contagious viral disease infecting goats and sheep. Symptoms include fever, discharge from eyes and nose, mouth sores, disturbed breathing, pneumonia, and death. It was first described in Côte d'Ivoire in 1942. In recent years it has been diagnosed in most countries in the Sahel region of Africa, the Middle East, and as far east as India. The symptoms are similar to rinderpest. A vaccine is available and can protect animals for as long as 3 years. Control of an outbreak is done by isolation of infected animals and vaccination of all surrounding animals. A low rate of infection exists in most affected countries. Occasional outbreaks occur and may result in nearly 100 percent fatality.

Rift Valley Fever. This is the same as described for cattle in Section 7.5.3.

Bluetongue. Bluetongue is a viral disease of sheep transmitted by mosquitoes. Goats may also be infected but do not typically show outward symptoms. There are over 20 serotypes (strains) of bluetongue, making positive identification especially difficult. Symptoms include fever, frothing at the mouth, and a bluish discoloration of parts of the tongue and regions of the hooves. It will also cause abortion in pregnant females. Mortality in adult sheep ranges from 10 to 50 percent. Testing is still being done, but current data indicate that this virus is worldwide. Vaccination of nearby animals is effective only if the serotype is identified. Control of outbreaks is by elimination of mosquito breeding places.

Sheep and Goat Pox. Sheep and goat pox are caused by a virus transmitted by contact or by airborne particles from infected animals. There is only one serotype, but several strains produce the disease in either goats or sheep. Symptoms include development of pox lesions on the skin and inside the mouth. It is present in Africa, the Middle East, India, and most of Southeast Asia. Infected adults may suffer up to 50 percent mortality while infected lambs and kids may have as much as 95 percent mortality. Control of outbreaks is primarily by destruction of infected animals and vaccination of nearby animals. Strict control on movements of animals and animal products is required.

7.6.4 Marketing

Sheep and goats provide four types of products for marketing: meat, milk, wool, and hides. Milk and hides are handled in much the same ways as for cattle and buffaloes and will not be covered in detail here. Meat and wool marketing have characteristics unique to sheep and goats and these will be described in more detail. Meat from sheep is further differentiated into meat from a mature sheep (mutton) and meat from young sheep (lamb).

Meat products of both sheep and goats are subject to heavy seasonal consumer demand. This demand is often linked to religious or cultural holidays. Each holiday has slightly different preferences for the age and weight of lambs or goats. A compilation of major religious and national holidays is shown in Table 7.2. Each holiday

TABLE 7.2. Holiday Markets for Sheep and Goat Meat

Religious Holidays	
Christian (eastern)	Easter: variable date
Christian (western)	Easter: variable date
	Christmas: December 25
	Navadurgara or Navratra
Hindu	Dashara: variable date
Jewish	Passover: variable date
	Rosh Hashanah: variable date
Muslim	Ramadan: variable date
	Eid-ul-Fitr: variable date
	Eid-ul-Adha: variable date

Nonreligious Holidays	
July fourth	United States
Carnival	Most of Latin America
National Independence days	Different for each country

has specific traditions calling for serving of goat or lamb meat as part of a meal. In some cases there is even a prescribed method of harvest such as the Halal and Kosher methods for Muslim and Jewish markets. In China, the demand for goat meat is stronger in the 6 colder months of the year. Most parts of Africa have a steady year-round demand with greater sales around religious holidays.

Under natural conditions, sheep and goats will produce lambs and kids during the spring months of a year. They will be ready for marketing 2 to 4 months later. This is not a problem for holidays with a relatively fixed date such as Easter, Passover, or Navadurgara. The Muslim calendar poses a complication for the supply of goat and lamb meats for religious holidays. Muslim holidays follow the lunar calendar, which shifts the holiday about 10 to 11 days each year relative to the Gregorian calendar. This means that these holidays often fall in periods when the natural breeding cycles of sheep and goats do not provide young animals for market. Modern husbandry practices can shift the production cycle as much as 2 or 3 months to compensate for this market shift.

Marketing channels from producers to consumers are fairly well established in the industrialized countries. The animals first go from the producer to local auctions or to marketing cooperatives. The animals then go either to local processors or regional auctions and then to processors. The packaged meat products go through the grocery wholesale and retail chain before they reach consumers. In a few cases producers may sell directly to individual consumers.

In the United States these marketing channels are still being established in many parts of the country. Prior to 1991 the United States was a net exporter of goat meat. Beginning in this year, imports were higher than exports and in 1994 there were no exports of goat meat. The United States must now import goat meat from Australia and New Zealand. This increase in demand within the United States is linked to the rapid rise of foreign-born residents, both naturalized and noncitizen shown in Figure 7.24. Their desire for familiar meats has led to a corresponding increase in goat and lamb production in the United States.

Sheep and goat production in other countries suffer from the same lack of refrigeration as beef products. Goat and sheep farmers have established markets, but

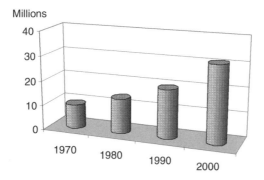

Figure 7.24. Foreign-born United States residents.

producers often fail to keep up with increased demand due to population growth. There are many reasons, but the majority are classified as marketing problems. A study on the constraints to adoption of improved technologies conducted in North West Province of Cameroon showed some of the problems blocking greater production. A basic factor is the size of individual farmer herds. Most farms are small, and the average herd size for goats is nine and for sheep eight animals.[29] This herd size means the capital for improvements is limited. Related to the herd size is the limited amount of available pasture land. The current market prices are perceived as too low to merit increased investment. The result is that farmers are making little additional investment to increase herd productivity and animal numbers.

Wool products are an income source for most sheep farmers and some goat farmers. Most wool from meat-type sheep is of ordinary quality and only yields a relatively small income for the farmer. Specialized wool-producing breeds of sheep and goats can give a good income if managed properly. Wool is used largely for clothing because of its insulating properties and moisture absorbing qualities. Each application requires a specific diameter, length of fiber, and quality as shown in Figure 7.25.[30]

The key to marketing is the quality of the wool. Wool quality is determined by the diameter of the individual fibers, their length, and amount of contamination by foreign matter, such as dirt, burrs, and seeds. Within the United States three systems are currently in use.

The American system (sometimes called the blood system) was developed in the 1800s. It classifies wool according to the percentage of merino blood in the breed of sheep providing the wool. Full-blood merino is classed as fine, then $\frac{1}{2}$ blood, $\frac{3}{8}$ blood, $\frac{1}{4}$ blood, low $\frac{1}{4}$ blood, and so on. This only gives an approximate classification of wool fineness and quality.

The spinning count system (or English system) is more exact. It is based on the number of "hanks" (560 yards) of yarn that can be spun from a pound of clean wool. Finer wool fibers will produce more hanks than coarse fibers.

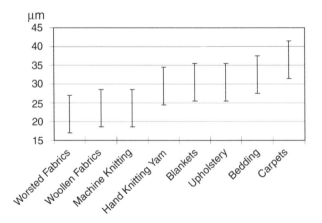

Figure 7.25. Fiber diameter and end uses of wool. (Adapted from *Harvesting of Textile Animal Fibers*, FAO.)

The micron system is based on actual measurements of individual fibers. Very fine fibers of wool measure less than 17.7 μm while very coarse fibers are 36 to 40 μm in average diameter. Because wool is a natural fiber, not all fibers are the same diameter. A batch of wool that is more uniform is more valuable than a highly variable batch. The uniformity of a batch of wool is expressed by its coefficient of variation. Lower coefficients of variation indicate more uniform wool.

The U.S. government has established standards for all wool products in commercial trade. Strict observance of these standards makes selling and buying more risk free.

Similar standards exist for "grease mohair," which is the fiber from Angora goats. Fine mohair measures less than 23 μm while the most coarse mohair is over 43 μm in average diameter.

7.7 SWINE PRODUCTION

> *Latin Name*: *Sus scrofa*
>
> *Other English Names*: Pig, sow, boar, gilt, barrow, hog

There are four subspecies of swine, each one occupying distinct geographic regions of the world. *Sus scrofa scrofa* is found in western Africa and Europe. *Sus scrofa ussuricus* is in northern Asia and Japan. *Sus scrofa cristatus* is common in Asia Minor and India while *Sus scrofa vittatus* is limited to Indonesia.

The countries with significant swine populations are shown in Figure 7.26. The country with the most swine is China with 488 billion swine. The second country in order of swine population is the United States with only 60 billion animals, followed by Brazil, Germany, and Spain. The countries with the lowest swine populations are predominantly Muslim or Jewish.

Swine are raised entirely for meat purposes. Other products obtained from swine are incidental to meat processing. When raised under sanitary conditions, swine are relatively disease free. However, swine left to forage around a village or that have become wild can be infected with parasites such as Trichinella. This causes the disease known as Trichinellosis (also known as trichinosis). These parasites infest muscle and blood vessel tissues causing pain and in severe cases may lead to death. Thorough cooking of the meat of infected animals will destroy the larvae. However, any portion not well cooked can harbor living larvae that will subsequently infect the person who consumes the meat.

Swine are often considered "dirty" because they use their snouts to root in the earth and mud. When left to roam freely, swine tend to wallow in wet earth or mud, mostly for cooling purposes on hot days. A coating of mud is used by many animals to dislodge ticks and smother lice and fleas. In cool weather, swine tend to be clean animals.

Another swine-related phrase in the English language is "to eat like a pig." Swine have the reputation for eating noisily and to spread bits of food around. There is some truth to this since swine are rather noisy when eating and, in their haste to eat, often drop bits of food.

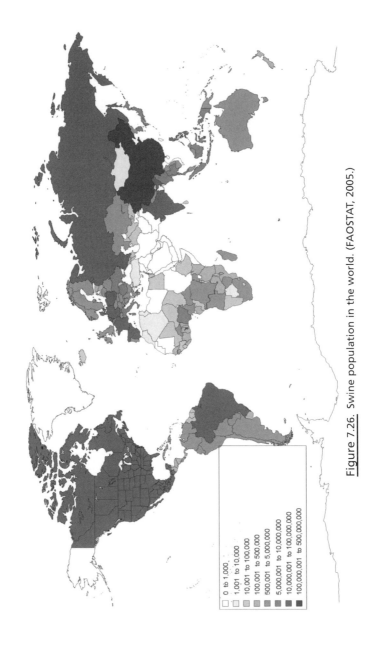

Figure 7.26. Swine population in the world. (FAOSTAT, 2005.)

☐	0 to 1,000
☐	1,001 to 10,000
☐	10,001 to 100,000
☐	100,001 to 500,000
☐	500,001 to 5,000,000
☐	5,000,001 to 10,000,000
☐	10,000,001 to 100,000,000
■	100,000,001 to 500,000,000

The saying "fat as a pig" has lost some of its meaning over the past 30 years in the United States. Historically, swine were raised as a source of fat to provide lard for cooking. With the increasing use of vegetable oils for cooking, there is less demand for lard and farmers are raising extremely lean swine for the marketplace.

7.7.1 Biology

Swine are monogastric animals. Their digestive system is similar to a human, consisting of a stomach, duodenum, ileum, and colon. They are omnivorous scavengers eating a diet consisting of grains, roots, and a few greens, plus insects and will also consume meat, even becoming cannibalistic.

Genetic studies indicate that domesticated swine are descended from the Eurasian wild boar, *Sus scrofa*. Domestication occurred nearly simultaneously in Europe and Asia about 9000 years ago. Wild swine were first recorded in cave paintings and carvings over 25,000 years old. The first records of domesticated pigs are from China, over 7000 years ago. Historical records show that pigs from China were brought to Europe in the eighteenth century for purposes of improving the European breeds. Swine have greater intelligence than most farm animals and have been used to replace dogs for hunting.[31] Domesticated swine often become wild if not kept in confinement, and it is likely that most wild pigs hunted in the United States are descendants of domesticated pigs.

Distinct breeds of swine have evolved over the years. To maintain the characteristics that distinguish a particular breed, farmers band together to form a registry that lists names and owners of livestock meeting the ideal characteristics for color, size, and the like. It is interesting to note that the first swine entered on a registry of purebred livestock was a boar owned by Queen Victoria of England in 1875. The major breeds of swine in the United States are Duroc, Landrace, Chester White, Poland China, Berkshire, Spots, Yorkshire, and Hampshire.

The meat of swine (pork) is considered "unclean" and not fit for human consumption in Jewish and Muslim religions. It is most likely this prohibition is related to the presence of parasites in many swine herds of the Middle East. Avoidance of foods that could carry disease makes good medical sense. Including it in religious doctrine makes for more healthy followers. There is no such prohibition in other religions.

"No man should be allowed to be president who does not understand hogs."
President Harry Truman

7.7.2 Feeding Requirements

The nutritional requirements of swine are similar to that of humans. As a result, we would expect them to compete with humans for the same types of food. In reality, swine will eat a wide variety of foods that humans do not prefer to eat. Historically, swine have eaten kitchen wastes that humans refuse to eat such as potato skins, meat scraps, maize husks, stale bread, and the like. Large cities have significant amounts of kitchen wastes from homes and restaurants requiring disposal. Often farmers utilize this kitchen waste to feed swine.

A problem with using kitchen wastes, and even slaughterhouse wastes, is disease transmission. The 2001 outbreak of foot-and-mouth disease in England was traced to a single infected salami casing used as feed on one farm. Trichinella is also transmitted primarily from eating meat scraps infected with the eggs. As a result of these problems, kitchen wastes are either cooked to destroy pathogenic organisms or mixed with compost to be used as crop fertilizer. In some countries, no animal by-products are permitted in livestock feed.

Swine are also fed many waste, or low value, products from the food-processing industry. This includes soybean hulls, distillers mash, maize-cob meal, beet pulp, sugar-cane juice, cottonseed meal, ripe bananas, cassava, and wheat middlings. Even cane sugar can be mixed with grains and fed to swine. A disadvantage is that pure sugars have only 80 percent of the gross energy value (kilojoules/gram of dry matter) of maize.[32] Reject bananas contain only about 5 percent protein, requiring other sources of protein to balance the diet.

For swine in an intensive management system, a well-balanced diet providing energy and proteins for growth is composed of a mixture of grains, green materials, and supplements. The mixture of ingredients is adjusted for each growth stage of the animals. Generally, three types of mixtures will be fed to swine: starter rations, finishing rations, and sow rations. A starter ration will consist of maize, sprouted oats, soybean meal, dried whey, salt, and vitamin supplements. The finishing (or growing) ration will contain more maize, but without oats or whey. The ration for sows will contain similar ingredients but with larger amounts of some vitamin supplements. Feeding swine in tropical areas under intensive management is similar around the world. The major differences relate to types of products used to attain the desired ration.

Starter rations are fed to the young pigs from weaning time (12 to 28 days age) until they are 10 to 20 weeks of age. These are similar to the young pigs shown in Figure 7.27. The finishing ration is fed for 15 to 40 weeks until the pigs reach marketing weight of about 110 kg. The sow ration is fed to pregnant and nursing sows. Pregnancy lasts 115 days and nursing of the piglets lasts 12 to 28 days.

Feeding swine under less intensive management systems usually increases the time needed to bring animals to market weight. When raised in pens, swine are fed a mixture of home-grown or purchased grains with kitchen scraps. They may be permitted to graze and forage for roots and nuts in a fenced field. The farmer may also purchase waste from food-processing plants such as cassava trimmings. Cassava is useful if dried or cooked to reduce the hydrocyanic acid content. If not treated, cassava roots will cause reduced weight gain and possible toxicity reactions.

In Vietnam, swine are seen as a form of savings. Management of the small herds concentrates on maintaining animal health and producing several pigs for sale each year. The overall return on farmer investment exceeds the interest payments for savings in commercial banks.

Figure 7.27. Young pigs after weaning.

7.7.3 Pests and Diseases

The OIE list A diseases affecting swine are described here. Other diseases affect swine but are considered to be a lesser economic threat on the international level.

Foot-and-Mouth Disease. This is the same as described for cattle in Section 7.5.3.

Vesicular Stomatitis. This is the same as described for cattle in Section 7.5.3.

Swine Vesicular Disease. Swine vesicular disease is a serious viral disease of swine. It is unique because researchers believe this is a disease transmitted from humans to swine. The virus is especially persistent, surviving up to 2 years in salted, dried, or smoked meats. Symptoms include fever, sores in the mouth, snout, feet, and teats. It has been diagnosed in Hong Kong and most of Europe. The disease is spread by feeding infected garbage or by infected animals showing no signs of the disease. It is not a fatal disease for swine but results in greatly reduced gain. The symptoms are nearly identical to foot-and-mouth disease. There is no vaccine. If a herd is infected, the only control consists of slaughtering and burying infected animals and all nearby animals.

African Swine Fever. African swine fever is caused by a virus spread by soft ticks and infected garbage fed to swine. In some cases the disease was caused by feeding waste from the processing of birds that had fed on the ticks, thereby becoming carriers. Once it has infected a herd, it spreads by contact among the animals. It is present in Africa and the island of Sardinia. The virus is relatively resistant, surviving nearly 5 months in salted meats. Infected animals have a fever and reddish skin often developing reddish spots. Pregnant sows will abort. The virulence of the strain influences the mortality. Mild strains may cause only 10 percent mortality, but more virulent strains can kill all animals in a herd. Death usually occurs within a few days. There is no vaccination available. Infected herds are slaughtered and buried. Attention to visitors and general herd sanitation are required.

Classical Swine Fever. This disease was previously known as hog cholera but the preferred name is now classical swine fever.[33] It is caused by a highly contagious virus that can persist for years in frozen meat. The disease appears in various forms from acute (with obvious symptoms) to chronic (with few symptoms). Symptoms include high fever, depression, and superficial and internal bleeding. In the acute forms there is near 100 percent mortality within 10 to 15 days after symptoms first appear. Swine with a chronic, less severe form of the disease will also eventually die. It is present in 36 countries around the world, being eradicated only in Australia, Canada, and the United States. Infection is through contact with other infected swine. This occurs mainly through auctions, livestock shows, and visitors to swine buildings. Preventive vaccines are available and can significantly reduce the prevalence of the disease. Infected herds must be slaughtered and buried. Normal sanitation procedures help to prevent infection of clean herds.

7.7.4 Marketing

The primary market product from swine is meat. Incidental products are made from the skin. Pork is usually marketed through the same marketing channels as beef or mutton. Similar requirements exist for sanitary harvest, handling, packaging, and transport.

7.8 AQUACULTURE

For the purposes of this section, we will define aquaculture as the rearing of aquatic organisms under controlled conditions.[34]

Within aquaculture there are many systems for controlling or capturing and rearing aquatic animals. We will cover only two systems closely related to land agriculture: integrated rice fields and ponds.

TABLE 7.3. Cultivated Aquatic Animals

Finfish	Crustaceans	Mollusks
Low value	Lobsters	Clam
Carp	Crab	Mussel
Tilapia	Shrimp	Oyster
High value	Prawn	
Trout		

Figure 7.28. European carp (*Cyprinus carpio*).

7.8.1 Aquatic Species Cultivated

Many types of aquatic animals and even plants are cultivated by humans to eventually harvest for food. The most popular types are shown in Table 7.3. Within each type of aquatic animal there are several general classes of animals and many species of each are cultivated.

The most widely cultivated aquatic animals are carp and tilapia. There are many species of each. Only a few can be covered in this chapter. Among the more common carp species cultivated are silver carp (*Hypophthalmichtys molitrix*), big head (*Aristichthys nobilis*), grass carp (*Ctenopharyngoden idellus*), black carp (*Mylopharyngodon piceus*), and common carp (*Cyprinus carpio*) shown in Figure 7.28.

The common tilapia species cultivated are *Tilapia mossambica* and *Tilapia nilotica*. The most common species cultivated in the United States is *Tilapia mossambica* shown in Figure 7.29.

Figure 7.29. *Tilapia mossambica* ready for market.

7.8.2 Biology

Records in China indicate that common carp were raised in rice paddies during the Wei dynasty (AD 220–265).[35] Many other Asian countries have recorded histories of cultivation of fish with rice. Before 1930, tilapia was cultivated almost exclusively in its continent of origin, Africa. It was introduced into most countries of the world starting in the 1940s.

Each species of fish has unique feeding habits and growth stages. Bighead and silver carp both feed on plankton. Silver carp in the larval stage feed mainly on zooplankton but turn to phytoplankton as adults, while bighead feed on zooplankton all their lives. This difference is due to anatomical differences in their gill rakers. Gill rakers are a series of projections located along the front edge of the gill arch. These act to filter zooplankton from the water. The gill rakers of bighead are short and sparse and unable to retain the smaller phytoplankton that the more numerous gill rakers of the silver carp can retain.

Grass carp are herbivorous consuming most types of aquatic and land grasses with their saw-toothed teeth. This species is often used to control aquatic weeds in waterways and canals. Dense populations of grass carp in rice paddies can cause damage to the rice crop but sparse populations are useful for weed control.

Black carp are carnivorous, feeding on snails and clams. Their teeth and callous pad act together to crush shells and eat the meat. Under cultivated conditions, they feed on silkworm pupae, earthworms, and animal entrails.

Common carp feed on rotifers, insect larvae, snails, young clams, shrimp, some plants, and seeds. Their mouth is shaped to permit them to dig in mud for organic matter.

Tilapia are omnivorous. At the larval stage they prefer zooplankton, but as they grow larger they will eat a large variety of plants, algae, and animals. The dense gill rakers of *Tilapia nilotica* makes it more efficient at feeding on the green algae, Chlorophyta, and the blue algae, Cyanophyceae. This gives it an advantage over other species that cannot digest these algae.

Growth rate is often a major consideration for choosing an aquacultured species. Most of the carp species have high growth rates. The optimum temperature for growth and food intake is 25 to 32°C with death occurring above 40°C. The best pH range of water for growth is between pH 7.5 to 8.5. Water below pH 6.0 or above pH 10.0 will retard growth. At higher levels of dissolved oxygen (4 to 5 mg/L) feeding is intense. Levels of dissolved oxygen below 1 mg/L will stop feeding and cause fish to gasp for air.

7.8.3 Management Systems

Integrated rice fields and pond aquaculture will be discussed here. Both systems are present in China, the largest producer of aquaculture species. It should be mentioned that a large part of the commercial production in China is done on large operations similar to that shown in Figure 7.30.

Integrated rice fields use the flooded conditions of rice paddies to nurture fish for sale. Some modification of the pure rice planting scheme is needed to accommodate

Figure 7.30. Southern China shrimp pond and shrimp.

fish. A variation on this method is a type of alternating pure rice and pure aquaculture in the same paddy.

Pond aquaculture uses existing or constructed ponds to culture fish. Normally these ponds are in association with other land animal or crop production enterprises.

Integrated Rice Field Aquaculture. This system involves using standing water in rice paddies to cultivate fish, freshwater prawns, crabs, turtles, and frogs. This description will only cover fish, the most common animal raised in rice paddies.

Normally, rice paddies are intentionally stocked with fingerlings that may be sold as larger fish for stocking or raised for market. The most suitable rice production ecosystem for aquaculture is the lowland irrigated rice system. This system has a guaranteed source of water to maintain adequate levels for both rice and fish production. In an irrigated rice ecosystem, the water level in the fields ranges from 2.5 to 15 cm deep. This is not suitable for fish production since water temperatures can exceed 40°C at midday. Rice cultivation requires lowering the water levels for weeding and harvest, making the field unsuitable for fish. Several modifications to the rice paddy permit both fish and rice to coexist. The height of the paddy bunds (embankments) are raised to 40 to 50 cm. This permits a greater depth of water at some stages and prevents fish from jumping out of the paddy. Screens are placed at the inlets and drains of the paddy to prevent fish from escaping. Fish refuges are constructed within the paddy. These consist of trenches 25 to 30 cm deep as shown in Figure 7.31, or simple holes 1 to 2 m deep throughout the paddy. To permit weeding, the water level is slowly

Figure 7.31. Fish refuges in a rice paddy.

lowered, making sure all fish are in the refuge trenches. After 1 to 2 days of weeding, the water level is again raised to let the fish roam freely in the paddy. The same system may be used for harvest, giving the fish a longer growth period. Harvesting of fish is done by hand when the water level is low.

Integrated Pond Aquaculture. A more common method of aquaculture uses existing or constructed ponds. These are normally rather small ponds to permit draining or manual harvest of mature fish. In the Philippines a small pond will cover less than 1000 m^2. Average sized ponds are 5000 m^2 in area with a depth of 1.5 to 2.0 m.[36]

Pond rearing of fish in China is widespread. Fish are commonly raised in association with a land animal such as swine, poultry, or silkworms. The wastes from the land animals are used to feed or fertilize the pond fish. Mulberry trees are often planted on pond dikes, permitting mulberry leaves, silkworm frass, and silkworm pupae to be added to the pond as they become available. Swine and cattle manure is also added to fertilize the pond. Swine are the most popular animals raised in association with pond aquaculture in China. Pigsties are built directly on the dikes to permit easy flushing of manure to the pond.

Animal manures have two functions in aquaculture. The first is to provide feed to the fish. Most animals do not digest all grain fed, passing some undigested grains through the gastrointestinal tract. These are available for fish to take up and digest. The second function is to provide nutrients for growth of phyto- and zooplankton in the pond. As the manures decompose, they release soluble nutrients that are used by other organisms to grow and reproduce. Fish feed directly on these plankton.

7.8.4   Feeding Requirements

The management systems described previously will produce fish for marketing with minimal effort. To realize the potential feed conversion efficiency possible with fish, it is necessary to use prepared fish feeds. As an example, minimal management practices in China will produce 375 to 450 kg of common carp per hectare. Intensive management practices including feeding prepared rations will yield 3000 to 3750 kg of carp per hectare. Much of this increase is from the direct intake of feed by "feed eating" species, but plankton eaters also benefit from the increased plankton production.

A typical prepared fish food will contain ground soybeans, wheat, or maize perhaps mixed with ground meal of soybean, peanut, or cottonseed left after oil pressing. Other ingredients may include wheat or rice bran, and minced plant materials such as grasses, sorghum leaves, melon vines, and vegetable leaves or roots.

Most cultivated fish species are omnivorous (eating both plant and animal matter), but the black carp is carnivorous. For these carnivorous species, feed will contain fish meal, small wild fish, silkworm pupae, shellfish, fish entrails, and earthworms. Sample rations for black carp and grass carp are shown in Table 7.4. The black carp ration contains animal protein while that of the herbivorous grass carp is of only vegetable origin. It is important to note that feeding fish only prepared feed will not give maximum gain. Natural sources of food must be present for optimum growth of the fish.

TABLE 7.4. Feed Composition for Carnivorous and Herbivorous Species

	Black Carp (%)	Grass Carp (%)
Straw powder	50	25
Soybean cake powder	10	25
Fish meal powder	5	
Sesame stem powder		25
Barley/wheat flour	15	1
Wheat/rice bran	10	24
Rapeseed cake	10	

Source: Training Manual: Integrated Fish Farming in China.

7.8.5 Pests and Diseases

The OIE list A diseases does not contain any diseases affecting aquatic animals. Most efforts are directed toward disease prevention since diagnosis, treatment, and cure are especially difficult with fish. Some of the more notable diseases are hemorrhagic septicemia, erythroderma, enteritis, bacterial gill rot, vertical scale disease, and saprolegniasis. In addition, many parasites infest aquatic animals, causing long-term damage. In most cases treatment consists of disposing of the infected fish and cleaning the pond or rearing enclosures.

7.8.6 Marketing

Marketing of fish is a bit more akin to that of poultry than other animals. Many countries in the tropics lack widespread refrigeration so that transport of live fish has been developed as an alternative. In the Philippines, small farmers sell to middlemen who come to their ponds with aerated tanks.[36] They are paid in cash for fish that are then transported to wholesalers in a nearby city. The fish may then be cleaned and packaged for sale in large fish stores or sold to individual buyers from the live tanks. An alternate market is to kill the fish on the farm and for the buyer to transport the whole fish on ice to a market where they are cleaned.

Fish are not fed for 24 hours before harvest to avoid off-flavors, a well-known problem in farmed fish. This is due to feeding large amounts of commercial feed and to uncontrolled fertilization resulting in shifts in the types of plankton in the pond.

In parts of eastern Europe, carp is the traditional Christmas dish much like turkey or ham in the United States.

Small farmers frequently produce fish for sale and for use in their homes. A farm family with a fish pond will normally eat fish 3 times per week. The reasons are low cost, freshness, and convenience. Given the average size of families and frequency of consumption, the average farm family will consume 21 to 31 kg per person each year. In the Philippines, about 90 percent of the fish produced in ponds will be sold. The remaining 10 percent are consumed by the family or given to neighbors. Average yields for tilapia in the Philippines are around 8 tons per hectare for a single harvest. Two harvests per year can yield nearly four times more net income than two crops of rice. Note that these yields are for a pond 1 ha in size. Most ponds are 0.1 to 0.5 ha and the family income must be reduced accordingly.

Larger fish farmers utilize more intensive management methods and have a lower cash cost of production. They also harvest large amounts of fish, giving them access to higher paying markets. In most cases these producers also sell to middlemen who then resell the fish to wholesalers in large cities. In industrialized countries, the fish are harvested, cleaned, and packaged on the farm. This decreases chances for development of off flavors due to spoilage.

7.9 OTHER FARM ANIMALS

A wide variety of animals have been domesticated by humans. The preceding descriptions of farm animals only cover the most numerous species cultivated on two or more continents. Following is a short description of species less numerous but still domesticated by farmers.

7.9.1 Alpaca (*Vicugna Pacos*) and Llama (*Lama glama*)

Both the alpaca and the llama are related to the camel. They are native to the Andean region of South America but have spread to North America and Australia. The alpaca is primarily kept for its very fine wool. An alpaca sweater rivals the famous cashmere sweater in warmth and light weight. The llama is used more for a beast of burden, carrying loads along narrow trails at high altitudes. They are well adapted to altitudes of 4000 m and more. Llamas are very territorial. When threatened, they become very nervous and will discharge nasal fluids and, eventually, gastric juices upon the unlucky animal or person. This is described as the "spitting" of the llama.

7.9.2 Camel (*Camelus dromedarius* and *Camelus bactrianus*)

There are two distinct species of camel used by farmers in the world. The most common is the Arabian, or dromedary, camel. This is often described as the "one-humped" camel, though in fact it actually has two humps. The front hump is much less developed than the large rear hump. The "two-humped" camel is the bactrian camel. The Arabian camel is widespread through North Africa and the Middle East. The bactrian camel is found mainly around the Gobi Desert in western China. The humps of both types are fat storage tissues to enable them to survive for 2 to 3 weeks without water. Farmers use them as draft animals, to carry loads long distances, for milk, meat, and even to race.

7.9.3 Dog (*Canis familiaris*)

Farmers have used domesticated dogs for thousands of years. Dogs are pack animals, and this seems to blend very well with human group behavior. Dogs are used to guard livestock, herd sheep and cattle, protect buildings, as a source of meat, and as pets. In some countries they are used as food. In most countries, however, dogs are kept as pets and serve little function other than protection.

7.9.4 Horse (*Equus caballus*) and Donkey (*Equus asinus*)

Both the horse and donkey are used by farmers to carry loads and pull carts or wagons. Horses are more widespread than donkeys, perhaps because of their multiple uses. Donkeys are useful for carrying or pulling heavy loads, while horses offer draft functions as well as riding, milk, meat, and even racing. In general, donkeys live longer than horses, are cheaper to maintain, and can carry loads longer distances than horses. The hybrid of a male donkey with a female horse is a mule, usually sterile.

7.9.5 Guinea Pig (*Cavia porcellus*)

Guinea pigs are not related to pigs nor do they originate in Guinea. They are small rodents, frequently sold for pets in the United States but used for food through much of the Andean region of South America. Large breeds have been developed weighing up to 2 kg for meat purposes. They frequently are raised in large boxes outside rural kitchens where they feed on vegetable scraps and grasses. In parts of Bolivia and Peru, they are a major source of protein for subsistence farmers.

7.9.6 Ostrich (*Struthio camelus*)

The ostrich occurs in the wild in the Sahel region of Africa but has been distributed throughout the world for meat purposes. Adults weigh from 90 to 130 kg and are up to 2.5 m tall. Their primary uses are for meat and eggs with some industries utilizing the feathers. Their eggs can weigh up to 1.3 kg. They are difficult to manage, being classified as a dangerous animal in Australia.

7.9.7 Turkey (*Melleagris gallopavo*)

The turkey was probably domesticated from the wild North American wild turkey. In contrast to the wild turkey, the domesticated turkey is often considered to be easily confused and foolish. The primary use of domesticated turkey is for meat purposes. Many industry-specific breeds have been developed with meat characteristics desired by commercial markets, such as "double breasted."

7.9.8 Yak (*Bos grunniens*)

Yaks are used by farmers in the high-altitude areas of Tibet and western China. They are related to cattle but are larger, weighing up to 1000 kg. Yaks are multipurpose animals for farmers in these areas. They serve as carriers of heavy loads and provide

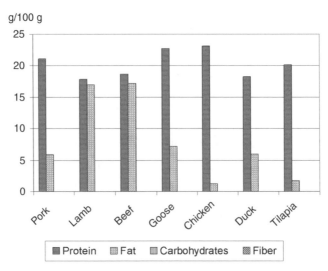

Figure 7.32. Nutrition components of meats. (From http://www.nal.usda.gov/fnic/ foodcomp/search/index.html.)

milk, wool, and meat. In some parts of Asia they are even used to pull plows. Their milk is formed into a cheese called *chhurpi* in Tibetan and Nepali.

7.10 NUTRITION

The nutritional value of animal products for human diets is shown in Figures 7.32 and 7.33. All meat products range from 16 to 24 percent protein. Among the grains, only

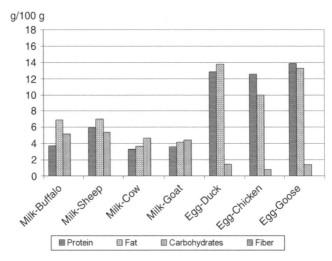

Figure 7.33. Nutrition components of milk and eggs. (From http://www.nal.usda.gov/fnic/ foodcomp/search/index.html.)

soybean has higher protein content. The fat content of meats varies with the specie. Chicken and tilapia have the lowest fat content while beef and lamb have the highest. Egg and milk products also vary with the specie. Milk from cows and goats have similar nutritional components while milk from Indian buffalo and sheep tend to have much higher fat and carbohydrate contents. The values shown in Figure 7.32 for pork, lamb, and beef are for a composite of trimmed retail cuts. The values for duck, goose, and tilapia are for whole, fresh raw meat. Chicken is represented by values for raw breast meat of broilers. The values for eggs in Figure 7.33 are for whole, fresh eggs, and for milk only values for whole, fluid milk are given.[37]

7.11 CONCLUSIONS

Farm animals provide food, clothing, and power to humans. In the most direct form, we consume various parts and products such as meat, eggs, and milk to provide high-quality protein in our diets. Many animal products provide articles of clothing made from wool and leather. Farmers use many types of animals to till the soil and carry produce from the farm to markets. Animals often utilize food materials not directly usable by humans such as grass or coarse grains, therefore avoiding direct competition for food. Some animal diseases can be passed to humans and sanitation is essential at all stages of production.

QUESTIONS

1. Which is the most numerous farm animal in the world?
2. Describe a landless livestock production system.
3. What is the difference between the diet of a ruminant and a monogastric animal?
4. What products do humans utilize from cattle?
5. What is the difference between a bullock and a cow?
6. Which animal provides mohair?
7. Why are fish and swine associated on farms in China?
8. How has immigration to the United States affected the demand for goat meat?
9. Which country has the largest population of domesticated ducks?
10. How much does an ostrich egg weigh?

REFERENCES

1. Available at: http://www.fao.org/documents/show_cdr.asp?url_file = //docrep/W0613t/w0613T01.htm.
2. Available at: http://www.cgiar.org/impact/research/fisheries.html.
3. Available at: http://water.usgs.gov/watuse/wuglossary.html.
4. *Global Livestock Production and Health Atlas*, FAO, Rome. Available at: http://www.fao.org/ag/aga/glipha/index.jsp.

5. H. Steinfeld and J. Mäki-Hokkonen, *A Classification of Livestock Production Systems*, FAO, Rome, 1995.

6. Available at: http://www.ithaca.edu/staff/jhenderson/chooks/chooks.html.

7. *Small Scale Poultry Production, FAO Animal Production and Health Technical Guide*, FAO, Rome, 2004.

8. Available at: http://www.epa.gov/oecaagct/ag101/poultryphases.html.

9. Goose Production, FAO Animal Production and Health Paper 154, FAO, Rome, 2002.

10. Feeding Chickens, Leaflet 2919, University of California, Cooperative. Extension Service, Yuba City, CA, 1983.

11. Agfact A5.0.1 Duck Raising, Part A, NSW Dept. of Primary Industries, New South Wales, 2004.

12. A. G. Cagauan, R. D. S. Branckaert, and C. Van Hove, *Rice–duck farming in Asia*, INFPD/FAO Electronic Conference on Family Poultry, July 2001.

13. Terrestrial Animal Health Code (2005), OIE, Paris, 2005.

14. Available at: http://www.state.nj.us/agriculture/avianinfluenza.htm.

15. Available at: http://www.bisoncentral.com/index.php?c=63&d=73&a=1022&w=2&r=Y.

16. Available at: http://ww2.netnitco.net/users/djligda/wbfacts.htm.

17. Available at: http://www.traill.uiuc.edu/dairynet/paperDisplay.cfm?ContentID = 1190.

18. Milk Production, National Agricultural Statistics Service, USDA, Washington, DC, April, 2006.

19. P. N. de Leeuw, A. Omore, S. Staal, and W. Thorpe, *Dairy Production Systems in the Tropics: A Review*, ILRI, Nairobi, Kenya, 1998.

20. A. Speedy and Rene Sansoucy, *Feeding Dairy Cows in the Tropics*, FAO, Bangkok, Thailand, 1989.

21. P. Auriol, Intensive Feeding Systems for Beef Production in Developing Countries, in *World Animal Review*, FAO, Rome, 1978.

22. Available at: http://www.fao.org/DOCREP/003/X6511E/X6511E00.htm#TOC.

23. Available at: http://ask.metafilter.com/mefi/28427.

24. *Manual on Simple Methods of Meat Preservation*, FAO Animal Production and Health, Paper 79, FAO, Rome, 1990.

25. Available at: http://www.ansi.okstate.edu/breeds/sheep/.

26. Sándor Kakovics, *Sheep and Goat Production in Central and Eastern European Countries*, REU Technical Series 50, FAO, Rome, 1998.

27. D. E. MacHugh and D. G. Bradley, Livestock Genetic Origins: Goats Buck the Trend, *PNAS*, Vol. **98**(10), 5382–5384, 2001.

28. S. M. J. Hossain, M. R. Alam, N. Sultana, M. R. Amin, and M. M. Rashid, Milk Production from Indigenous Black Bengal Goat in Bangladesh, *J. Biol. Sci.*, **4**(3), 262–265, 2004.

29. African Research Network for Agricultural By-products (ARNAB). *Overcoming constraints to the efficient utilization of agricultural by-products as animal feed.* Proceedings of the Fourth Annual Workshop held at the Institute of Animal Research, Mankon Station, Bamenda, Cameroun, 20–27 October 1987, ARNAB, Addis Ababa, Ethiopia.

30. O. J. Petrie, *Harvesting of Textile Animal Fibers*, FAO Agricultural Services Bulletin 122, Rome, 1995.

31. Swine History, http://www.cyberspaceag.com/farmanimals/swine/swinehistory.htm, in www.cyberspaceag.com, by the *High Plains Journal*, Kansas.

32. Pérez, Rena, *Feeding Pigs in the Tropics*, FAO Animal Production and Health Paper 132, FAO, Rome, 1997.

33. *Foreign Animal Diseases*: "The Gray Book," Committee on Foreign Animal Diseases of the United States Animal Health Association, 1998, Richmond, Virginia.

34. Aquaculture in the Classroom, Available at: ag.arizona.edu/azaqua/extension/Classroom/home.htm, Univ. of Arizona, Phoenix, 2006.

35. M. Halwart and M. V. Gupta, *Culture of Fish in Rice Fields*, World Fish Center Contribution No. 1718, FAO, Rome, 2004.

36. *An Evaluation of Small-Scale Freshwater Rural Aquaculture Development for Poverty Reduction*, Asian Development Bank, Manila, Philippines, 2005.

37. USDA National Nutrient Database for Standard Reference. Available at: http://www.nal.usda.gov/fnic/foodcomp/search/index.html.

BIBLIOGRAPHY

An Impact Evaluation of the Development of Genetically Improved Farmed Tilapia, Asian Development Bank, Manila, Philippines, 2005.

Aquaculture in the Classroom, ag.arizona.edu/azaqua/extension/Classroom/home.htm, Univ. of Arizona, Phoenix, 2006.

Buckeye Meat Goat Newsletter, **4**(1), January 2006.

A. Coché and D. Edwards, *Selected Aspects of Warmwater Fish Culture*, Training Courses on Aquaculture, FAO, Rome, 1988.

Ethnic Holiday Calendar, at http://sheepgoatmarketing.info.

Foreign Animal Diseases: "*The Gray Book*," Committee on Foreign Animal Diseases of the U.S. Animal Health Association, Richmond, VA, 1998.

E. Giuffra, J. M. H. Kijas, V. Amarger, O. Carlborg, J. T. Jeon, and L. Andersson, The Origin of the Domestic Pig: Independent Domestication and Subsequent Introgression, *Genetics*, **154**, 1785–1791, April 2000.

Halwart, Matthias, and M. V. Gupta, *Culture of Fish in Rice Fields*, World Fish Center Contribution No. 1718, FAO, Rome, 2004.

Handling and Marketing Wool, University of Maine Cooperative Extension Bulletin #2070, March 2006.

How Pigs are Raised, at www.manitobapork.com, Manitoba Pork Council, Winnipeg, Canada, 2006.

J. P. Hunter and N. L. Mokitimi, The Evolution of the Wool and Mohair Marketing System in Lesotho: Implications for Policy and Institutional Reform, Network Paper No. 20, International Livestock Centre for Africa, Addis Ababa, 1989.

R. Kott, *Wool Grading*, Bulletin MT 8380, Montana State University, Bozeman, 1993.

Livestock as Food Security in Viet Nam, Poverty Alleviation and Food Security in Asia, RAP Publication 1999/4, FAO, Rome, 1999.

D. Machin and S. Nyvoid, *Roots, Tubers, Plantains and Bananas in Animal Feeding*, FAO Expert Consultation, FAO, Rome, 1991.

R. O. Myer and J. H. Brendemuhl, 4H Project Guide: Swine Nutrition, 4H225, IFAS Extension, University of Florida, Gainesville, FL, 2003.

O. J. Petrie, *Harvesting of Textile Animal Fibres*, FAO Agricultural Services Bulletin No. 122, FAO, Rome, 1995.

R. Pérez, *Feeding Pigs in the Tropics*, FAO Animal Production and Health Paper 132, FAO, Rome, 1997.

F. Pinkerton, *Meat Goat Marketing in Greater New York City*, Project Report for The Center for Agricultural Development and Entrepreneurship, State Univ. of NY, Oneonta, 1995.

Recognizing Peste des Petits Ruminants: A Field Manual, EMPRES-FAO, Rome, 1999.

Rendered Animal Products for Swine, at www.renderers.org/links/swine.htm, National Renderers Association, Washington, D.C. 2006.

Smallholder Livestock Production: Constraints on the Adoption of Improved Technologies, in *African Research Network for Agricultural By-products (ARNAB)*, 1989. Overcoming Constraints to the Efficient Utilization of Agricultural By-products as Animal Feed, Proceedings of the Fourth Annual Workshop Held at the Institute of Animal Research, Mankon Station, Bamenda, Cameroun, 1987.

Swine History, at http://www.cyberspaceag.com/farmanimals/swine/swinehistory.htm, in www.cyberspaceag.com, by the *High Plains Journal*, Kansas.

Training Manual Integrated Fish Farming in China, Training Centre for Integrated Fish Farming, Wuxi, China, 1985.

United States Standards for Grades of Wool, USDA, Washington, D.C., 1968.

United States Standards for Grades of Grease Mohair, USDA, Washington, D.C., 1973.

8

CLIMATE AND FOOD PRODUCTION

8.1 CLIMATE AT THE THREE FARMS

Weather characteristics of the areas of the three farms are given in the pictures depicting the sky during the December through February time period 2005–2006. The Philippine sky (Fig. 8.1) and the Ecuador sky (Fig. 8.2) are similar to the Ohio farm sky (Fig. 8.3), and yet the Philippines and Ecuador are green and the Ohio countryside is brown, the trees all being without leaves. Heavy thick clouds make the day dark in both the Philippines and Ohio. However, in the Philippines the clouds will deposit rain and dissipate while those in Ohio will not provide any precipitation on this particular day and will continue to cover the area for perhaps several days. When the conditions are right, these types of clouds in Ohio will provide precipitation in the form of snow or drizzle at this time of the year.

The Ecuador farm is on the equator while the Philippine and U.S. farms are in the Northern Hemisphere. Ecuador and the Philippines have no summer/winter temperature difference, but the United States farm does. For this reason the surroundings in the United States are very different, no plants growing, from those at the other locations where plants grow all year round.

World Food: Production and Use. By Alfred R. Conklin, Jr. and Thomas Stilwell
Copyright © 2007 John Wiley & Sons, Inc.

Figure 8.1. Sky at the Philippine farm during rainy season. Leaves in the foreground are mango and in the background are bananas, coconut, and other fruit-bearing bushes and trees. (Courtesy of Henry Goltiano.)

The Ecuador countryside and sky (Fig. 8.2) indicate a partly cloudy day with consequent shadows on fields. Cloud cover is consistent with the lower rainfall in Ecuador compared with Ohio or the Philippines, as shown in Figure 8.4. The verdant green is a result of a relatively constant rainfall and temperature regime at Quito, Ecuador.

As seen in Figure 8.5 the temperature at Leyte and Quito are both very constant, varying less than 2°C (1.1 for Ecuador and 1.8 for Leyte) throughout the year. However, the temperature at Quito hovers between 18.9 and 20°C while at Leyte it

Figure 8.2. Clouds and countryside during January in Pinantura, Ecuador. The fields are outlined by fence rows. (Courtesy of David Cerón.)

Figure 8.3. Winter clouds at the farm in Wilmington, Ohio. Field in the foreground is used as pasture.

hovers between 33 and 35°C, and thus Leyte is more than 10°C warmer than Quito. At Wilmington, Ohio, the temperature goes from an average low of 1.7°C in January to an average high of 29°C in July and then back to 4.3°C in December. Thus there is a dramatic, cyclic change in temperature, which is not seen in either Quito or Leyte.

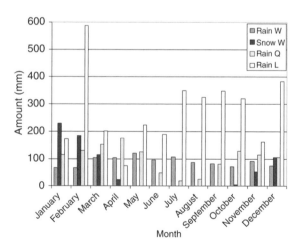

Figure 8.4. Shows the average precipitation for the sites of the three farms. Rain W and Snow W are for Wilmington, Ohio; and Rain Q and L are for Quito, Ecuador, and Leyte, Philippines, respectively. Rain and snow for Wilmington are 10-year averages while those for Leyte are 2-year averages and 20-year averages for Quito. [Data from Clinton Country Soil Survey, Leyte State University Climatological Station and Qwikcast (http://qwikcast.weatherbase. com), respectively.]

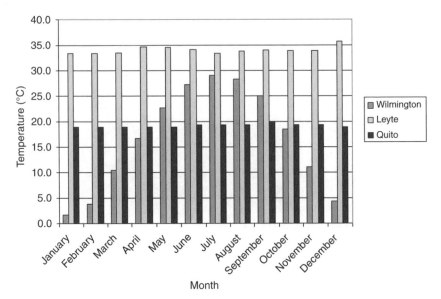

Figure 8.5. Average daytime temperatures throughout the year at the three farms. The temperatures for Wilmington are 10-year averages for Leyte, 2-year averages and 20-year averages for Quito. [Data from Clinton County Soil Survery, Leyte State University Climatological Station and Qwikcast (http://qwikcast.weatherbase.com), respectively.]

In central Ohio temperatures suitable for crop production occur from May through September, although some farmers plant in April and delay harvest until November to take advantage of growth during frost-free days during these months. In January through March and October through December, there will be killing frost, which is called this because it is cold enough to kill all annual crops except those specifically adapted for winter growth such as winter wheat.

As seen in Figure 8.4 Leyte averages around 3000 mm (3335 for the 2 years reported here) of rain a year and has the highest rainfall throughout the year of the three farms with particularly high rainfall in February. During this month, there are mud slides in Leyte, particularly the southern part, which are often quite destructive and cause significant numbers of deaths. Rainfall in Quito is lower than Leyte but slightly higher than Wilmington, except for the months of June, July, and August. This includes an estimated snow contribution of 25.4 mm of water for each 254 mm of snow. This estimate is subject to wide variation.

Another difference is day length. The differences in day length between the Philippines, Ecuador, and Ohio in the months of December and June are given in Figure 8.6. There is no difference in day length in Ecuador while the greatest difference is in Ohio and an intermediate difference in the Philippines. More detailed discussion of day length differences and its importance to food production is given below.

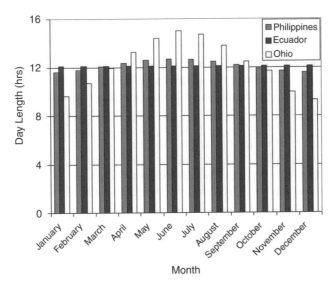

Figure 8.6. Day-length differences between summer and winter at the three farms.[3]

In both the Philippines and Ecuador the relatively constant temperature and rainfall means that food can be grown all year around. On Leyte rice paddies are seen in all stages of production at all times of the year. Some paddies will be plowed and ready for planting while others will be newly planted, some with rice plants at midstage of development and some heading (forming grain heads) and some being harvested. Thus, the climate provides for constant food production.

However, high rainfall and high temperatures, as in Leyte, are not conductive to growth of all plants, and so in some cases some crop plants are preferentially grown in months with less rainfall such as April through June (Fig. 8.4). In other cases crops may be protected from excess rainfall by putting a "roof" of transparent plastic over the planted area.

In Quito, Ecuador, the relatively constant temperature also allows the production of food all year around. As noted, the temperature is lower so crops that do not do well in the Philippines will do well in Ecuador. Tomatoes do better in climates cooler than the Philippines and thus do better in Ecuador.

As noted, crop production on the Ohio farm is limited to those months that are frost free. However, some garden crops such as radishes and peas do better during the cooler months while maize and watermelon do better during the warmer months. Also any crop that takes a longer warmer growing season such as rice will typically not be grown, although special conditions, not suitable for commercial or even household use or production, can be arranged and thus allow the plants to grow and even produce. This might mean starting and growing the plants in a greenhouse for a significant period of time before planting in the field.

8.2 CLIMATIC ZONES OF THE THREE FARMS

The climatic zones in each of the areas of the three farms are described very differently. In the Philippines climatic zones are largely determined by rainfall patterns. Type II, which is the type experienced by the Philippine farm, as shown in Figure 8.7, has no dry season and has high rainfall in the winter months, particularly February as can be seen in Figure 8.4. Type IV zone on the lower eastern side of Leyte is similar to type II in that rainfall is evenly distributed during the year, but it does have a short dry spell at some time during the summer months.

In Ecuador the climatic zones are primarily determined by the altitude of the area, as shown in Figure 8.8. There are thus three climatic zones: the costal area, which tends to be tropical with high humidity and high temperatures, the sierra or central highlands, which is where Pinantura, Quito, is located, are variable in temperature, depending on altitude, and the Eastern lowlands are the rainforest type of climate.

In Ohio the lowest temperatures during the winter are the most important consideration in terms of crop production, particularly with reference to the ability of fruit and nut tree to survive the winter (Fig. 8.9). These temperatures also serve to determine what types of protection animals need during the winter months, particularly in terms of making sure their water source is not frozen.

In addition to the lowest expected temperatures, the average first killing frost in the fall and the last in the spring are also important dates for determining safe planting and heating requirements for animals. For Clinton County, Ohio, the last killing frost in

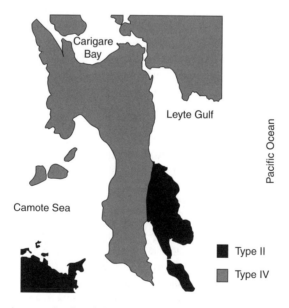

Figure 8.7. Climatic zones in the Philippines determined largely by rainfall patterns. (Data from http://www.fao.org/ag/AGL/swlwpnr/reports/y_ta/z_ph/phmp131.htm.)

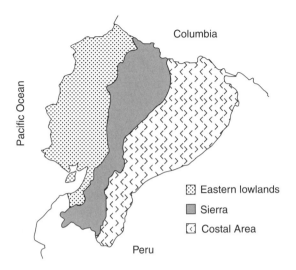

Figure 8.8. Climiatic zones in Ecuador. (Data from http://www.fao.org/ag/agp/agpc/doc/counprof/ecuador.htm.p8.)

the spring is on May 15th with the first killing frost in the fall on September 27th. The growing season, with temperatures above $0°C$ is approximately 145 days long[1]. The values given are the most conservative in that both the killing frost and growing season length are variable from year to year.

Each of these types of climatic characteristics and how they affect crop and animal production are discussed in more detail below.

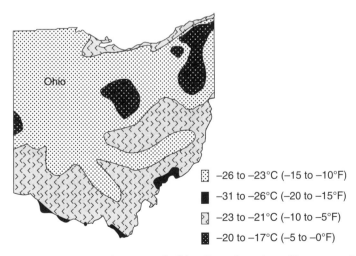

Figure 8.9. Climatic zones in the state of Ohio. (Data from http://www.growit.com/bin/USDAZoneMaps.exe?MyState = OH.)

8.3 CLASSIFICATION OF CLIMATIC ZONES

As a first approach to world climatic zones, one could look at the general climatic zones of the earth, as shown in Figure 8.10. They can be divided into tropical (between 15° north and south latitude), subtropical (between 15 and 30° in both the Northern and Southern Hemispheres), temperate (between 30 and 60° in both the Northern and Southern Hemispheres), and the Arctic and Antarctic regions (from 60 to 90° north and south latitude, respectively). Agriculture, both plant and animal production, is generally restricted to the first three of these because of temperature.

Two other major ways of classifying the world's climate are the Köppen and the Thornthwaite classification. In the Köppen system three climate groups based on latitude and designated by roman numerals, as shown below, and five major climatic zones, designation by A, B, C, D, and E; see Table 8.1, based on temperature and precipitation are described. The three climatic groups based on latitude are determined by the dominance of combinations of air masses as follows:

Group I: Low latitude climates that are controlled by tropical and equatorial air masses

Group II: Midlatitude affected by tropical air masses moving north and polar air masses moving south

Group III: High latitude climates dominated by polar and artic air masses

The A regions are simply those tropical regions experiencing high rainfall throughout the year. The B designation is subdivided into desert regions that have the subdesignation W and thus would be given the full designation BW. The other region is semiarid, also called steppe, and is given the subdesignation S and thus would be given the full designation BS. All the zones excepting large parts of E are used for agriculture. Keep in mind that region BS and some parts of region BW cannot be farmed without irrigation.

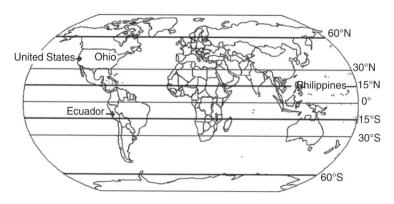

Figure 8.10. Climatic zones based on latitude.

TABLE 8.1. Basic Köppen Climate Zones

Zone Designation[a]	Characteristics
A	Evenly distributed rain year round and constant high temperatures year round, i.e., a tropical climate
B	Little rain and wide variations in temperature. This zone is subdivide into two subgroups
C	Rain in winter with dry summers typically of midlatitude areas
D	Moderate rainfall with wide variation in temperature. This is the climate of large continents
E	Cold climates with short periods above freezing

[a]Letters correspond to those in Figure 8.11.

Further subdivisions are made using lowercase letters for their designation as shown in Table 8.2. Thus, f designates adequate rainfall in all months with no dry season (used with A, C, and D). High rainfall areas, with short dry period, such as rainforest, are designated by m (used only with A). Dry summers are designated by s and dry winters by w. Further distinctions can be made by using additional lowercase letters as needed.

Knowing the climate classification and knowing the types of crops grown in that zone gives a good indication of crop adaptation to an area. A generalized world map showing Köppen classifications is shown in Figure 8.11. While this classification system gives good information relative to agriculture, it does not provide specific information, such as total rainfall or soil moisture available for crop production. Also local conditions, such as proximity to mountains, rivers, and ocean currents may significantly change localized weather conditions that would allow production of crops that would not normally grow in the region.

However, to understand crop and animal production, it is essential to know not only about temperature, rainfall patterns, and the altitude of a particular area but also

TABLE 8.2. Further Subdivisions to Köppen Climate Zones

Designation	Characteristic	Major Zone Used with
a	Hot summers over 22°C	C, D
b	Warm summer but less than 22°C	C, D
c	Cool summer with >4 months over 10°C	C, D
d	Very cold winters, i.e., temperature below −38°C	D
h	Hot and dry mean annual temperature <18°C	B
k	Cold and dry mean annual temperature >18°C	B

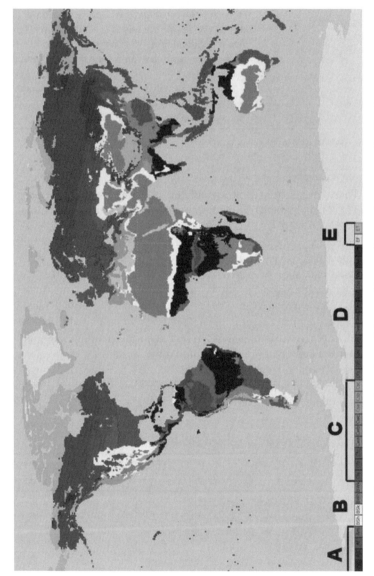

Figure 8.11. Climatic regions of the world according to the Köppen classification.

evaporation, water loss from the soil's surface and transpiration, water loss from the plant leaves. These latter two, along with the former three, determine how much water is available for crop and animal production. It is primarily these five that account for variations in crop and animal production in various parts of the world.

The Thornthwaite climate classification takes into account evaporation from soil surface and transpiration through leaves of plants. These two components are hard to measure independently, and so agriculturalists generally combine the two into the amount of water lost from soil surface and plants and term this combined water loss as evapotranspiration. This is the amount of water needed to produce a crop. Thus, one needs to know both the rainfall and the evapotranspiration to know how much water is available for growing a crop or for watering animals.

Measurement of evapotranspiration is extremely difficult in that, although moisture in the air can be measured, it is difficult to differentiate between the moisture already in air and that which is added by the plant and that evaporated from the soil surface. In addition the water vapor arising from evapotranspiration is rapidly diluted in the atmosphere. In addition other factors such as dew formation can cause additional difficulties in determining evapotranspiration.

A way around some of these problems is the open-pan evaporation method of estimating evapotranspiration. Here the amount of water lost from a standardized open pan of water is related to evapotranspiration by various conversion factors.

Figures 8.12 and 8.13 show graphs of rainfall verses evapotranspiration for a year in a dry and a wet climate. Wet climates are those where the rainfall exceeds evapotranspiration, in dry or desert climates rainfall is less than evapotranspiration. These two factors, that is, rainfall and evapotranspiration, can both be expressed as centimeters (cm) or millimeters (mm) of water and graphed together on a month by month bases. In this way shortfalls in soil water can be determined and irrigation plans developed

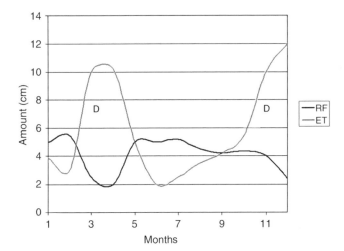

Figure 8.12. Rainfall (RF) verses evapotranspiration (ET) for a dry climate; D represents times of water deficit.

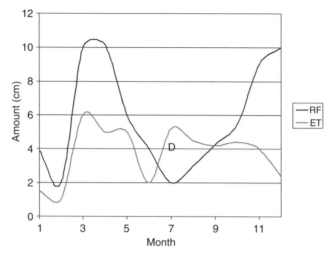

Figure 8.13. Rainfall (RF) verses evapotranspiration (ET) for a wet climate; D represents times of water deficit.

if water for irrigation is available. This method thus gives a ready, visual, method of determining the availability of water for crop production.

Figure 8.12 is an example of a dry or desert-type climate. There is a deficit in soil water throughout most of the year. Excess moisture at some times is not enough for crop production during the remainder of the year. Under these conditions there is insufficient water for recharge of groundwater, and irrigation from groundwater is not advisable because it will not be sustainable.

Figure 8.13 is an example of a wet climate, where there is a deficit during short periods that is more than made up for the rest of the year. Soil storage of moisture is sufficient to provide water for crops during this time. Excess water during the remainder of the year either runs off into streams and rivers or percolates into groundwater. Under these conditions excess water can then be safely used for irrigation (Chapter 9 for further information about irrigation).

Note that this method gives great emphasis on rainfall and evaporation. However, other climatic factors needed for crop production, that is, temperature, day length, and the like must also be appropriate for the crop to be grown.

8.4 CROP PRODUCTION AND CLIMATE

People live and raise both crops and animals in almost all climatic zones, exceptions being very high mountains, the permanently frozen parts of both the north and south polar regions, and those very dry regions without sources of water for drinking and irrigation. Different climatic conditions usually mean that crops and animals adapted to that climate are grown and raised. There are several exceptions to this. One is that plant and animal breeders are constantly breeding new varieties of crops and animals adapted to areas where they are not naturally grown or raised. Second, maize is

TABLE 8.3. Weather Events Important in Food Production

Event	Average	High	Low	Effect
Temperature °C (°F)	21 (70)	41 (105)	5 (41)	Plants cannot grow at low and do not do well at very high temperatures
Rainfall[a] (mm)	1000	2000	500	Low rainfall results in insufficient water for crop growth. High rainfall results in high humidity and many diseases[b]
Wind	NA	50 km/h and higher	NA	High wind can knock over crops and prevent pollination

[a]Ice, hale, and snow may also be problems in some areas.
[b]Irrigation may be used to produce crops in low rainfall areas if there is a sufficient water source.

typically grown in all areas of the world where the warm season is long enough to allow it to mature and there is sufficient water from either rain or irrigation.

Temperature, altitude, and rainfall are three climate factors that dramatically affect crop and animal production. First and perhaps foremost is temperature, second is altitude above sea level, which also controls temperature, and the third is rainfall (the effects of these on crop production are summarized in Table 8.3). The length of time during the year when the temperature is high enough to allow crop growth will, in part, determine what crops can be grown and when. As altitude increases into the troposphere, the temperature decreases, affecting not only temperature but can also be important in determining rainfall pattern and thus the crops and animals raised.

8.4.1 Temperature

Although some areas of the world experience high temperatures, generally high temperature alone will not preclude crop growth. However, low temperatures, especially those below freezing, will prevent crop growth. However, there are numerous food plants that are temperature hardy* and thus can survive and even grow at quite cold temperatures. In some cases crops may require low temperatures for initiation of flowering, such as apples, or the color associated with the fruit, such as oranges.

As noted above the Earth can be divided into artic, subartic, temperate, semitropical, and tropical regions (Fig. 8.10) or into regions according to other schemes. It might seem that simply knowing the latitude, above or below, the equator would be enough to explain the temperature regime of an area. This is not the case for numerous reasons. Ocean currents, prevailing winds, and altitude are three of the most important factors in addition to latitude that help to determine the temperature regime of an area.

Weather along costal areas is strongly affected by ocean currents. In areas where warm currents occur, such as those along the coast of Europe, these keep areas near the sea warmer than might be expected. On the other hand in areas where cold currents prevail, temperatures will be cooler than might otherwise be expected. The northern

*The word *hardy* is used to indicate a plant or animal able to withstand harsh environmental conditions.

coast of California would be an example of this type of situation. In either case the agriculture of the area will be profoundly affected by the temperature regimes controlled by these ocean currents.

Likewise prevailing winds blowing from a warm area or across a warm ocean current will cause temperatures to be higher than expected in those areas affected by the wind and similarly for wind blowing from areas of low temperature. For winds to have a significant effect, the winds must be prevailing, which means that they always or almost always blow from the same direction.

Large bodies of water also have a moderating effect on the temperature of an area. The heat capacity of water is much higher than soil, and so land generally heats and cools faster than water. Land areas near water cool and heat slower than areas far from large bodies of water. The Philippines is surrounded by ocean. The consequence is that the temperature is never very cold and never very hot, usually around 35°C and because of the abundance of water the humidity is always high.

Although not a variable factor in agricultural weather, rain shadows caused by mountains can have dramatic effects on rainfall. Wind blowing off the ocean, if it encounters a high mountain, will be pushed up high enough to cause clouds to form and rain to fall. As the wind passes the mountains, all the water will be removed and the area beyond is a desert or semidesert. This happens on the coast of northwestern United States in the states of Washington and Oregon.

8.4.2 Altitude

Altitude is a very different phenomenon. As the altitude in the troposphere increases, the temperature decreases. Perhaps one of the most dramatic effects of increasing altitude on temperature is in Kenya, which is split by the equator, and Tanzania where Mount Kenya and Mount Kilimanjaro both have snow on them all year around even though both mountains are very near the equator. Likewise the highlands of Kenya, Tanzania, and Zimbabwe have relatively moderate temperatures considering their closeness to the equator. In and around the capital of Zimbabwe, Harare, the temperature gets low enough that one can see frost early in the morning during the winter months, that is, June, July, and August. In addition temperatures are low enough to turn oranges orange (Chapter 6). In the Philippines, which has a definite tropical climate, mountainous areas, such as in the north central portion of Luzon, allow production of such cool-weather crops as strawberries and snow peas.

Baguio, Philippines

The Philippines lies between 5° and 21° north latitude and thus most of the country has a tropical climate. In spite of this, in the central mountains of Luzon is a valley in an area generally referred to as Baguio, which because of its altitude is much cooler that the remainder of the country. During the winter months of December through February night time temperatures will approach freezing, and occasionally it will snow. In the summer months it is also cooler than most of the Philippines. It is cool enough that cool-weather crops such as strawberries (Fig. 8.14) for which the valley is famous, can be successfully and economically produced.

Figure 8.14. Strawberries and flower production in the Banget Valley near Baguio, Philippines.

This illustrates that the altitude at which the agricultural activity is taking place has a strong effect on the type of agriculture practiced and the types of crops grown. Cold-weather crops such as cabbage, snow peas, and strawberries are grown both at high latitudes and at high altitudes. Many high-altitude areas are steeply sloping, have little soil and are thus not suitable for most types of agricultural production. However, such areas may be planted to various types of tree or bush-type plants for fruit production.

8.4.3 Precipitation

Rain can be heavy as in tropical areas, which get 2000+ mm of rain a year, or very little such as deserts with 200 mm or less of rain a year. In addition to the amount of rainfall, the time of year when it occurs and its intensity are also important. Rain can fall during months when crops are not planted and stored in the soil for plant use later in the growing season provided there is enough rain to allow significant soil storage. On the other hand small amounts of rain can come only during the growing season and thus be a limiting factor in crop production. The amount of grain, fruit, and vegetables produced are being largely determined by this rainfall. This, of course, assumes all rain-fed agriculture with no irrigation.

In addition to the quantity, the fate of precipitation after it falls is extremely important. It can infiltrate into the soil and just wet the soil profile. It can be greater than that required to wet the profile and drain or percolate below the soil and be stored in the subsoil; it may also be so little that all precipitation that infiltrates into the soil is subsequently lost from the soil surface when the soil dries. As mentioned above, water is lost from soil by both evaporation from the soil surface and by transpiration through plant leaves. The combined losses are termed evapotranspiraion. For successful crop production it is essential that there be enough water in the soil to provide for both the crop needs and the water lost by evaporation. If insufficient water is available, crop growth will decrease and yield will be lost.

8.5 PRECIPITATION PATTERNS AND CROP PRODUCTION

In some regions total precipitation, including snow and sometimes ice, may be as important or more important than rainfall alone in determining the amount of water available for crop production. In the great plains of the United States snowfall may contribute to total water available for crop growth. Snowfall in mountains may provide water in streams and rivers coming out of mountainous regions and subsequently being used as irrigation water. When considering water availability for crop production all sources—rain, snow, and ice—must be considered.

In addition to simple total rain or precipitation during the year or growing season, the intensity of the storms producing the rain must be taken into account. The reason is twofold. First, very intense short-duration rainfall events may produce mostly runoff and little infiltration of water into the soil. Second, such rainfall events may cause extensive and severe erosion, which degrades the productivity of the soil, not only water-holding capacity but also ability to support plant growth.

Sahelian Rainfall Event

In the western Sahelian region, the region of intermediate rainfall immediately surrounding the Sahara Desert of Africa, is an area of course sandy soils and short intense rainfall. I was constantly thwarted by intense rains while installing erosion control structures near villages situated north east of Filingué, Niger. This continued for some time until I decided that I must watch a rainfall event, as it occurred, in one of our experimental fields, so that I could see what was happening and thus determine what steps to take in making more effective erosion control structures.

One day a dark threatening cloud developed over an experimental area, shown in Figure 8.15, that was about a kilometer away. This particular area had a tall rock, from the top of which one could look down on the experimental plot. I rushed to the plot and climbed to the top of the rock just as the storm broke. Rain fell in large intense drops, the wind blew in gusts, there was lightening and thunder, and a great deal of sand in the air. In addition to being in a very dangerous place, I was unable to see anything because the rain was so intense. The storm lasted 30 minutes and deposited 25 mm of rain. When the rain, fog, and sand cleared, some of the installed erosion control structures, the furrows in the picture, had indeed failed; however, I was no closer to finding the answer for having tried to witness the event.

For agricultural purposes one can designate areas where rainfall occurs during the growing season or in some cases two seasons. In the central highlands of Kenya farmers talk about the long and short rainy seasons. In these situations the rain occurring in a period of time determines the crop growth and productivity because there is little or no rainfall at other times during the year, and thus there is no chance for storage of large amounts of water in the soil.

Another situation common in the humid tropics is where rainfall is spread more or less evenly throughout the year. Here the rain is such that there is sufficient or excess

Figure 8.15. Experimental plot in Niger. Arrow shows direction of water flow and furrows designed to stop the flow.

water during all months of the year. This means that as long as there is no other limiting factor such as photoperiod sensitivity of the crop or insect or microbial infection, crops can be planted and grown at any time during the year. This is the case with rice. In the area between Baybay and Ormoc, near the Philippine farm of Donio, on the island of Leyte one observes rice paddies in all stages of growth and production at all times of the year.

In some areas rainfall occurs in the nongrowing season and water stored in the soil is used during the growing season. In these cases it is often beneficial to supplement soil-stored water with irrigation. In the wheat-growing midwestern part of the United States precipitation falls as snow during the winter. While winter wheat may be planted and seen in the fields, it is, for the most part, not actively growing. Active and productive growth occurs in the late spring and often depends on soil-stored water. In many places rivers fed by melting snow in the surrounding mountains are used for irrigation in these areas.

In those areas where there is insufficient rain for rainfed crop production, extensive irrigation will be practiced if there is a sufficient source of water. This may be a stream or river, which originates in mountains, where there is higher rain or snowfall that melts to provide water for the rivers. In some cases this may be done on a small scale and be called water or rain harvesting.

Another source of water for irrigation is wells. Here water is pumped out of the ground and used to irrigate crops. In this case extreme caution must be exercised. First, the recharge, water coming into the aquifer being used, must be high enough to replace all the water used for irrigation in a year. If this is not the case, irrigation will not be sustainable, and the water will be depleted and future irrigation will fail. Second, wells may be dug or drilled into water sources that are contaminated. This has happened in California with selenium and in Bangladesh with arsenic. Third, the water may be salty such as to preclude its use as irrigation water. Thus, extensive testing, both geological and chemical, of waters to be used in irrigation must be undertaken.

Water for irrigation from all sources, that is, rivers, streams, lakes, and aquifers, contains salts. The salt content may be relatively low or may be a little high, but the water is still suitable for irrigation. In either case buildup of salt in the soil will occur if this water is used for irrigation and steps must be taken, usually by leaching and drainage, to remove excess salts before they buildup to levels that will inhibit plant growth.

Less common but still important in some areas is precipitation other than rain such as snow, ice, and hail. In some areas snow is a positive weather factor in that it provides water to plants either directly by melting into the ground or, when it falls in mountains, it melts providing water for streams and rivers and is then available for irrigation in valleys or other lower lying areas. Wet snow, which sticks to plants, can on the other hand cause damage by breaking leaves and limbs, thus causing a loss of productivity.

Precipitation in the form of hail can occur in areas that do not get cold enough to have snow or ice storms and often occurs in warm seasons when crops are actively growing. Hail can rip leaves and damage both vegetative and fruit portions of crops. In some extreme cases it may even destroy the whole plant.

Wind, usually accompanying precipitation, can cause two types of damage. For crop plants nearing maturity, it can cause the plants to fall to the ground. This is called lodging (Fig. 8.16) and usually either prevents harvest or results in a significant decreased yield.

Very windy conditions during pollination can cause pollen to be dispersed such that it does not reach the proper receptors, resulting in very poor yields. Fortunately, the time period over which crops are susceptible to this type of damage is limited and the damage does not often occur. A summation of wind conditions affecting crop production is given in Table 8.3.

Data given in Table 8.3 are general in nature. Many areas of the world will have either higher or lower average temperatures and rainfall. Also these factors can

Figure 8.16. Lodging of nearly mature wheat.

change dramatically from year to year. Some years may be droughty while others may have excessive rainfall. Also wind events can be highly variable, although there are some areas of the world where there are monsoon seasons with high rainfall and typhoon seasons with high wind. In the northern Philippines there is no, or very low, rainfall December through March while May through August, the monsoon season, will have very high rainfall.

8.5.1 Excess Precipitation

Excess precipitation just before, during, or after the growing season can be very detrimental to crop production. On a large scale excess precipitation can cause excessive soil erosion and mud slides and prevent farmers from preparing fields for planting. It can also saturate soil, filling all the pore space with water and thus depriving plant roots of oxygen.[†] In addition it can provide an environment conducive to enhanced growth of fungi and various plant pathogens and deleterious insects. At harvest heavy rains may knock down plants and lead to spoilage of grain in the field. It will also lead to a need for increased drying time once the grain is harvested. All may lead to a decrease in yield.

8.5.2 Deficient Precipitation

Deficient rainfall during the growing season, in areas not suitable for irrigation, is commonly referred to as a drought. Typically, drought is thought of as insufficient water, either as rainfall or as stored water in the ground, at any stage of the growing season. If there are droughty conditions before planting, lack of soil water may delay germination and the emergence of plants. Thus farmers may delay planting, hoping that precipitation will come. Dry conditions during the growing season will decrease yields and will decrease fertilizer efficiency. It may also interfere with pollination and the size of the grain heads. However, once the grain has fully filled the head, an extended dry spell will be beneficial to harvesting and storage of grain.

Drought and Alfalfa
Alfalfa is a deep-rooted, 150+ cm deep, legume commonly grown as a pasture or hay for animals. An alfalfa field that had a deep soil with a clay subsoil experienced a prolonged dry season. During this drought, the alfalfa was able to draw water from deep in the soil profile while the common weeds in the field were not. The consequence of this drought was that the alfalfa field was almost completely free of weeds.

[†]This is true of most crop plants except lowland rice and taro, which are commonly grown with roots submerged in water.

8.6   DAY LENGTH

The length of sunlight is extremely important in crop production not only in terms of the amount of light available for photosynthesis but also in that some plants are photoperiod sensitive. This means that they respond in their growth and reproductive habits to the length of the day. The day length of each month for the approximate location of each of the three farms is given in Figure 8.6. Ecuador, being on the equator, has no difference in day length between winter and summer. On the other hand there is almost $5\frac{3}{4}$ hours difference in day length between winter and summer in Ohio. The Philippines being above, but near, the equator have a day-length difference of approximately $1\frac{1}{4}$ hours.

These same variations in day length December to June occur in the Southern Hemisphere except that December day length is longer and that of June shorter. In either case the effect of day length on crop production is expected to be similar.

It would be logical to assume that the amount of sunlight and its intensity might be a controlling factor in crop production. Two situations occur that might affect the amount of light plants receive: (1) areas of the world where there are constant clouds, fog, rain, or other overcast conditions and (2) areas where the day length changes from summer to winter. Vegetative production might be expected to be adversely affected by clouds and high rainfall because clouds obscure the sun. Likewise areas of mist and fog would be expected to have low productivity. Sunlight obscured by these conditions does not usually adversely affect crop production. However, other adverse effects such as lower temperatures or high relative humidity may adversely affect crop production because of increased disease pressure. Of these, generally, changes in day length are more important in crop production.

Areas with very low temperatures might also be expected to be unsuited to any agricultural production. However, in northern latitudes where there are very long days during the summer, short-season cold-tolerant crops can be grown and produce large yields.

Day length can dramatically affect crop growth and production in plants that are photoperiod sensitive. Many plants respond to day length, for instance, growing faster when the day length is shorter or not flowering unless the day length is correct. Certain varities of rice, when planted after the day length needed for flowering, will grow a whole year before flowering and producing grain. However, if they are planted 4 months before the proper day length, they will grow, flower, and produce grain in 4 months. To avoid these types of problems, most agricultural crops, particularly the grains, have been bred to remove or reduce photoperiod sensitivity.

8.7   LENGTH OF GROWING SEASON

Each crop needs an appropriate length of suitable weather for it to grow and produce successfully. Generally, growing season length is considered the time between killing frost events. However, some crops such as tomatoes are more sensitive to low temperatures and some crops such as snow peas are less sensitive. Some crops require low temperatures in some parts of the year or growing season to bear fruits, and some simply show increased productivity when the temperatures are low at certain stages of growth.

If the combination of day length and temperature are not sufficient, poor or no crop may be produced. There are some interesting exceptions to this rule. Some crops have a wide tolerance for day length and temperature. Cabbage can be grown in temperate and even some tropical climates. However, it can also be gown in cold climates with very long day lengths such as Alaska. Under these conditions the cabbage heads are very large. Common sugarcane varieties require from 18 to 24 months to mature. Because of the long growing time, sugarcane is typically grown where there are 365 frost-free days during the year (there are some cold-hardy sugarcane varieties as well as varieties that have shorter growing periods).

8.8 GROWING DEGREE DAYS

Absolute high and low temperatures may have a dramatic effect on the crop grown, however, the number of days with temperatures needed for growth is also important and is ascertained by determining the degree days. The degrees above some base temperature each day are added to get a total degree day for that length of time the temperature is above some base temperature. The optimum temperature for a variety of crops is given in Table 8.4. For a crop having a temperature of 10°C, below which it stops growing or is damaged, this is the lower temperature limit for this crop. A day with a temperature of 20° would count as 10° days because it is 10° above the base temperature. For each crop and each variety there will be a different number of degree days needed to produce a mature crop.

The first date in a year with a temperature above the cutoff is the beginning of the count of degree days. Likewise the last day of a growing season with a temperature above the minimum is the last day counted in determining degree days. For planning purposes the average first and last days having the cutoff temperatures can be used in estimating the degree days available for crop production.

TABLE 8.4. Optimum Growth Temperatures for Selected Crops

Crop	Optimum Temperature (°C)	Temperature (°C) for Maximum Yield
Wheat	15–20	15–20 (grain yield and grain weight)
Potato	18–20	5–20 (tuber production)
Pea	14–18	15–25 (flowering)
Sugar beet	17–22	20–22 (sugar production)
Barley	17–22	22
Cabbage	15–24	10 (hasten flowering)
Apple	20–24	15–20 (pollination)
Strawberry	5–26	14 (favor flowering)

For countries such as the Philippines and Ecuador where two of our three farm examples are located the temperature never goes below 10°C, and thus the number of degree days is generally a mute point. For crops that require a certain number of cool or cold days to produce a crop, such as apples, the number of cool days and the coolness can be calculated in a similar way. Thus, while the Philippines are not suitable for apple production, Ohio in the United States is ideal.

8.9 HUMIDITY

Humidity refers to the amount of water in air and is commonly reported as the relative humidity. Dew point is also related to humidity, being the temperature at which water condenses out of air. The lower the relative humidity and dew point the dryer the air. High humidity and high dew point both indicate conditions favorable to precipitation.

The relative humidity has a pronounced effect on crop water needs and drying before harvest. Hot, dry, low-humidity climates will require more water for crop production, usually in the form of irrigation, than will humid, cool, or hot climates. Drying and water loss will be even more pronounced if there is air movement as wind or even a slight breeze, and this may necessitate irrigation. Drying of a mature crop will be faster in hot low-humidity climates.

Some crops will have higher pollination rates and grain production under humid conditions. The humidity of an area can thus have a significant effect of the success of certain crops. In low-humidity regions this may partially be offset by irrigation, particularly by sprinkler irrigation, at the time of flowering. However, humidity is a factor in crop production in all areas of the world.

8.10 ADVERSE WEATHER CONDITIONS

All plants and animals are susceptible to adverse weather conditions, as illustrated in Table 8.5. Adverse weather conditions include intense high rainfall, most often associated with high wind. High-intensity rainfall will cause soil erosion and scarring of land, as seen in Figure 8.17 (learn more about erosion in Chapter 9). Erosion removes the most productive, high-fertility soil, leaving the field or area less productive. Keeping the soil covered with either plant residues or growing plants controls water erosion. High rainfall can also decrease crop production, as noted in Table 8.5.

High winds such as those in typhoons, hurricanes, and tornados can do both physical and production harm to crops. Physical damage might be caused by breakage of limbs or stems. Breakage can be caused by the winds themselves and by material carried in the wind. In some cases this may be as simple as sand scarring of leaves. In other cases it may be due to flying objects hitting the plants. Breakage of plant parts may cause loss of productivity because of loss of fruiting plant parts. However, for many crops lodging is a primary cause of harvesting problems associated with high winds (Fig. 8.16).

TABLE 8.5. Effect of Adverse Weather Conditions on Agriculture Productivity

Plant or Animal	Adverse Condition	Effect
Sugar beet	Excessive rain	Decreases sugar content
Onion	Low temperature and low humidity	Decrease tuber production
Pear	Low night temperatures <10°	Premature ripening
Plum	Cool spring 7–8°	Embryo death
Strawberry	High temperatures 30°+	No flower bud production
Pigs	High >40°C temperatures	General decrease in productivity
Cattle	Low <10°C and high >40°C temperatures	General decrease in productivity

As noted above a different problem can also occur when high winds interfere with pollination. Knocking plants down by any mechanism but most often by water and wind is called lodging. The occurrence of lodging any time during plant growth causes plants to yield less. However, it is more common for rain or wind to knock over crops when they are heading out and the tops are heavy with grain. When the grain heads come in contact with the ground, several things can happen. They may begin to rot and thus be unusable. They may sprout and again not be useful as food. More often the bigger problem is that the grain is hard to harvest and thus large amounts of yield are lost.

All low-lying areas have fertile soil and so are prized agricultural lands. However, these same lands are subject to flooding. Flooding can wash away crops. However, it

Figure 8.17. Soil erosion caused by intense rainfall on bare soil. (Courtesy of Matthew Deaton.)

can also cause lodging and leave the ground so wet that equipment cannot be used to harvest the crop. Generally, hand harvesting before grain deterioration is also precluded, although if the ground freezes, it may become hard enough to allow for mechanical harvesting.

Hail is an important consideration in some locations because it occurs during the growing and fruiting season of crops. It can be extremely destructive both to edible and nonedible portions of crop plants including grains, vegetables, and fruits. In some locations, such as the northern highlands of Kenya farmers often plant beans under corm. The theory is that the maize stalks and leaves will protect the beans during hail storms.

In Sections 8.4.3 and 8.5 we discussed the adverse effects of both excess and deficient precipitation. These, of course, also represent adverse weather conditions.

8.11   CLIMATE CHANGE AND POSSIBLE EFFECTS ON FOOD PRODUCTION

Four aspects of climate change can have a predictable effect on food production: increasing carbon dioxide in the air, increasing temperature, changes in rainfall, and increasing winds. The first two changes should have a positive effect on crop production. The third could be neutral or negative, particularly in relationship to animal production. The last will have variable negative, neutral, and positive effects. Changes in storm and rainfall patterns will dramatically affect crop production, however, predicting the locality, number, and intensity of storms is not possible.

Carbon is an essential nutrient for all plants and is obtained from the atmosphere in the form of carbon dioxide. Thus, increasing carbon dioxide concentrations in the atmosphere are expected to lead to increases in plant growth, productivity, and yield and thus food availability. To a certain extent increased plant productivity tends to balance or counteract the increase of carbon dioxide in air.

Increases in temperature would lead to an increase in the land available for crop production. Northern Hemisphere land, which is presently too cold or the growing season too short because of low temperatures, would be available for farming. There are large areas in North America and northern Russia that would become productive by increased temperatures. Thus, one can make the argument that increasing world temperatures would increase food availability.

To a certain extent this would be offset by loss of land along the oceans and other bodies of water as their levels rise from melting polar ice. Flooding of these lowland areas would provide high-quality fish sanctuaries and thus an increase in fish productivity that could also be exploited to increase food production. Thus, this loss of farmland would be, at least partially, offset by increased fish production.

Changes in rainfall can dramatically change agriculture and agricultural productivity. Increases in rain can provide water needed for crop production and could potentially allow crop production or increased crop production in areas where there is little crop production today. On the other hand decreases in rainfall will decrease agricultural productivity and can have long-range effects even out of the immediate rainfall area. Some models show increasing temperatures decreasing water in major rivers in Africa. This would be important because these provide direct food as fish and indirect food as the water is used for irrigation for crop production.[2]

The Niger River (Fig. 8.18) starts in Guinea and runs through Mali, Niger, and finally Nigeria. The river provides direct food in the form of fish and indirect food in providing water for irrigation. Any dramatic change in water flow in the river would affect both. The figure does not show the numerous tributaries of these rivers, which also provide food and water.

In addition to changes in water availability, climate changes will also change wind patterns. Increased windy conditions would be either neutral or negative depending on the wind, its speed, humidity, and when it occurred. Strong winds during the early parts of the growing season would have little effect on crop growth. As crops mature, strong winds can adversely affect pollination, and, when grain formation begins, wind can knock over crops, causing lodging, making harvest difficult and causing a loss of yield.

In dry climates strong dry winds can increase evapotranspiration and thus increase the need for water in soil or for additional irrigation. They would also increase the

Figure 8.18. Major rivers of Africa—the four longest and most important; the Nile, Niger, Congo, and Zambezi are shown.

amount of water animals needed to stay healthy. Strong wet winds would have an adverse effect on animals, particularly in cold climates; however, they would have little effect on pant growth.

An indirect effect of increased windiness would be an increase in soil erosion by wind. This would mean that additional soil conservation steps would need to be taken by farmers to protect soil from erosion and to protect plants from blowing soil, which can damage plant leaves and decrease yield as noted above.

The potential effect of increased storm intensity is hard to predict. Increased storm activity could increase rainfall in dryer climates. This could increase crop productivity. On the other hand increased rainfall could lead to prolonged saturation of soil and to flooding, both of which would lead to destruction of crops. The overall effect on food availability would depend on the time and nature of the rainfall or storm event, that is, in a food producing time frame, a highly unpredictable situation.

One other possible adverse effect of increased storminess is hail. This would be hail stones, frozen water at times 2 to 3 cm in diameter and sometimes larger, falling from the sky during relatively warm periods, that is, when crops are growing. Hale will destroy leaves and stems and beat down crops, thus effectively destroying crops. Only certain storm events are likely to produce hail; however, as storminess increases, it can be assumed that there will be a corresponding increase in hail storms.

All forms of storms can have adverse effects on transport and storage of farm produce and thus lead to a loss of food. Interruption of transport by destruction of roads and rail lines and the downing of power lines will all adversely affect food availability. Both dryer and wetter conditions lead to a decrease shelf life of agricultural produce.

On the one hand global warming and increased carbon dioxide in the atmosphere will lead to increased plant growth and productivity and thus to increase food production. On the other hand increased storminess could offset these increases and result in no increase in food availability or a decrease in food production.

8.12   CONCLUSIONS

Both the Köppen and Thornthwaite climate classifications are useful in understanding conditions that affect the crops that can be grown productively in a locality. Rainfall, temperature, and day length are three of the most important climatic conditions that control plant growth. In addition altitude can have a significant effect in that the temperature decreases as altitude increases. Humidity can also affect crop production in that it can control viability of pollen and the growth of destructive microorganisms. The heat degree days concept is one important way of finding if there are enough warm days to allow a crop to come to maturity in a given locality. Adverse weather conditions, high winds, heavy rainfall, and hail can cause yield decreases or even total crop destruction. Global climate change will have positive, neutral, and negative effects on food production and availability depending on the types of changes and where the changes occur.

QUESTIONS

1. The various climatic zones are delineated in Figure 8.1. Explain some reasons why these definite boundaries might not always hold.
2. Explain why irrigation is not advisable in all areas even though there may be water in wells.
3. Describe the various rainfall patterns in relationship to crop production.
4. Explain some effects of day length and frost-free days on the crops that may be grown in an area.
5. Cold-weather crops can successfully be grown in some tropical countries. Describe the places, generally, where this is possible.
6. Explain the possible effects of increasing atmospheric carbon dioxide on crop production.
7. Describe evapotranspiration and explain why it is important in determining an area's suitability for crop production or needing irrigation.
8. An area is determined to have just enough rainfall to satisfy crop production needs, but crops cannot be grown in the area. Explain how this might occur.
9. Describe the adverse effects high wind can have on both soil and crops.
10. Define growing degree days and explain why they are important.

REFERENCES

1. T. E. Lucht, S. J. Hamilton, M. H. Deaton, D. B. Dotson, and J. Allen, Soil Survey of Clinton County, Ohio, found at http://soils.usda.gov/survey/online_surveys/ohio/index.html.
2. Africa faces big water shortage, 3rd March 2006, found at: http://news.bbc.co.uk/cbbcnews/hi/newsid_4770000/newsid_4771500/4771592.stm.
3. Calculated from data from http://www.qpais.co.uk/modb-iec/dayleng.htm.

BIBLIOGRAPHY

C. D. Ahrens, *Meteorology Today: An Introduction to Weather, Climate and the Environment*, 8th ed., Brooks/Cole, New York, 2006.

C. D. Ahrens, *Essentials of Meteorology: An Introduction to the Atmosphere*, 4th ed., Brooks/Cole, New York, 2005.

J. Chang, *Climate and Agriculture: An Ecological Survey*, Aldine, Chicago, 1968.

N. J. Doesken and A. Judson, *The Snow Booklet: A Guide to the Science, Climatology, and Measurement of Snow in the United States*, 2nd ed., Colorado State University Publications, Fort Collins, Colorado, 1997.

J. H. Griffeths, *Handbook of Agricultural Meteorology*, Oxford University Press, New York, 1994.

D. L. Hartmann, *Global Physical Climatology*, Academic Press, New York, 1994.

J. Horel and J. Geisler, *Global Environmental Change: An Atmospheric Perspective*, Wiley, New York, 2004.

L. R. Kump, J. F. Kasting, and R. G. Crane, *The Earth System*, 2nd ed., Prentice-Hall, Upper Saddle River, NJ, 2003.

W. Rudioff, *World Climates*, Wissenschaftliche Verlagsgesellschaft, Berlin, 1981.

C. W. Thronthwaite, An Approach Toward a Rational Classification of Climate, *Geographical Rev.* **38**, 55–94, 1948.

J. M. Wallace and P. V. Hobbs, *Atmospheric Science*, 2nd ed., Academic Press, New York, 2006.

9

SOILS AND WATER

9.1 SOIL ON THE THREE FARMS

Soil profiles at the three farms, shown in Figures 9.1 to 9.3, are very different in both looks and composition. The Philippines (soil profile in Fig. 9.1) has high temperatures and rainfall leading to rapid decomposition of organic matter. This, coupled with the constant monocropping* of maize and the constant removal of organic matter, means that this soil cannot develop an obvious A horizon (Fig. 9.1; see discussion later in chapter).

In Ecuador (Fig. 9.2) there are moderate temperatures and rainfall leading to the buildup of organic matter in the soil. This soil is dark is color, has a very strong structure, and, although it is hard to see in pictures, a very definite A horizon. The field is rotated and organic matter left on it so that there is a constant input of organic matter. In spite of the field having a steep slope, it shows no sign of water erosion, although there has been movement of soil under the pull of gravity, that is, colluvial movement.

---

*Monocropping is when only one crop is grown in a field.

World Food: Production and Use. By Alfred R. Conklin, Jr. and Thomas Stilwell
Copyright © 2007 John Wiley & Sons, Inc.

Figure 9.1. Soil profile on the larger of Donio's corn farms. Because all organic matter is removed after harvest, there is no obvious A horizon in the soil on this farm. The profile is highly uniform at the dept of 1 m. (Courtesy of Henry Goltiano.)

In the United States incorporation of organic matter coupled with lower rainfall and cooler temperatures during the spring, fall, and winter are conducive to buildup of organic matter and the development of a prominent A horizon; see Figure 9.3. Variations in color of the B horizons shows that there is poor water drainage in this soil.

Figure 9.2. Soil profile on farm of Aída and Octavio.

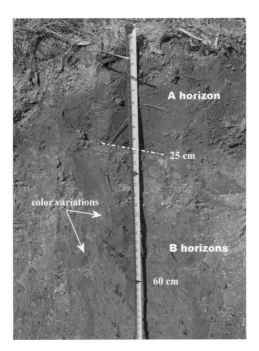

Figure 9.3. Soil profile on Steve's farm. Note the distinctive A horizon that extends to a depth of 25 m. The lower horizons show variations in color associated with the soil being saturated with water for a significant period of time during the year. When pit was dug in June, water was encountered at a depth of 1 m.

9.2 INTRODUCTION

All food production depends on soil. It is the medium in which the vast majority of our food is produced. This is true for both cultivated plants and food obtained from wild plants and animals. For those people that depend on animal products for their sustenance, the animals they depend on eat plants growing in soil. Even fish food used in raising fish on fish farms depends on crops grown in soil. Soil is even involved in fish growth in the wild because dissolved soil minerals and organic matter carried into bodies of water, including oceans, are part of the food of marine plants and animals.

Soil is at the same time absolutely common and extremely complex.

Soil is common in that almost every person sees it every day or at least sees plants growing in it every day. Excepting places where bare rock is exposed or where there are bodies of water, it is the common covering of Earth's surface. Most people have plants growing around and in the home. In many cases plants inside the home are grown in artificial soil since the inside of a house is not the natural environment for soil and it does not do well under these conditions. It is also in the air as dust and in water as suspended material, for example, muddy water.

Soil can be discussed on a large, medium, small, or molecular scale. On the large scale, soil covers the Earth and as such it makes up the topography, or lay of the land. It has slope, aspect, the direction a slope is facing (i.e., north, south, east, or west), and depth.

The slope of land is important in determining its suitability for crops and its tendency to erode along with measures needed to control erosion. Its aspect will determine when and how sunlight is received. In northern climates this may be very important in how fast it warms up in the spring, which in turn determines when planting can commence.

Water is also essential to agricultural production and cannot be separated from soil. Its continuing source as precipitation as rain, snow, or ice is essential, and the use of water by people must not exceed these inputs if sustainable human population and agriculture is to occur. Watersheds, upland, hilly, or mountainous areas planted to trees are an especially productive way of capturing and storing precipitation.

Water passing through soil will be filtered and purified by destruction of organic matter, but it will also dissolve salts and in some cases these salts may contain toxic components. Water to be used for drinking by humans and animals and for crop irrigation must contain few salts and be free of toxic components. Because of these constraints, the construction of wells, which make use of stored water, must be carried out with care.

9.3 AVAILABLE LAND

Three Farms

Table 9.1 gives a comparison of the land available for cropping in the Philippines, Ecuador, and Clinton County in Ohio. While Clinton County in the United States has less total land and can only be cropped once a year or with relay cropping perhaps 1.5 crops per year, it has the highest percentage of farmland compared to the Philippines and Ecuador. The Philippines has the second highest percentage of cropland while Ecuador has the lowest. However, three crops per year are possible in both Ecuador and the Philippines.

Table 9.2 gives the total land area and percent of arable land in the countries of the world. Total land areas are taken from FAO data (Food and Agriculture organization of the United Nations, http://faostat.fao.org/default.aspx), and the percentages of arable land are calculated from this data. It includes all arable (farmable) land from excellent to marginal. Not included is land planted to trees nor land suitable for agriculture if irrigated. Different sources will give different amounts of total and arable land for

TABLE 9.1. Agricultural Land (ha) in Areas of Three Farms

Country, County, or Island	Total Area		Farm Land		Percent[a]	
	ha	acres	ha	acres	Cropland	Comments
Clinton County, OH USA	106,873	264,088	82,405	203,627	77	Prime farmland
Ecuador	28,365,000	68,408,653	8,075,000	19,953,759	28	Cropland
Philippines	30,000,000	73,679,411	12,200,000	30,146,856	41	Cropland

[a]Percentages are rounded up.

Source: Data for Clinton County is from the Clinton County Soil Survey and for Ecuador and the Philippines from data abstracted from http://faosta.fao.org/default.aspx 2003.

TABLE 9.2. Percent of World Arable Land

TABLE 9.2. *Continued*

Country	Area^a	Percent Arable	Country	Area^a	Percent Arable
Afghanistan	63,088	5	Estonia	4,493	57
Albania	2,829	31	Ethiopia	112,895	34
Algeria	230,452	2	Fed. Rep. of Russia	1,674,146	17
Angola	123,775	52	Fiji	1,815	10
Argentina	277,685	33	Finland	33,078	22
Armenia	2,970	22	France	54,550	65
Azerbaijan	8,558	49	Gabon	26,486	53
Bahamas	1,336	57	Gambia	1,092	56
Bangladesh	14,003	64	Gautemala	11,045	30
Belarus	206,150	74	Georgia	7,013	33
Belgium	3,005	65	Germany	35,436	69
Benin	11,790	79	Ghana	24,181	73
Bhutan	4,036	0	Greece	13,243	43
Bolivia	108,903	57	Guinea	24,602	39
Bosnia and Herzegovina	5,125	51	Guinea-Bissau	3,369	46
Botswana	57,485	12	Guyana	20,907	53
Brazil	853,637	55	Haiti	2,723	30
Bulgaria	11,057	69	Honduras	11,490	24
Burkina Faso	27,514	69	Hungary	9,213	79
Burundi	2,815	41	Iceland	10,109	0
Cambodia	181,970	59	India	306,140	59
Cameroon	46,274	68	Indonesia	189,220	20
Canada	978,404	10	Iran	161,601	6
Central African Republic	61,857	73	Iraq	43,041	7
Chad	128,075	23	Ireland	6,933	45
Chile	75,202	9	Israel	2,179	25
China	934,949	17	Italy	30,314	50
Côte d'Ivoire	32,465	73	Jamaica	1,132	15
Columbia	112,184	43	Japan	37,178	33
Congo	34,366	45	Jordan	9,006	2
Costa Rica	5,200	18	Kazakhstan	270,282	9
Croatia	5,678	62	Kenya	59,440	30
Cuba	11,068	63	Korea Dem.	12,443	26
Czech Republic	7,833,000	78	Korea Rep.	9,685	32
Denmark	4,408,000	69	Kuwait	1,657	0
Dijbouti	2,299,000	0	Kyrgyzstan	19,840	8
Dominican Republic	4,879	37	Lao	23,001	21
East Timor	1,448	52	Latvia	634	83
Ecuador	28,684	40	Lebanon	1,030	15
Egypt	98,786	0	Lesotho	3,010	13
El Salvador	2,015	22	Liberia	9,870	43
Equatorial Guinea	2,681	8	Libya	160,950	1
Eritrea	12,116	5	Lithuania	6,449	81

(Continued)

(Continued)

TABLE 9.2. *Continued*

Country	Area[a]	Percent Arable	Country	Area[a]	Percent Arable
Macedonia	2,509	39	Saudi Arabia	1,951	0
Madagascar	59,159	43	Senegal	19,510	47
Malawi	11,959	52	Sierra Leone	7,203	50
Malaysia	32,669	20	Slovakia	4,815	59
Mali	124,852	18	Slovenia	2,004	50
Mauritania	104,383	1	Solomon Islands	2,802	9
Mexico	96,062	23	Somalia	64,448	7
Moldova Republic	3,331	84	South Africa	122,230	20
Mongolia	155,100	0	Spain	50,909	35
Morocco	39,658	21	Sri Lanka	6,574	47
Mozambique	79,894	70	Sudan	248,828	32
Myanmar	66,490	30	Suriname	14,429	54
Namibia	81,966	10	Swaziland	1,770	44
Nepal	14,725	11	Sweden	44,358	18
Netherlands	3,825	41	Switzerland	4,111	27
New Zealand	26,617	25	Syria	18,603	26
Nicaragua	12,909	32	Tajikistan	14,277	14
Niger	118,254	5	Tanzania	93,819	64
Nigeria	91,207	62	Thailand	51,603	52
Norway	31,808	2	Togo	5,720	72
Oman	31,009	0	Tunisia	5,720	72
Pakistan	79,847	6	Turkey	77,823	43
Panama	7,569	29	Turkmenistan	469,004	1
Papua New Guinea	466,480	19	Uganda	24,219	51
Paraguay	19,905	56	Ukraine	59,512	83
Peru	128,922	56	United Arab Emirates	7,613	0
Philippines	30,000	41	United Kingdom	24,418	52
Poland	30,983	70	United States	946,837	39
Portugal	8,780	44	Uruguay	17,907	77
Quatar	1,107	0	Uzbekistan	44,519	13
Romania	23,593	70	Venezuela	92,388	53
Rwanda	2,450	19	Vietnam	33,391	27

Source: (Abstracted from Food and Agriculture Organization of the United Nations; online databases, FAOASTAT, http://www.fao.org/es/ess/top/country.html).
[a]In thousands of hectares.

the same countries; however, these percentages give an idea of the land available for food production.

Some countries, notably desert countries in the Middle East, are listed as having no arable land because they have no rain-fed agriculture, and all agricultural production depends on irrigation. These countries can and do, in some cases, have extensive land devoted to agriculture and food production. Most of Egypt receives less than an inch of rain per year and may receive no rain for many years. However, extensive agriculture is carried out along the Nile River, which is used as a source of irrigation water. In other cases water for irrigation may be produced by desalination, that is, removal of salt from seawater, or obtained from wells.

As seen in Table 9.2, there is little indication of a relationship between percentages of arable land and the well-being of the inhabitants of a country. For some countries this is because they can go from food surplus to deficit without any change in the agriculture situation, arable land, rainfall, and so forth. In other cases it may be due to agricultural practices, government policies, or land tenure systems. However, the amount of arable land does put an upper limit on the potential agriculture productivity of a country.

9.4 SOIL TYPES

Different soils are grouped into similar observable characteristics. There are many ways of grouping soils. Many countries, France, the former Soviet Union, and the United States, are just a few examples of countries that have developed different classifications of soils. FAO, under the auspices of the United Nations, also classifies soils into 26 different groups (http://www.fao.org/AG/agl/agll/key2soil.stm). In the United States soils are classified as being in one of 12 soil groups, given in Table 9.3. Figure 9.4 shows the extent of each soil type in the world. Of the 12 soil orders defined by the USDA (U.S. Department of Agriculture), the Entisols, Inceptisols, Aridisols, Mollisols, and Alfisols are most widely distributed and used for agricultural production. Of the soil orders, the Gelesols are probably the least likely to be use for food production because they are frozen much of the time.

Table 9.4 gives an indication of soil types designated in the FAO nomenclature. Although the FAO system contains many more primary soil designations, they can be related back to the USDA system and to the soil's suitability for agricultural production by reference to Table 9.4.

Each type of soil occupies a certain extent of the landscape but may also be found in many different localities, as shown in Figure 9.4. On a medium scale the observable characteristics of individual soils and soil components such as horizons, structure, and color are used to differentiate one soil from another. These same characteristics, such as the thickness and condition of the A horizon, which is the surface horizon of a soil, are very important in determining the management of a soil for maximum food productivity. A deep dark uncompacted A horizon will be more conducive to crop production than will other types of A horizons.

The geographic and other large features of soil are easy to observe but give little hint as to the complexity of soil. When looking at soil on a small to molecular level, its complexity becomes very evident. On the small scale we can observe soil components, that is, sand, silt, clay, organic matter, air, and water. Each of these components is complex in its own way and contributes to the overall molecular complexity of soil.

The complexity occurs even within soil texture, which is the relative proportions of sand, silt, and clay in a soil. Sand size particles in soil can be divided into many different subsizes from very fine to very coarse. Clay components may be very well defined crystalline structures or they may be amorphous, that is, without definite form. In

TABLE 9.3. Soil Orders According to U.S. Department of Agriculture (USDA)

Soil Order	Characteristics	Major Uses
Inceptisol	Soil that has little or no development	When topography is suitable, may be used for crop production
Entisol	More developed than Inceptisol although development is slight	When topography is suitable, may be used from crop production
Alfisol	Light surface horizon with well-developed horizons	Widely used for crop production
Mollisol	Deep dark surface horizon. Develops under grass vegetation	Widely used for crop production
Aridisol	Soils that occur in arid regions	May be used for crop and animal production if water is available
Vertisol	Soils with high activity clays that develop wide (20 cm) and deep (100 cm) cracks when dry	Can be farmed, although water infiltration and seedbed preparation are difficult
Andisols	Soils containing material from volcanoes. Often has high gravel content and fast drainage	Can be used for agriculture if sufficient water is available
Gelisols	Soils with frozen horizons	Generally not used for agriculture
Oxisols	Common in tropical climates have had many components leached out of the soil	Are commonly used for agricultural production particularly pineapple production
Ultisols	Ultimate in soil development have a full complement of horizols	Are generally suitable of crop production
Spodosols	Sandy soils developed under acid-producing vegetation, has horizon of deposited aluminum and organic matter often with accompanying iron	Used for both tree and agricultural crops
Histosols	Organic soils	Are generally high productive and often used for production of vegetable crops

addition different clays have different characteristics such as shrink–swell and their cation exchange capacity that holds cation nutrients (elements with a positive charge) in a form available to plants. Organic matter, air, and water also add their own forms of complexity to soil. It is the complexity of soil that makes it particularly suitable for plant growth, and understanding and managing this complexity can lead to increased and improved food production.

In many cases it is not necessary to be able to identify a soil in terms of its official name. Soils can be sandy, silty, or clayey. It is relatively simple to feel a soil's texture by wetting it and working the wetted soil between thumb and forefinger (this is often referred to as the feel method). A sandy soil will have a gritty, sandlike, feel. A silty soil will have a flourlike feel, and a clay soil will be plastic, can form shapes it can retain, and clays are usually sticky. Basically, sandy soils have high infiltration and percolation rates, that is, water moves through the soil easily. Silty soils will have slightly lower infiltration and percolation rates and hold the most water for plant use. Clayey soils will have low infiltration and percolation rates.

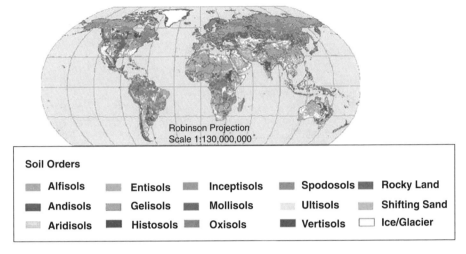

Soil Orders

Alfisols	Entisols	Inceptisols	Spodosols	Rocky Land
Andisols	Gelisols	Mollisols	Ultisols	Shifting Sand
Aridisols	Histosols	Oxisols	Vertisols	Ice/Glacier

Figure 9.4. USDA world soil regions. (Courtesy of U.S. Department of Agriculture, Natural Resources Conservation Service, Soil Survey Division, World Soil Resources, Washington, D.C. http://soils.usda.gov/use/worldsoils.mapindex/.)

If the textural name, percentage of sand, silt, and clay in a soil is known, it can be related back to specific percentages of sand, silt, and clay. These are given in a textural triangle shown in Figure 9.5. The names are easy to follow in that the word *sand* or *sandy* indicates a large percentage of sand in the soil. The same is true for silt and

TABLE 9.4. Partial Comparison of USDA and FAO Soil Classification Schemes

USDA Classification	FAO Classification
Inceptisols	Cambisols, Andisols, Rankers
Entisol	Arenosols, Fluvisols, Regosols
Ultisols	Acrisols
Histosols	Histosols
Mollisols	Chernozems, Greyzems, Kastanozems, Phaeozems, Rendzinas
Oxisols	Ferrasols
Alfisols	Luvisols
Ultisols/Alfisols	Nitrosols
Vertisols	Vertisols
Aridisols	Xerososls, Yermosols
Spodosols	Podzols
Lithic subgroups	Lithosols, Solonchaks (salt), Solonetz (sodium)
No correspondence	Planosols[a]

[a]Have a bleached, temporarily water-saturated topsoil on a slowly permeable subsoil.
Sources: FAO-UNESCO. *Report on the Agro-Ecological Zone Project*, Food and Agriculture Organization and United Nations Educational, Scientific and Cultural Organization, World Soil Resources Report 48, Rome, Italy, 1978. FAO-UNESCO. *Soils of the World*, Food and Agriculture Organization and United Nations Educational, Scientific and Cultural Organization, Elsevier Science, New York, 1987. See also p. 321.

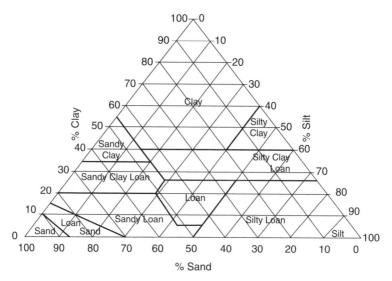

Figure 9.5. Textural triangle showing the textural name for soils with various amounts of sand, silt, and clay.

clay. The term *loam*, found in the lower middle of the textural triangle, is a little different. This term indicates that all three soil fractions—sand, silt, and clay—contribute equally to the characteristics of this texture soil. The textural names of soils are encountered frequently in all agricultural work and food production.

Some soils contain a large amount of gravel, rocks, or stones. In these cases the textural name will be prefixed with the appropriate designator, for example, stony silt loam. Gravel, rocks, and stones can be important in crop production particularly in carrying out cultivation and because such soils often have limited water and nutrient holding or supplying capacities. Stones and rocks can be removed from a field to make soil and water retaining structures (Fig. 9.6); however, it is usually impossible to remove all larger rocks from a field.

A second characteristic of soil important in food production is its profile. That is, what horizons it has and their characteristics; see Figures 9.1, 9.2, and 9.3. For instance, a soil that is deep and sandy, with some clay in its lower horizons, will be good for crops because it will provide deep rooting and provide water so that the crop can withstand drought. Such soils will be good for root crops, allowing for easy development of tubers. A soil with a lower layer that is high is clay might be good as a rice paddy since the clay will slow or in some cases prevent the downward movement of water. Again one can determine the characteristics of each horizon by the feel method mentioned above.

Another important aspect is any change in the soil profile with depth. Any change in texture or compaction will tend to impede the movement of roots and water down through the soil profile. This can be highly detrimental to rooting of plants and can prevent them from developing a rooting system that will fully explore the soil. This in turn will limit the ability of the plants to take full advantage of the nutrients and water contained therein.

Figure 9.6. Rock collected from badly eroded fields on a hillside in the Philippines used to construct "dams" or bunds to prevent water eroding the soil. Ocean is visible in the background.

9.5 SOIL COMPLEXITY

Soil is complex in that it is made up of air, water, and solids. Making it even more complex is that soil air is different from atmospheric air, soil water is a solution of inorganic and organic ions and molecules, and the solid portion is made up of both inorganic and organic components. The inorganic components have different sizes, composition, and activities in soil. The organic components represent all compounds of biological origin in addition to humus synthesized during the decomposition of organic matter. It is organic but is nothing like the compounds from which it is made.

In addition soil components do not act independently of each other. Inorganic components, sand, silt, and clay along with organic matter combine into what could be called a secondary structure. These secondary structures, called peds, lead to a soil having improved crop-growing capabilities. In addition air and water are trapped in pores in soil and thus are an intricate part of this secondary structure.

In addition to these complexities, soils in the field are different depending on how they are formed and from what they are formed. If rock is pulverized, it forms a powder that can be any color but is commonly gray and thus is very different from soil. In the field soil has horizons, which differentiate one soil from another; see Figure 9.7. The surface horizon, called the A horizon, is darker in color and has more organic matter than lower horizons. The lower horizons have more clay and are often redder in color. Some soils are very young and so have little horizon development, while older soils have extensive horizon development. Generally, soils that are not too young or old are best for crop production, although all types of soils, except Gelesols as described above, are used for crop and animal production. Gelesols are used by grazing animals that are hunted for meat.

Organic matter has a dramatic effect on the surface horizon. It helps form the A horizon, and in the case of Figures 9.1 to 9.3 this A horizon is termed an Ap horizon because it has been plowed. Organic matter also aids in a soil's ability to produce plants and support animals. It slowly decomposes, releasing plant nutrients in forms that are available to plants. Organic matter also increases the amount of plant available water that a soil can hold. It does this by two mechanisms: one is its intrinsic ability to hold water and the second is its involvement in improving soil structure. Soil with good structure, enhanced by organic matter, can hold more water, especially water available to plants. Organic-matter-enhanced peds are larger than soil particles (sand, silt, and clay), making them harder to move by water or air. Thus, they make soil less susceptible to wind or water erosion. Organic matter can also adsorb materials that might be toxic to plants or animals and render them nontoxic.

On the other hand the higher clay content in lower horizons is also beneficial. This increased clay layer is called the B horizon or in many cases a series of B horizons. These can often be identified by simply feeling the difference in texture between the surface A horizon and the lower B horizons, which would be the clayey subsoil in Figure 9.3. Clay retards the downward movement of water, thus holding it in the

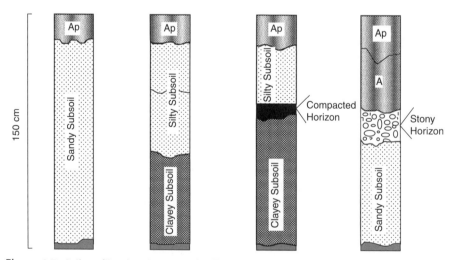

Figure 9.7. Soil profiles showing several different types of conditions. The Ap horizon is an A horizon that is plowed. The lowest horizon in each is the water table. Compacted and stony horizons restrict the downward extension of roots.

upper horizons for plant use. Clay minerals themselves hold more water than other common inorganic soil components and thus store water for plant use. In addition through their cation exchange capacity (CEC) clays hold plant nutrients in forms that are available to plants.

There are other horizons in soil and horizon subdivisions that are used by soil scientists to identify different soils. Some of these may have a dramatic effect on plant and animal production. This is the case when the surface horizon is very low in organic matter and there is a buildup of salts in these and other horizons. In some types of environments soils can become very acidic and in others very basic, and either of these conditions will also adversely affect productivity.

Another aspect of soil is its depth. We think of food crops as having roots that explore the soil to a depth of 150 cm ($\sim$5 ft). This means that shallow soils may not have enough rooting depth for profitable crop production. Although areas having shallow soils may produce grass for pasturing animals, they may not be suitable for crop production. Also compacted and stony horizons (Fig. 9.7) will restrict root growth and thus may result in decreased yields of various crops. In Figure 9.8 there is a layer of sand with a layer of gravel over it and a layer of sand on top of this. Colored water added to the top layer of sand will stop at the gravel layer. This illustrates the fact that any abrupt change in any characteristic of any horizon will tend to impede the downward movement of water and roots through the soil profile.

9.6 SOIL INORGANIC COMPONENTS

Soil inorganic components come from rock. Rock is broken down into sand and silt-size particles by physical forces such as tumbling in a river, grinding in ice such as glaciers,

Figure 9.8. Water moving down through layers of various coarseness.

by freezing and thawing, wetting and drying, and heating and cooling. In addition chemical weathering such as oxidation, hydration, dissolution, and being attacked by chemicals given off by roots occurs and also results in the breakdown of rock. Clay can be released from rock during its decomposition and undergo chemical changes in the soil environment to form other clays. Clay can also be synthesized in the soil as rocks disintegrate and components dissolve and recombine.

The material resulting from these activities is called the soil parent material, and it must undergo soil formation to have a soil with horizons. The soil factors involved in soil forming are time, climate, biota, parent material, and topography, which interact over time to produce a fully developed soil. A fully developed soil would have many horizons and be perhaps 150 cm deep. To develop this soil would take on the order of 1000 years. This is why protecting soil from accelerated erosion is so important.

Should a person be able to discern the characteristics of a soil simply by looking at the composition of the rock from which it is forming? Unfortunately, the answer is no. This is never the case. To cover all the reasons why this is so would require a book of some significant length. Here we will discuss only some of the more important reasons why soil is different from its underlying rock. The three chief reasons are leaching, by which components are lost from the forming soil, deposition of soil or soil parent material from water or air, and the loss of soil from its surface due to erosion.

Soil and soil components are constantly in motion, being carried by air, water, ice, and gravity. This means that a soil in one area is always losing material while other soils are gaining. Thus material is constantly being lost and gained. In the case of deposition, soil has deposited on it material from other areas, which is not like the rock from which the soil is forming. This means that the composition, particularly of plant nutrients, will be different than would be expected in soil forming only from one type of rock.

Because soil is different from underlying rock, it is not possible to know about the fertility of a soil, an area, or landscape without actually doing a chemical analysis of the soils or the area in general.

Of the inorganic components in soil the clays are the most important because of two important characteristics. Clays have a large surface area and thus have high adsorptive and absorptive capacities. Because it is often hard to distinguish between absorptive and adsorptive processes in soil, the process is often called sorption. Both inorganic and organic components can be sorbed by soil clays.

A second important characteristic of clay is cation exchange capacity (CEC). Clays carry negative charges at edges where bonds are broken. Many clays, including those commonly found in soil, also have isomorphic substitution that gives rise to additional negative charges. Both of these sources of negative charge result in positively charged species, mostly metal cations, being attracted to the surfaces of clays. This attraction is called the cation exchange capacity. The word *exchange* is included because these cations can readily exchange with cations in the soil solution. One of the main plant nutrients is potassium that occurs as a positive cation, K^+, in soil and on the CEC. Because this cation is easily exchangeable, it is readily available to plants.

Other important cations, such as calcium Ca^{2+} and magnesium Mg^{2+}, are also attracted to the exchange sites and are thus also readily available to plants. Calcium and magnesium are essential nutrients for plants but are in high concentration in most soils and so are not generally added as fertilizer. Both calcium and magnesium are instrumental in controlling soil pH, for which purpose they are added to soil, and so they have a dual role in plant production.

9.7 SOIL ORGANIC COMPONENTS

Soil organic components come from living and dead plants, animals and microorganisms, and humus, which is synthesized when organic matter is decomposed. Plants add organic matter from both tops and roots. Typically, soil that has grass, which in some areas dies in the fall each year, will have a deeper darker surface horizon than other soils. Animals add organic mater in the form of excreta, hair, and skin while they are alive. In death they add the organic matter of their body to the soil. Microorganisms add organic matter that they cannot digest and waste products from decomposition of organic matter (OM). In addition they add their bodies when they die.

$$OM + O_2 \xrightarrow[\text{degradation}]{\text{Aerobic microbial}} CO_2 + H_2O + Energy + Released\ plant\ nutrient + Humus$$

$$(9.1)$$

$$OM \xrightarrow[\text{degradation}]{\text{Anaerobic microbial}} CO_2 + H_2O + CH_4 + Energy + Released\ plant\ nutrient + Humus$$

$$(9.2)$$

Organic matter added to soil undergoes rapid decomposition, especially if the soil is moist and at a temperature above $15°C$. Decomposition is generally faster under oxidizing conditions, illustrated in reaction (9.1), as opposed to reducing (anaerobic) conditions, illustrated in reaction (9.2). Under either set of conditions small pieces of organic matter undergo the fastest decomposition.

A great number of changes occur during the organic matter decomposition process. First, the complex organic and biomolecules are broken down. This releases plant nutrients, commonly abbreviated by capital letters, that is, nitrogen compounds (N), phosphorus (P), potassium (K), sulfur (S), and potentially all the other plant nutrients that happen to be present in the organic matter. In all cases these are intended to be forms available to plants. Humus, which has high absorptive capacity for water, plant nutrients, organic molecules, and other organic material, is produced, or synthesized, during this decomposition process. Humus also readily releases sorbed nutrients for plant use.

Soil organic matter is highly variable from soil to soil and region to region. Organic matter can be less than 1 percent in many African soils, around 1 percent in many soils in the United States, and 2 percent in some soils in the central Philippines. However, it is not uncommon to find soils containing 10 percent organic matter. There is even a soil order, the Histosols, that are primarily organic matter; see Table 9.3.

Generally speaking, organic matter is not taken directly into plants. However, organic matter has dramatic beneficial effects on soil, especially in relationship to plant growth. In addition to being a source of plant nutrients, it increases soil water-holding capacity and improves soil structure. It does this by helping to bind soil particles together to form peds. All benefits of organic matter in soil are hard to elucidate, and even small increases of organic matter are beneficial to plant growth and crop production.

One general way in which crop production throughout the world can be increased is by increasing the organic matter in soil. This can be done in a number of different ways. Plowing all plant residues into the soil, applying animal manure, composting, and green manure crops are the four most common. In many places in the world and many traditional cropping systems involve burning crop residue before plowing or otherwise cultivating a field. Burning crop residue certainly reduces the residue to a form (ash) that is easier to handle. However, at the same time, it also results in the loss of most nitrogen and sulfur and some phosphorus and potassium plus other nutrients.

The addition of animal manures to soil is an excellent way to increase both the organic matter of soil and also its nutrient content. Indeed in organic farming, manure, from any type of animal, often plays a central role in maintaining soil fertility. The exception to this would be untreated human waste because disease can be spread in this way. In all other cases the manure should be maintained in good condition before addition to soil. For instance, exposure to rain will lead to leaching and loss of nutrients and thus degrade the usefulness of the manure as a soil amendment.

Manures contain easily decomposed organic matter and so should be composted for some time before plowing in soil. Fresh organic matter will decompose aerobically, and this process, which requires oxygen, will lead to an anaerobic (without oxygen) state in soil. Because plant roots need oxygen, such a condition will be detrimental to plants if too much undecomposed organic matter is incorporated into the root zone.

Composting involves taking organic matter from any source and placing in a pile to decompose. Frequently, this pile would consist of organic kitchen waste, weeds pulled from the garden or other sources, and any other organic waste. The compost pile (Fig. 9.9) may be bounded with wooden or wire fence and organic matter added as it becomes available. Once decomposed the organic matter, which is now composted, is removed from the bottom and put on the garden or field.

Care must be taken during composting to keep pest animals—mice, rats, raccoons, and the like—out of the compost pile. However, inclusion of worms will hasten composting and improve the quality of the final compost. In this case both compost and worms are then added as a soil amendment.

Green manures are also an excellent way to improve a soil's organic matter content. In the strictest since this involves growing a crop—wheat, rye, rye grass, soybeans, or sweet clover, for example—and plowing it under before it matures; usually during the flowering stage. In many respects a field allowed to grow weeds that, once tall and near maturing, are plowed under is also like a green manure crop and will add organic matter to soil. However, it is not a green manure crop if the grain is harvested and then the plant residues are plowed under.

Some organic matter, plant residues such as wheat straw and other high-carbon plant materials such as sawdust, has little nitrogen. When these are incorporated into the soil, microorganisms take nitrogen from the soil to carry out the decomposition

Figure 9.9. Compost pile with fresh organic matter on top and composted organic matter at bottom.

of this organic matter. The soil becomes deficient in nitrogen, and plants cannot obtain enough of this nutrient and their growth is decreased. Thus, these types of materials are best composted or the decomposition allowed to continue for some time before crops are planted. This does not apply to legumes because they fix their own nitrogen.

Like fresh manure, a freshly plowed under green manure crop contains a great amount of readily decomposed organic components. The decomposition of these can lead to oxygen depletion in the A horizon, and thus the soil will not be suitable for immediate crop production. However, green manure crops can be very beneficial in increasing a soil's organic matter and its structure.

9.8 SOIL BASIC FERTILITY

In this section we will not discuss fertilizers because they are discussed in detail in Chapter 11. Basic fertility is the innate ability of soil to supply plant nutrients to crop plants. Some soils will have naturally high levels of some plant nutrients; others will have low levels. Because of its chemistry nitrogen is never likely to be at high levels in soil except where large amounts have been applied as fertilizer or if the soil is near some natural source, such as a soil that receives the drainage from a bat cave, a livestock holding area, or is a nesting area for thousands of birds. Legumes fix their own nitrogen from the air and are thus independent of other nitrogen in the soil. However, high nitrogen levels in soil reduce the amount of nitrogen legumes fix.

Unlike nitrogen, some soils can have naturally high levels of phosphorus, and others can have high levels of potassium. It is not common, however, to find soils with both of these nutrients at high levels. High levels of phosphorus can occur around phosphate deposits such as those in the midwestern part of Florida. In soils developing from parent material containing large amounts of potassium-bearing minerals can have large amounts of potassium such as the soils in northwestern Luzon in the Philippines.

Calcium, magnesium, and sulfur are often in sufficient supply in soil. However, micronutrients may be lacking. Calcium, as ground limestone, is added to soil to increase pH, while sulfur is added to decrease pH. Magnesium accompanies calcium as it is a common constituent of limestone. Micronutrients are very different in that they are needed in very small amounts and are thus added with other fertilizer elements. They are of particular concern because in high concentration they can be toxic to plants and in some cases animals and humans if excessive amounts are applied. Thus, great care must be exercised in applying micronutrients to soil.

It is rare to find a soil that is equally deficient in all plant nutrients. It is more common to find a soil particularly deficient in one nutrient and somewhat lacking in other nutrients.

With continued cultivation and crop production on the same piece of land or field without fertilizer, manure, or organic matter application, all nutrients become deficient and crop production is dependent on the nutrients released from parent material during the growing season. This leads to very low crop yields and a degradation of the soil and its ability to support crop production. However, crop yield will never go to zero, as there is always some decomposition of soil parent material that releases crop nutrients into the soil, and nutrients from decomposing plant materials.

An exception is rice production. In South East Asia rice is often irrigated by water from surrounding mountains. This irrigation water contains plant nutrients dissolved as the water moves through soil and down rivers. Plants irrigated using this water can make use of these nutrients. Therefore, these rice paddies have more natural plant nutrients than might be expected. The result is that these rice paddies can sustain higher yields without fertilization.

This type of irrigated rice production takes place in areas that have rainfall as high as 2000 mm per year. This high rainfall means that there is no problem with salt buildup in these areas.

In desert areas there is insufficient rainfall and so crops must be irrigated. Desert areas generally have low humidities and high evaporation and evapotranspiration[†] rates. Therefore, salts in irrigation water are deposited in and on the soil when evaporation of irrigation water occurs. These salts will have two effects on crop production. First, the pH of the soil will be above 7.5. Second, high salt levels will prevent plant growth when they exceed a certain level depending on the crop. In all cases irrigation can result in increased salt in soil. Therefore, irrigation needs to be carried out with caution and requires special management to keep these soils productive.

Because of the importance of plant nutrients in crop production, it is common to take soil samples from a field and have them analyzed; see Figure 9.10. A representative soil fertility report is given in Table 9.5. The farmer generally gives the sample number, and the report may also have a laboratory number assigned to each sample that may also be on the report. Generally, it is desirable to have soil pH in the range of 6.5 to 7.0 for most crops, although some specialized crops such as blueberries may require lower pHs. For acid soils a buffer pH is given and can be related to tables that give the appropriate levels of lime needed to bring acid soils to the

---

[†]Evapotranspiration is the combined loss of water due to evaporation and transpiration.

Figure 9.10. White spots in field show where soil sample was taken.

desired pH. Soils with pH above 7.5 can be acidified by the addition of sulfur to bring the pH down to that needed for the crop to be grown.

Organic matter (OM), cation exchange capacity (CEC), percent base saturation, (%BS), and all other important characteristics used to make recommendations about the amount and types of fertilizer needed are given in the soil analysis report. Although nitrogen level is not given, knowledge of general soil nitrogen levels and previous cropping are used to modify nitrogen recommendations based on crop needs for maximum production.

Fixed nitrogen, that is, nitrogen combined with other elements (not elemental nitrogen as occurs in air), can occur in a number of different forms in soil and changes rapidly, is highly labile, is easily lost, and thus its concentration changes rapidly over time. For this reason it is not generally profitable to do a nitrogen analysis. However, in some lower rainfall areas it is useful to analyze the soil profile for nitrogen compounds and take their availability into consideration when making fertilizer recommendations.

Phosphate and potassium soil analysis are straightforward and are very important in making fertilizer recommendations. Local conditions, the crop being grown, and soil conditions will dictate the amounts and types of these fertilizers needed.

TABLE 9.5. Soil Fertility Report

Sample No.	pH	Buffer pH	Percent OM	CEC	Percent B S	N	P	K	Ca	Mg	Other[b]
AR25a	5.5	6.0	1.1	12	75	[a]	12	150	3520	290	
AR30b	6.3	6.7	0.5	9	80	[a]	15	200	4250	300	
AR10c											

[a]Nitrogen may be reported in many forms.
[b]Other might be any of the micronutrients or other components.

Calcium and magnesium are both important in crop production; the amounts available are related to both the pH and the CEC. In addition, the ratio of calcium to magnesium needs to be within a certain range for optimum crop growth. Both of these elements are found in limestone, which is generally ground to produce lime for correction of acid soil conditions.

The other column in Table 9.5 refers to micronutrients, such as zinc (Zn), copper (Cu), iron (Fe), selenium (Se), manganese (Mn), molybdenum (Mo), and so forth and may include and specify any of them. In general, micronutrient analysis is relatively more expensive than the analysis for the other nutrient elements, and so it is done sparingly and only when observation of plants leads one to believe that there is a micronutrient deficiency.

9.9 SOIL EROSION

Soil erosion is a worldwide problem that negatively impacts crop production. There are two types of soil, erosion, geologic and accelerated. Geologic erosion is the natural loss of soil that occurs even on the best vegetated soils and is a natural phenomenon. Accelerated erosion occurs most often on sloping bare soil and is caused by either wind or water passing over a bare field; see Figure 9.11. Soil loss is greater than geologic erosion and is most important and most damaging to crop production if not controlled.

Accelerated erosion occurs as a result of agricultural practices and to a lesser extent as a result of construction. Any soil without vegetative, plant residue or other cover is subject to accelerated erosion. During traditional cultivation, plowing, disking, and planting, the soil is most subject to accelerated erosion. Once the crop has canopy closure, the time when the leafy tops of crops completely shade the ground, the threat of erosion is greatly reduced.

Raindrops striking a bare soil surface dislodge soil particles, break up soil peds, and thus allow these particles to be removed from the field. The process of raindrop

Figure 9.11. Gully erosion caused by rain on a field not protected by a growing crop or crop residue. (Courtesy of Mathew Deaton.)

dispersion is typically called splash erosion. Water flowing over the surface of the soil moves these dispersed particles and removes them from the field. More plant nutrients are contained in the surface soil, and so as surface layers are removed so is fertility, which decreases the soil's productive capacity.

If the water is moving as a sheet across a field and down a slope, this is called sheet erosion. This will result in a more or less even removal of the surface layer of soil. If water concentrates in a low area and runs off the field, then depressions will be cut in the field. This is be called rill erosion; see Figure 9.11. Evidence of these types of erosion are removed by cultivation, and so it may not be evident or observable that erosion has taken place. However, when rills become too large to drive across, they are not easily covered up and are called gullies.

To stop water erosion, first protect the soil surface from being directly hit by rain-drops. Second, prevent water from running across the field. This can be accomplished by mulches, plant residues, growing plants, and performing cultivation perpendicular to the slope of the field.

Reduced-till and no-till practices are additional methods used in controlling soil erosion. Reduced tillage, as the name suggests, involves using the minimum tillage necessary to plant a crop, thereby limiting the exposure of soil to erosion. There is no clear or finite definition of reduced tillage, it could mean plowing and disking at the same time or disking and planting without plowing, however, the central concept is to reduce the number of tillage operations necessary and to get seed growing as quickly as possible. No-tillage or no-till involves killing all the plants growing in a field with a herbicide and then planting directly into the soil without any further soil preparation. This keeps the soil covered and protected from both wind and rain erosion while the crop is growing.

What Farmers Say About No-Till

Farmers who do not use no-till have a large number of reasons why it will not work. The field looks trashy in the spring when it is first planted. Indeed some ren-ters have been told that they can no longer rent because they are not taking care of the land because it looked so trashy in the spring and fall. Another complaint is that it leaves all crop residue on the field, and this makes an ideal breeding ground for both animal and microbial pests, for example, rabbits that eat the young crop plants and diseases that attack crops.

Farmers who use no-till have different complaints. First, they do not see any of the problems that the farmers who do not use no-till see as problems. They like that there are less trips over the field; they can use tractors with less horsepower and use less fuel, and their soil is always protected against erosion. The complaint is that when it rains there are so many worms coming out of the ground that it is not pleasant to walk around. This is interesting because one often hears that agricultural chemicals kill all worms in soil. This is apparently not the case with chemicals used in no-till farming. It is true, however, that residues on the soil surface produce an environment conducive to worms.

Other methods that conserve both soil and water are farming on the contour, bunds, and terraces. Crop rows on the contour tend to slow or stop water movement down the slope, which means that soil does not move down the slope. Berms or bunds are like small dams across the field that stop and hold water, thus preventing it from leaving the field. In some cases soil eroded from higher in the field will be caught on the up side of the berm, and thus eventually it can be expected that a terrace will form. Figure 9.6 shows a badly eroded field and rock "dams," or berms, made from rocks collected form the field and placed on the contour to prevent erosion.

Terraces are structures on a hillside where the soil has been removed from the upper side of a slope and moved to the down side to make a relatively flat piece of land that can be farmed and that has a reduced risk of erosion. At the down side of the terrace a steep bank connects with the next lower terrace. This steep bank is kept protected by permanent grass vegetation and not farmed or tilled.

Note that in all these cases the objective is to prevent soil erosion by preventing as much water as possible from running down and off a hillside. Water that does not run off a hillside has only two other options. It can infiltrate into the soil, thereby increasing soil water, or it can evaporate. In most cases infiltration will be faster than evaporation. Infiltration must be fast enough to prevent saturation of the soil profile for a period of time that will otherwise kill the crops planted there. In most situations this is not to be a particular problem. However, if it is a problem a drainage tile can be used to remove excess water.

Soil can also be eroded by wind, called wind erosion. It occurs any time wind dislodges soil particles and carries them away or out of a field. In this case keeping the soil covered and preventing the wind from contacting the soil surface can reduce or eliminate erosion. Keep in mind that wind has to be strong and continuous to effectively dislodge and carry soil away. Generally, the soil must also be dry because water causes soil particles to stick together, and it has significant weight, thus making it difficult for wind to carry wet soil.

Prevention of wind erosion is similar to preventing water erosion. Keeping soil covered with plant residues and growing plants prevent wind erosion. Structures, usually living trees, called windbreaks used to slow wind speed and keep it off the ground are effective in stopping wind erosion. It may seem strange but a porous windbreak does more to prevent wind erosion than does a solid windbreak. A porous windbreak allows some wind to go through, thus slowing it and allowing suspended soil particles to fall out. Thus, in this case leaving stubble (plant stalks below the grain head) in the field after harvest will help prevent wind erosion. Each row of stubble acts as a windbreak. All windbreaks work best when they are at right angles to the prevailing wind.

9.10   WATER ON THE THREE FARMS

Water for both Donio and Aída and Octavio come from springs in the mountains some distance from their farms. Water is piped to the houses by a local authority, and they pay a monthly fee for the water. Steve gets his water from a well near his barn and pays only for the electricity needed to pump the water.

Figure 9.12 shows the water pipe and faucet outside of Donio's house. The water's source, a spring in the mountains, is enclosed in a cement structure that protects the water from contamination and from which the water pipes emanate. Because the spring is higher than the house, water moves by gravity from the spring to the house and no pumping is necessary. Doino or a member of his family must carry water, usually in a plastic container, to the house where it is used for drinking, cooking, and cleaning. Often it is the children's job to obtain water. Water costs P80 (80 pesos) per month, which is equivalent to about $1.52 USD/month.[‡] Donio does not use water from the faucet to irrigate any plants. He relies entirely on rain to water his plants.

In Ecuador the farm water also comes from a spring; see Figure 9.13. However, in this case a pipe with a screen over its mouth is used to collect the water and move it from the spring to a holding tank. If the screen comes loose, the water system can become plugged with leaves and other debris. Water moves from the holding tank to the house via gravity; no pump is needed, and the water is piped directly to an outside washing facility, into the kitchen, the bathroom, and then to the garden near the house where it is used for drip irrigation. Water costs Aída and Octavio $1.50 USD/month (Ecuador uses U.S. dollars as it currency).

Of our three farms Aída and Octavio are the only ones that irrigate. Water, from the same source used for the house is piped to the garden area, filtered, and then fed to commercial drip irrigation pipes that are placed either on top or on the side of the raised beds used for crop production; see Figure 9.14.

Figure 9.12. Pipe and faucet, about 1 m above the ground, that serves as the water source for Donio's house. (Courtesy of Henry Goltiano.)

[‡]P is Philippine peso; USD is U.S. dollar.

Figure 9.13. Spring that is the water source for Aída and Zorro's farm in Ecuador.

In the United States Steve uses an electric pump to pump water from the well to the house and for the cattle and pigs; see Figure 9.15. The well is covered by a cement slab and a round access cover and surrounded by a small enclosure, called a pump house, which protects the well and pump. Steve paid for digging the well, putting in the covering, house, pump, and all the necessary piping to carry water to the house and animals. However, only a small amount of his monthly electric bill

Figure 9.14. Drip irrigation pipes on beds raised ∼25 cm high in Aída and Zorro's garden.

Figure 9.15. Murphy well house. The well, bottom left, round hole and lid, and associated pump, unseen in back, and pipes that serve as the water source for Steve's farm.

goes to pay for pumping water. Steve does not use water from his well to irrigate any plants. He relies entirely on rain for all his crops.

9.11 WATER AND SOIL

Soil cannot be separated from water and water cannot be separated from soil. Water is constantly being added to and removed from soil, and the moisture available for crops is different in different parts of the world, as seen in Figure 9.16. Rain falling on soil adds water that either runs off the surface or infiltrates and percolates through the soil profile. Some water falling on soil will be held in pores against the pull of gravity. Some of this water will be lost by evaporation from the surface of the soil, and some will be lost by transpiration from the leaves of plants. As discussed earlier the total process is called evapotranspiration.

Several things happen as water percolates through soil. Microorganisms decompose organic contaminants dissolved in the water, and solid components, both organic and inorganic, are filtered out because of the tortuosity of the path of the water through soil pores. However, as water passes through soil, it also dissolves both organic and inorganic components from soil to become part of the soil solution. When taken up by plants, many of these components are used as plant nutrients.

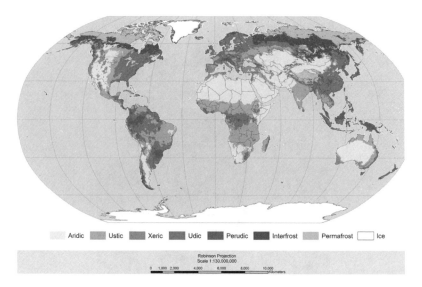

<u>Figure 9.16.</u> World soil moisture regimes where: Aridic is dry more than half the year. Ustic is between Aridic and Udic. Xeric in winter is cool and wet and in summer is warm and dry. Udic is not dry for more than 90 days. Perudic is always wet. Interfrost is frozen at some depth most of the year. Permafrost soil is frozen at some depth all the time. (Courtesy of U.S. Department of Agriculture, Natural Resources Conservation Service, Soil Survey Division, World Soil Resources, Washington, D.C. (http://soils.usda.gov/use/worldsoils.mapindex/.)

Niger and Water
The agriculture problem in Niger is the lack of water because the country lies on the edge of the Sahara Desert. Therefore, planting on the contour, contour furrowing, and increasing the organic matter in the A horizon will all result in slowing or preventing rain runoff as well as increasing infiltration and the water-holding capacity of the soil. Mulching and minimum tillage will also result in water conservation and an increase in the available water for crop production and thus food availability. Irrigation is not an option in many areas as there is no source of water for irrigation.

9.11.1 Water Sources

Water for agriculture comes from three main sources: rain, bodies of water (including streams, rivers, ponds, and lakes), and wells. These sources are not equal in quantity nor quality. Natural rain when available in sufficient quantity is the best source of water for crops. Worldwide well water is most frequently used for human and animal consumption. Bodies of water are used for both irrigation and drinking water for humans and animals.

Figure 9.17. Mountainous watershed, covered with trees, on the island of Leyte. In the foreground is the Camote Sea.

Rainwater can be conserved for agriculture in several ways. Cultivation can be carried out in such a way—with contouring, strip cropping, and conservation tillage—to maximize infiltration into the soil. This will both provide water for crops and allow more water to move through the soil profile and into the groundwater. It is also possible to store rainwater by the use of cisterns (sealed), ponds, or other catchments. However, such storage of rainwater is not as good overall as storage in soil. For example, Ferrie, a farmer in Zimbabwe, has a large granite rock outcrop[§] behind his house. He collects rainwater, which runs off this rock and channels it into cisterns around his house. The cisterns are not sealed and the water percolates horizontally down slope from these cisterns, providing subsurface water for a myriad of crops that are planted in plots situated down slope from the cisterns.

Another water storage system is called a watershed (Fig. 9.17). This is often a mountainous or sloping area that is heavily wooded. Trees and associated soil hold water that is slowly released into the surrounding environment, usually in the form of springs, streams, and rivers. When managed correctly, watersheds can be a source of continuing clean water for animal and human consumption. If the watershed is large enough, the water it releases may also be used for irrigation.

Great care must be exercised when using water from other sources, that is, other than watersheds and rainwater, for drinking, irrigation, or other agriculture purposes. There are two reasons for exercising caution: excessive use of well water may cause the well to go dry, that is, the water is being used faster than it is replaced. The second reason is that water from wells, streams, rivers, lakes, and ponds may contain salts or toxic compounds or elements. There are many places in the world where this happens; for example, Bangladesh has toxic arsenic levels in some well water, and

---

[§]Rock outcrops describe any situation where rocks stick out of the soil.

this has caused severe health problems for many people. This can also be caused by human activities such as mining, manufacturing, or waste systems.

All irrigation schemes must provide not only for sufficient water for the irrigation project but also for the people living in the area. It must also take into account the need to not only remove salts and salt buildup in soil but also the eventual fate of these salts. Although simple irrigation systems requiring only gravity are in operation in Asia, Africa, and South America, most systems require pumps, which require maintenance and spare parts, and, if a motor does the pumping, there must be fuel and lubricant for the motor.

Figure 9.18. Center pivot irrigation showing tubes directing irrigation water toward the ground. Corn is at the four-leaf stage. (Courtesy of Luke Baker.)

Figure 9.19. Irrigation by spraying water across fields.

9.11.2 Irrigation Methods Around the World

The Philippines
In the Philippines it is common for rice farmers to place rocks across a stream upstream from their rice paddies to impound water, which is then directed through a series of open canals to the paddies. No expensive infrastructure is involved once the canals are dug. Paddies are flooded with water covering the whole area to a uniform desired depth. Thus no water distribution system in the field is needed. Water from the river may pass through several paddies owned by several different farmers between the initial water takeoff and the water finally exiting into the ocean. The consequence is that many rivers have little water in them by the time they reach the ocean.

This system has a secondary benefit in that it can handle large rainfall events without causing flooding.

In spite of irrigation, soils in the Philippines generally do not have salt buildup problems because of the high rainfall. However, in northern Luzon there is a distinct dry season, and as a consequence there are concerns related to salt buildup from irrigation in this part of the country.

Ecuador
All types of irrigation practiced in the Philippines and the United States are used in Ecuador, although not to the extent used in the United States.

The United States
Much irrigation in the United States is done by overhead sprinkler irrigation, shown in Figure 9.18, via pipes carried over the field on large wheels. Initially, these irrigation systems sprayed water into the air to spread it out over the area to be irrigated. Spraying water in this fashion is a wasteful irrigation method because it maximizes evaporation of water, resulting in less water available to plants, and it encourages salt buildup in the soil. Often this type of irrigation is called center pivot because the pipe and wheels move in a circle around the field leaving a circular field of green in an otherwise brown environment. Today, these irrigation systems have hoses, which drop the water down closer to the soil surface before spraying, thus improving water use efficiency.

Other types of irrigation such as spray (shown in Fig. 9.19), which has much higher evaporation than other systems, and furrow and drip, which can also have mulch to further conserve water, is shown in Fig. 9.20, are also common. In furrow irrigation ditches are dug across a field and crops are planted on the ridges in between. The crops are irrigated by periodically filling the ditches with water from a main irrigation ditch. In drip irrigation small plastic tubing is run down the row of crops and there is a small "leak" at each plant such that each plant has a source of water. This is the most water-efficient irrigation method and is even more efficient when combined with any type of mulch.

These methods require large amounts of water and expensive equipment to implement. Because of this, they are not applicable to many parts of the world and are not available to most of the world's farmers.

Figure 9.20. Plastic mulch to control loss of water by evaporation in drip irrigation in straw-berry production in Florida. Plastic pipe in the left middle of the picture is part of the drip irriga-tion system.

9.12 CONCLUSIONS

Both soil and water are essential for food production and cannot be separated from one another. Soil texture determines how well it holds water and nutrients for plants. Organic matter improves soil's ability to support crop production and is an input available to all farmers. The arrangement of horizons will also affect the growth or crops and in some cases may dictate the types of crops that can be grown. The fertility status of soil must be analyzed to determine which and how must of each are needed. Conserving soil by preventing or controlling erosion by either water or wind is important to maintain a soil's productivity.

Water used in food production often comes from rain, and this is the best source. Excess water can be used for irrigation, provided it will be replaced and the water does not contain any limiting components such as high-salt content or toxic elements. Irrigation systems throughout the world are varied, but generally only the most water-efficient systems should be used, unless the area receives very large amounts of rain. Irrigation, no matter where it is used, must have technical support to successfully maintain it long term. This not only entails maintenance of the irrigation system but also testing both soil and water to determine salt contents.

QUESTIONS

1. Explain how soil is basic to all life on Earth.
2. What are the basic, observable characteristics of soil that are used to differentiate one soil type from another?

3. Describe the characteristics of an A soil horizon and describe what characteristics of an A horizon result in increased crop production.

4. Explain why a person should not be able to determine the natural (nonfertilizer) crop nutrients in soil by knowing the rock that underlies that soil. Explain why this is not correct.

5. Explain three reason why increasing the organic matter in soil should lead to an increase in crop yield in any soil.

6. List the basic crop nutrients found naturally in soil. Why might these be different from the same nutrients found in the soil's parent material?

7. Explain how soil is complex, giving the components it contains and how these components are related to each other.

8. Explain how soil erosion depletes soil of plant nutrients.

9. Reproduce the chemical equations that show how organic matter decomposes in soil and what the products of this decomposition are.

10. In conserving soil, water is also conserved. Explain.

BIBLIOGRAPHY

N. C. Brady and R. R. Weil, *The Nature and Properties of Soils*, 12th ed., Prentice-Hall, Upper Saddle River, NJ, 1999.

D. D. Frangmeier, W. J. Elliot, S. R. Workman, R. L. Huffman, and G. O. Schwab, *Soil and Water Conservation Engineering*, Thomson Delmar, New York, 2005.

V. C. Jamison and E. M. Kroth, Available Moisture Storage Capacity in Relation to Textural Composition and Organic Matter Content of Several Missouri Soils, *Soil Sc. Soc. Am. Proc.*, **22**, 189–192, 1958.

A. E. Olness and D. W. Archer, Effect of Organic Carbon on Available Water in Soil, *Soil Sci.*, **170**, 90–101, 2004.

E. M. Romney, A. Wallace, J. W. Cha, and J. D. Childress, Different Levels of Soil Organic Matter in Desert Soil and Nitrogen Fertilizer on Yields and Mineral Composition of Barley Grown in the Soil, *Commun. Soil Sci. Plant Anal.*, **7**, 51–55, 2001.

C. D. Tsadilas, I. K. Mitsios, and E. Golia, Influence of Biosolids Application on Some Soil Physical Properties, *Commun. Soil Sci. Plant Anal.*, **36**, 709–716, 2005.

A. Wild, *Russell's Soil Conditions and Plant Growth*, Longman Scientific & Technical, Essex, UK, 1988.

10

RAW MATERIALS OF AGRICULTURE

10.1 THREE FARMERS

Subsistence Farmer. Donio owns no land and has minimal farm equipment and buildings. An inventory of all his farm tools show only 13 tools used for his farming work. One of these shown in Figure 10.1 is a wooden frame with metal spikes used to remove weeds in maize. An estimated value for his entire set of farming tools is less than US$100. His only expenses for the crops he plants are for rice seed and fertilizer. He saves maize seed from each crop for seed. The fertilizer he purchases is applied at less than recommended rates because he cannot afford to buy very much at one time.

Family Farmer. Aída and Octavio own their land and have a few buildings for their family and livestock. Most of their farm production goes to feed their family, but a significant part is sold to meet family expenses. Most of the fertilizer for their farm is from composted manure and kitchen scraps. The most expensive materials purchased for their crops and animals are fertilizer for their potato crop (US$96/year) and agricultural-grade molasses for the animals (US$120/year). All of their farm tools are shown in Figure 10.2. They have no tractor or truck. They sell potatoes from their two

World Food: Production and Use. By Alfred R. Conklin, Jr. and Thomas Stilwell
Copyright © 2007 John Wiley & Sons, Inc.

Figure 10.1. Weeder used for maize.

harvests each year to have a monthly farm income of about US$250 to 300. The sale of pigs and chickens is irregular but brings in extra money for school and clothing.

Commercial Farmer. Steve Murphy and his sons own around 486 ha of land and rent about 814 ha more on a yearly basis. With so few people planting and harvesting the crops, they rely on machinery to keep costs down. They own three large tractors and nine smaller ones. They also own or share other large equipment with Steve's brother. They own several large barns for storage of equipment, hay, animals, and for equipment repair. The open-sided shed shown in Figure 10.3 is used to protect five small tractors from rain and snow. The large tractors are kept in another shed. Nearly all the materials used on their farm are purchased. Only some hay and maize are produced to feed their beef cattle during the winter. All seed, fertilizer, and chemicals are purchased. Even the

Figure 10.2. Tools used in Ecuador.

Figure 10.3. Open shed for tractors.

food for the family is purchased in a grocery store. Steve's youngest son, William, plants a small garden each year, but the greatest part of the family food comes from stores.

10.2 WHAT RAW MATERIALS DO FARMERS USE?

In agricultural terminology the raw materials used to produce crops and animals are "inputs." These are the things used to produce a product such as wheat or milk. For most farmers there are three types of inputs: free, variable, and fixed. Free inputs are often taken for granted because they are free. Examples are sunlight, air, and rain. Variable inputs are things such as fertilizer or seed. The cost depends on the quantity planted or applied. Fixed costs are expenses that the farmer has to pay even if no crop is planted. An example is the purchase of a tractor. It must be paid for even if the farmer is sick and cannot till the land. It costs the same if it used for 10 ha or for 100 ha.

A farm can be considered as a type of factory where raw materials are used to make products. A general list of the types of raw materials used in agriculture is shown in Table 10.1. Fixed materials are those that are purchased once and serve for many

TABLE 10.1. Types of Raw Materials in Agriculture

Free	Fixed	Variable
Sunlight	Roads	Labor
Oxygen and CO_2	Land	Seed
Rainfall	Buildings	Fertilizers
	Equipment	Weed control
		Electricity
		Vaccines and medicines
		Irrigation water
		Credit

years and crops. Variable materials are those that must be purchased repeatedly for each new cropping season or group of animals.

10.3   SUNLIGHT

Sunlight is taken for granted by most people but is a critical input for farmers. The amount of sunlight determines what types of crops can be grown and their quality. For example, the east coast of Spain has a Mediterranean-type climate with warm, dry spring weather. Dry weather means few clouds and lots of sunshine, perfect for crops with low moisture requirements to produce sweet fruits. Grapes and oranges are two crops that need this type of climate. Raising oranges in a rainforest climate will produce brightly colored fruits but with low sugar contents. The high rainfall of a rainforest requires clouds that block the sun. The result is less photosynthesis to make sugars for sweet oranges.

Crops such as sweet corn grow in direct relation to the amount of sunlight and warm days in the season. A calculation of *growing degree days* will predict when the sweet corn is ready to pick. A company that cans sweet corn will contract with farmers to supply sweet corn through the entire summer. Knowing how many calendar days a variety needs to accumulate the needed growing degree days, the company can tell a farmer when to plant. The company will then contract with farmers to plant on specific dates so their canning plant will have a continuous supply of sweet corn during the summer. If their calculations are off, or if the weather is bad, they will have to stop canning for 2 to 3 days, requiring a cleaning of the entire assembly line! A lot of expense can be avoided by accurate calculations of growing degree days.

10.4   OXYGEN AND CO_2

Carbon dioxide is needed by plants to perform photosynthesis. Oxygen is essential for animals and humans to exist. In most cases these gases are equally distributed around the Earth. A few commercial greenhouses practice CO_2 enrichment to increase yields of flowers or tomatoes. These greenhouses actually purchase CO_2 in tanks and carefully release it during the early daytime to speed up photosynthesis.

10.5   RAINFALL

Rainfall is also free but is a major limiting factor for farmers. Average rainfall is a useful guide to determine which crops may be grown in a specific area. The numbers shown in Table 10.2[1] are approximate but give an idea of differences between crops. The actual rainfall needed to raise a crop depends on when the rain falls during the crop growth cycle, the fertilizers applied, and temperature.

As a general rule, areas with less rainfall also are subject to wide variations in monthly rainfall. Areas with higher rainfall tend to have more predictable average

TABLE 10.2. Rainfall Needed for Crops

Crop	Minimum	Average
Chickpea	150 mm	150–200 mm
Sorghum	280 mm	600–1000 mm
Peanut	300 mm	500–600 mm
Barley	400 mm	400–450 mm
Maize	500 mm	1200–1500 mm
Wheat	500 mm	570–600 mm
Rice	500 mm rainfed	>1500 mm irrigated

Source: From Ref. 1.

monthly totals. This has a large effect on farming practices. A farmer in southwest India can expect the monsoon rains to start around June 1. It may be as much as 2 weeks earlier or later, but this is the average. This means that farmers must have their seed purchased, fields tilled, and rice ready to transplant before this date. If a farmer is not ready, tillage and transplanting rice become more difficult as the monsoon goes on.

Consider the calendar of farming operations that farmers must follow in Ohio, Zimbabwe, and India by consulting the rainfall charts in Figure 10.4. Even though the total yearly rainfall of each location is similar, the distribution of the rain dictates totally different farming calendars. A farmer in Bulawayo, Zimbabwe, will prepare planting materials in October, but an Indian farmer in Patna, India, will be harvesting rice in October. Planting for this farmer in north India will be in October for wheat and in late May for rice. The wheat will be planted just after rice harvest. A farmer in Ohio is blessed with precipitation all year long. Planting of summer crops will be done in May with harvest in October or November. Some crops, such as winter wheat, will be planted in late September but won't be harvested until after the winter passes and the crop matures in July the following year. The even distribution of precipitation throughout the year makes agriculture high yielding and less uncertain than in Zimbabwe or India.

Rainfall is considered separately from irrigation in this chapter because most farmers depend solely on rain for their crops. Installation of an irrigation system and use of irrigation water requires a very different set of conditions, and this topic is considered in a separate section.

10.6 ROADS

Roads are possibly the most important input for a commercial farmer. If there are no good roads to an area, farmers cannot get their product to markets. They also cannot get needed fertilizer and seeds to their farm. In areas where the government is not able to adequately maintain roads, it is common for farmers to collectively do periodic maintenance before planting and harvest.

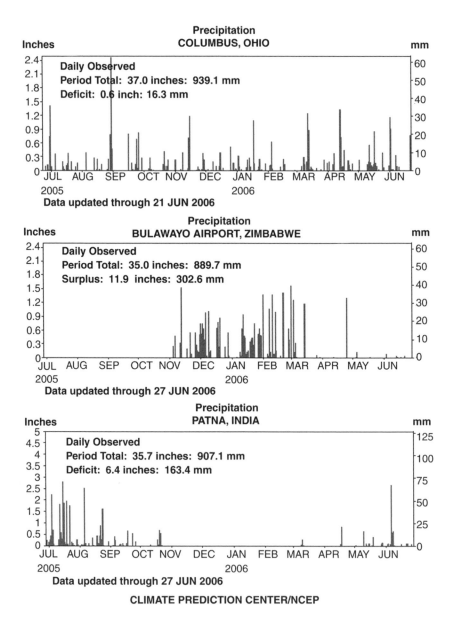

Figure 10.4. Contrasting precipitation patterns. (From Climate Prediction Center, NCEP.)

The types of roads our three farmers must use every day are seen in Figure 10.5. The road in the Philippines has potholes and may not be passable after heavy rains. The road to the farm of Aída and Octavio in Ecuador is narrow and winding, sometimes discouraging to a truck driver. The road leading from the farm of Steve Murphy is

Figure 10.5. Road to markets in the Philippines, Ecuador, and the United States.

smooth and clear all year round. He has no problems driving large trucks to market. The good quality of the road also means he will have fewer repairs to his trucks and lower overall operating costs than other parts of the world.

10.7 LAND

For crop production, land is the essential ingredient. It is possible to raise crops in a building without soil (hydroponics), but it is hard to beat soil as a readily available, cheap growing medium. The food of most land animals comes from land-based plants. Humans eat primarily land-based plants and animals. In most societies, the wealth of a person is based on the amount of land he or she owns or controls. The quality of the land for agriculture is also fundamental to its value. As you can see in Figure 10.6, the size of fields and their topography vary greatly. In some cases, the slope of the land is so steep that it cannot be used for agriculture. For the purposes of this chapter, we will only consider land used for agricultural purposes.

The agricultural value of land depends on many factors: organic matter content, depth of topsoil, drainage, ease of tillage, distance to markets, road access, and slope. In the United States all agricultural soils have been classified according to their technical characteristics and productive value. The taxes paid on land are determined by the soil classification. Value of land for sale depends in large measure on its productivity. Two general classes of agricultural land values are watched by the Department of Agriculture: cropland and pasture land. The trend of United States cropland prices from 1982 to 2005 are shown in Figure 10.7.[2] Since 1987 there has been a steady increase in average cropland prices. One hectare purchased in 1982 would now cost nearly twice as much. Of course, this is an average. Prices on pasture land are lower and prices for highly productive land are higher.

We often think that in developing countries, land prices must be much lower. The opposite is often the case. Take the case of northern India. Because of the population pressure, there is competition for land. In 2002, irrigated land in Punjab Province was reported as having dropped in price due to a drought. Land prices had dropped as much as 20 percent to give an average selling price of US$10,869 per hectare. Compare this to an average price of less than US$4000 per hectare in the United States.

Another factor that reduces the value of land for agriculture is the question of clear title. In the United States we take it for granted that after purchasing a parcel of land the land is ours. In some countries there are several causes of land tenure insecurity. Red tape and dishonest officials are causes for delays and confusion in registering a land purchase. Simple registration of title for a parcel of land can take years.

Other countries have to deal with different classes of land ownership that sometimes conflict. In much of southern Africa, there are three recognized land tenure systems: private, state owned, and customary. Customary land tenure is based on unwritten traditional rules with the land being administered by traditional leaders. There may be no written proof that the land belongs to a tribe or group of tribes. Tribe members do have rights to use of the land subject to approval of the council. Rights to use of land may change as family or tribal conditions change leading to some uncertainty about

Figure 10.6. Land in the Philippines, Ecuador, and the United States.

Figure 10.7. Historical U.S. farm cropland value. (Adapted from USDA–NASS, August 2005.)

investments in the land. The amount of land managed under customary land tenure ranges from 14 percent in Botswana and South Africa to 80 percent in Mozambique and 81 percent in Swaziland.[3] This leads to a large degree of uncertainty for the individual farmer about investing in land. It is difficult for an individual to purchase land.

A related problem is caused by HIV/AIDS. Many families have lost one parent, often the male, leaving the mother to care for the children and cultivate the land. In Lesotho, Malawi, and South Africa it sometimes occurs that upon the death of the male head of household, the patrilineal kin assign rights of cultivation to another family to avoid underutilization of the land.[3] Movements are underway in these countries to guarantee usage rights to women and even older children.

The land tenure problem can be summed up best by Figure 10.8. Greater land tenure security is associated with higher agricultural production and more sustainable natural resource use. Higher agricultural production results in long-term investment such as irrigation systems, machinery, and buildings. Sustainable natural resource use is reflected in reduced soil erosion and maintained, or improved, soil fertility.

In an attempt to improve agricultural production and security of land tenure, some countries undertake what is commonly known as *agrarian reform*. The ultimate goal is

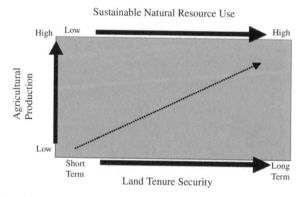

Figure 10.8. Relationship between land tenure security and agricultural production.[3]

to distribute unused, or unproductive, land to landless rural laborers. The theory is that people will invest much more effort in their own land than that of an employer. In practice, the reverse often happens. Because most countries have little frontier land that can be brought under cultivation, the large plantations of individual land holders are often seized and subdivided into small parcels. These are distributed to individuals that previously worked as laborers on the farm or to people from cities desiring land. This is politically popular but economically disastrous.

A case in point is Zimbabwe. In November 2002 the government seized 14 million hectares of land from large white farmers. This coincided with invasions and land grabs by laborers, often with the death of the landowners. To make matters worse, a cyclical dry period affected maize production. Large landowners were either forced off their land or sold out at reduced prices. The result was a dramatic fall in maize production, the staple food of many Zimbabweans. Before 2000, Zimbabwe was a net exporter of maize to neighboring countries. At the end of the 2004/2005 crop year (maize is planted in our fall and harvested in our spring) production had fallen to less than one-third of previous levels and Zimbabwe was importing maize.

Small farmers in Zimbabwe traditionally cultivated maize, but their yields were rarely greater than 2000 kg/ha. In contrast, large commercial farmer's yields rarely dropped below 3000 kg/ha, often rising to 5000 kg/ha.[4] The loss of large commercial farmers left most of the country's maize production in the hands of small farmers with few resources to purchase fertilizers and hybrid seed needed to achieve high yields. The net effect was an abrupt reduction in national maize production and a large increase in imports of maize.

10.8 BUILDINGS

Most agricultural farms need buildings of some type. They may be for storage of harvested grains, protection of livestock from weather or predators, to shield machinery from rain, or to perform sorting and packaging of produce. In many developing countries buildings constructed by subsistence farmers are made of locally available materials such as adobe, reeds, or bamboo, as in the left picture of Figure 10.9. Larger, commercial farmers often purchase complete buildings in a ready-to-construct package.

For any farmer, a building is an investment. For farmers trying to make a profit on an enterprise, a building should be adequate for its purpose, but no more. Spending large amounts of time and/or money on a structure that will outlive its usefulness is not economically sound.

In the examples shown in Figure 10.9, a separate building made with woven bamboo walls serves to store the maize harvest owned by the farmer in the Philippines. This was made entirely from materials picked up by Donio. The farmer in Ecuador is constructing a more solid swine building with concrete block to withstand the movements of the animals and to protect them from predators. It will be finished with a corrugated metal roof. One of the buildings on the commercial farm in the United States was specially built for swine production. It has provision for manure disposal and automatic watering and feeding. In each case the materials were

Figure 10.9. Farm buildings in the Philippines, Ecuador, and the United States.

easily available locally and were constructed at minimal cost to house the needed equipment or animals.

10.9 EQUIPMENT

The equipment used by farmers varies according to the climate, soil, crop, and economic condition of the farmer. For the Philippine subsistence farmer the most expensive items he owns is often the equipment he uses to farm the rented land. Even though the lining board used to mark rows for transplanting rice is a simple tool, it is essential for precisely spacing the transplanted rice plants. An estimate of the value of all tools used by the subsistence farmer is about US$130.

For the commercial farmer in the United States equipment is a major investment. Steve Murphy estimates that the cost of all his equipment used in farming approximates $750,000. It requires a specialized farm shop for maintenance and several buildings to protect the equipment from rain and snow.

The equipment used by the commercial farmer in the United States is designed to reduce labor to the minimum. The tractor used for planting in Figure 10.10 is equipped with a satellite-guided steering system to automatically guide the tractor and planter from one end of the field to the other. The planter uses suction to pick up individual seeds and drop them at precisely spaced intervals and preset depths with minimum disturbance of the soil. This combination will plant all day long without missing rows or seeds to give the optimum plant spacing. It does all this with only one person. Each item of equipment requires training in its use and calibration. For example, the satellite-guided self-steering tractor needs at least 1 hour of introductory training for a person to use it. Operation of the combine harvester needs several hours to learn the needed adjustments of the cutter-head and sieves. The integrated computer must also be calibrated and data downloaded after harvest.

A grain combine that costs over $230,000 has a 340-hp engine, harvests about 9 m wide of maize or wheat, has air conditioning, a stereo radio, and several computers to monitor functions and record grain yield, moisture, and protein content.

Figure 10.10. Farming tools used in the Philippines, Ecuador, and United States.

10.10 LABOR

Labor is a major cost in agricultural production. The work is also strenuous, often in dirty, hot conditions. As a result, farm labor is usually left to poorly trained workers who receive low pay. The information in Figure 10.11 shows that one worker can manage about 1 ha of crops in the Philippines where the work is largely accomplished without machinery. In Ecuador, the average farm size is larger but more farmers have access to tractors. One person can manage about 2.5 ha. In the United States the average size is about 195 ha. With more machinery, one person can manage many hectares. In the case of Steve Murphy, they have a very efficient operation where 4 people manage 1300 ha.

Another labor-related problem is the level of education needed in these countries. In the Philippines, Donio works with only an elementary education. In Ecuador, Aída has had training in farm management and specific technical enterprises. In the United

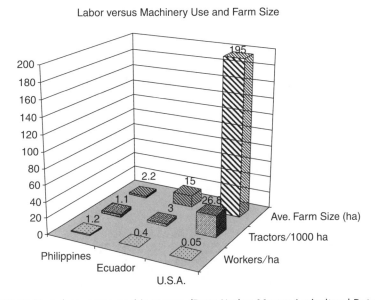

Figure 10.11. Labor versus machinery use. (From Nation Master Agricultural Database.)

States Steve has a college degree plus numerous short training courses and conferences. It is difficult to find labor with the knowledge to operate complicated machinery.

10.11   SEED

Seed is vital for all farmers (Fig. 10.12). In times of famine a farmer will guard seed until starvation approaches. Only then will the family sacrifice their seed. Seeds can be generally grouped in three broad categories: open pollinated, hybrid, and transgenic. There is no outward difference in appearance to indicate the type of seed. The important differences are contained within the genetic makeup of each seed.

10.11.1   Open Pollinated Seed

Before the 1900s all seed came from open pollinated plants. Farmers around the world saved some of their harvested seed to use for planting the next season. The wealth of a farmer often rested on his ability to carefully select the best seed for next year's crop. By selection a farmer could increase yields, improve insect resistance, improve disease resistance, change seed color, and even seed size. This method of selection and saving seed has been used by humans since we first domesticated plants. Farmers who were better at selecting good seed, or were just lucky, sometimes sold seed to neighbors. Other times they jealously guarded their seed to prevent their neighbors from gaining an advantage.

10.11.2   Hybrid Seed

The development of hybrid maize started in 1920 in the United States. By 1943 most of the maize in Iowa was hybrid. A unique feature of a hybrid crop is that new seed must be purchased each year. Replanting the seed harvested from a hybrid will only give 50

Figure 10.12. Maize saved for seed in the Philippines.

percent hybrids. The other 50 percent will be like the lower yielding inbreds used to form the hybrid. Making hybrid seed was at first very time consuming. Each plant had to be hand pollinated, as in Figure 10.13. Later, it became possible to make hybrids using special planting methods and male-sterile lines. The advantages of a hybrid are better yields and often improved insect and disease resistance. A disadvantage is that new seed must be purchased each year. This has another, less obvious, disadvantage. There must be an existing market structure to deliver and sell the improved seed to farmers when they need it without loss of quality or quantity. If the seed is shipped by rail during the hot season just before rains start, there will be loss of germination due to the heat. There will be broken bags because of improper handling. Delays in shipment will make these losses worse. If the shipment arrives even a few days late, farmers will plant other seed that is available and the store will be burdened with seed it cannot sell. The use of hybrid seed requires an efficient distribution system.

Many vegetables are hybrids. Some of the most popular tomato varieties are hybrids requiring that new seed be purchased each planting. In addition, tomato seeds tend to have poor viability. This means that even when treated well, only 85 to 90 percent of the seeds in the package will germinate. If the seed has been subjected to heat, moisture, or kept over a year, the number of viable seeds will be less than 50 percent. Tomato farmers keep seeds in a cool place or a refrigerator to maintain viability. Some seeds are very difficult to produce, and this may make them more

Figure 10.13. Making hybrid maize.

expensive. It is not uncommon to see hybrid specialty tomato seed sold for $US0.20 each seed!

10.11.3 Transgenic Seed

Many farmers are now planting transgenic seeds such as the varieties of maize shown in Figure 10.14. Most of these crops are hybrids and due to the nature of hybrids, new seed must be purchased each year. Because the development of these transgenic crops is so costly, the companies make each purchaser sign an agreement promising not to replant the harvested seed and not to sell to others for replanting. This is a way for the company to recover development costs.

For hybrid crops, seed saving is not a problem. Replanting the seed from a hybrid will result in lower yields. However, some other crops are largely self-pollinated. This means that a variety will reproduce itself each year and new seed does not need to be purchased. This is the case with soybeans. Seed can be selected each year and replanted to produce the same variety as the original. There is no loss in yield. The only thing stopping most farmers from planting harvested transgenic seed is the producer agreement they signed when they purchased the original seed.

For farmers in industrialized countries, the legal penalties for breaking a producer agreement are severe enough that few take the risk of replanting transgenic seed. However, a notable exception occurred in Brazil when herbicide-tolerant soybeans were introduced there. Many farmers in the state of Rio Grande do Sur signed agreements with the vendor not to replant their harvested seed. In spite of this, many farmers did replant their harvested seed. In 2004 over 5 million hectares (30 percent of the total soybean area) were planted to transgenic soybean, but none had been purchased from the company responsible for developing the variety. The company withdrew from Brazil. This prevented Brazilian farmers from purchasing transgenic hybrid maize and reduced their competitive advantage in the world market. As a result of this confrontation, the Brazilian congress passed a biosafety law in March, 2003.[5] When it was signed by the president in March 2005, it provided a legal framework for companies to enforce their agreements and opened the way for sales of new transgenic varieties.

Figure 10.14. Transgenic maize variety trial.

Farmers in India purchase transgenic cotton seeds each year because the new varieties cost less to produce a higher crop of cotton. However, there has been a reaction among parts of farm society to actively preserve nontransgenic open pollinated seed. The Protection of Plant Varieties and Farmers' Rights Act passed in 2001 gives individual farmers legal protection for the distribution of registered varieties of crops such as rice and maize. This means that a farmer who has developed an outstanding variety can give it a unique name and sell it with the same legal protection as large multinational companies.

10.12 FERTILIZERS: CHEMICAL OR ORGANIC

Fertilizers are an essential factor for increasing and maintaining yields of any crop. Continually harvesting grain or forage from a field without replacing any of the nutrients carried off is referred to as "mining the soil." Ancient civilizations were able to survive for many years because they developed their agriculture in floodplains along rivers. The yearly floods brought in a new layer of fertile soil from eroded hills upstream. Without this yearly addition of nutrients the crop yields decline and the fields eventually must be abandoned.

The relationship between cereal grain yields and use of chemical fertilizers is clearly shown in Figure 10.15. The regions of the world with lowest grain yields per hectare (Oceania, former Soviet Union, sub-Saharan Africa) also have the lowest use of chemical fertilizers. Western Europe and East Asia have higher grain yields and higher rates of application of chemical fertilizers. Of course, application of fertilizer is not the only factor that causes grain yields to increase. Variety, weed and insect control, density of planting, and time of planting are only a few of the other factors that contribute to higher grain yields.

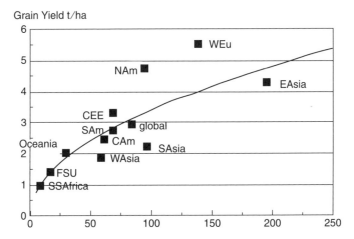

Figure 10.15. Relationship between grain yields and fertilizer use. (Chart courtesy of A. Krauss, International Potash Institute, Horgen, Switzerland.)

Studies have indicated that 40 to 60 percent of the yields for crops harvested in England and the United States are due primarily to fertilizer application.[6] In most areas of the tropics with a history of settled agriculture, the portion of grain yields due to fertilizers is probably similar. Areas of the tropics that are being cleared and put under cultivation for the first time are much more demanding. Fields cleared in the Amazon basin of Peru were cultivated for 15 years. Over 90 percent of the crop yields were attributed to chemical fertilizer application. In other words, without chemical fertilizers yields would have been reduced by 90 percent.[6]

In this chapter fertilizer is considered as anything added to the soil to increase crop production. Farmers in the United States frequently use the word *fertilizer* only for inorganic, or chemical, additives. Many other countries consider anything added to the soil as fertilizer.

For clarity we will use some more specific terms. Chemical fertilizers are compounds containing growth-promoting nutrients for plants that have been manufactured. Examples are urea, ammonium nitrate, and superphosphate. They may be applied as a gas (anhydrous ammonia), liquid (ammonium sulfate), or dry granules. Green manure is a term applied to green plants that are cut or plowed under before flowering for the purpose of adding both organic matter and nutrients to the soil for a crop to be planted. Examples are rye, crotalaria, and forage soybeans. Compost is a material made by decomposing organic materials. It is similar to green manure, but it has been decomposed, screened, and dried. Any plant or animal waste can be composted. Examples are sewage sludge, pig manure, and plant residues. Organic fertilizers may be either composted materials or certain chemical fertilizers. The concept is that they must occur in nature and do not come from factories. Examples are sodium nitrate (saltpeter), seaweed, and earthworm casts.

Dry chemical fertilizers are the most common form of nutrients applied to crops around the world. About 13 elements are considered essential for plant growth. The list of essential elements for plants in Table 10.3 shows that humans need more elements than plants. In fact, some of the elements, like arsenic, are poisons! The key factor is quantity. Like table salt, a very small quantity is beneficial, but more is deadly.

The nutrient element needed in largest quantities by plants is nitrogen used to form proteins, chlorophyll, and DNA. Most chemical nitrogen fertilizers are manufactured using natural gas. Their prices are dictated by the price of natural gas on the world market.

A hectare of maize producing 9500 kg of grain will typically require about 125 kg N/ha for adequate plant growth. However, plants cannot use the pure form of most elements. In the case of nitrogen, gaseous N_2 is inert and cannot be used by plants or humans as a nutrient. The most common form of nitrogen taken up by plants is either as ammonium (NH_4^+) or nitrate (NO_3^-). A hectare of maize producing 9500 kg of grain will typically require about 125 kg N/ha for adequate plant growth. If we want to apply 125 kg N/ha in the form of ammonium nitrate (34 percent N), we will need to apply 367 kg of ammonium nitrate fertilizer per hectare.

The second most important element for plant nutrition, in terms of quantity needed, is potassium. Potassium functions as a catalyst, permitting most plant enzymes to

TABLE 10.3. Essential Elements for Plants and Humans[a]

Essential for Plants	Essential for Humans
Nitrogen	Nitrogen
Potassium	Potassium
Calcium	Calcium
Magnesium	Magnesium
Phosphorus	Phosphorus
Sulfur	Sulfur
Manganese	Manganese
Iron	Iron
Chlorine	Chlorine
Zinc	Zinc
Copper	Copper
Molybdenum	Molybdenum
Boron	Sodium
	Fluorine
	Iodine
	Selenium
	Silicon
	Chromium
	Arsenic
	Vanadium
	Tin

[a] The order of the elements is arranged by decreasing quantities needed.

function. It also is needed to control the opening and closing of pores in leaves that regulate water movement in plants. Most of the potassium in maize plants is concentrated in the leaves and stalks. Potassium fertilizers are mostly composed of ground up potash rock. One of the largest potash rock mines in the world is in Saskatchewan, Canada.

One hectare of maize producing 9500 kg of grain will remove about 45 kg of potash (K_2O) per hectare. This assumes that only the grain is removed from the field, not the stalks and leaves. If the potassium is applied as pure potash rock (KCl equal to 60 percent K_2O), about 75 kg of potash rock must be applied per hectare.

The third most important element for plants is phosphorus. It is part of the ATP molecule (adenosine triphosphate) used to move energy around in plants and animals. In contrast to potassium, more of the phosphorus is concentrated in the grain than in the stalk and leaves. Phosphorus fertilizers are most commonly formed by treating the mineral, apatite, with acids to make the phosphorus more soluble. The largest apatite mines are located in Morocco.

A farmer harvesting one hectare of maize grain producing 9500 kg of grain will remove 60 kg of phosphorus. Phosphorus fertilizers are rated by the equivalent amount of soluble P_2O_5. A common phosphorus fertilizer is triple superphosphate containing about 45 percent P_2O_5. To replace the phosphorus removed in the grain, we will need to add 133 kg of phosphate fertilizer per hectare.

At this point, you begin to see that harvesting grain from one hectare removes nitrogen, potassium, and phosphorus from that hectare. If these nutrients are not replaced, next year's crop will yield less, and eventually the soil will become so poor that we will not be able to raise enough maize to replace the seed used in planting. Just to replace the nutrients carried off in the grain wagon, we must add 367 kg of ammonium nitrate, 75 kg of potash, and 133 kg of triple superphosphate. To simplify things, many farmers apply a fertilizer with these three elements combined. A fertilizer identified as 27–13–10 will contain the equivalent of 27 percent N, 13 percent P_2O_5, and 10 percent K_2O in the bag. The rest of the bag is sand or lime. To replace the 125 kg of nitrogen per hectare that our maize crop will remove, we will need to apply 463 kg of the mixed fertilizer. For 10 ha of maize we will need to buy 4.7 tons of mixed fertilizer, transport it to the field, and then spread it evenly on the soil. For the forage crop shown in Figure 10.16 it only takes a few minutes to drive a machine over the crop field, applying the correct ratio of nitrogen, phosphorus, and potassium fertilizers.

But what about all the other essential elements? Plants require these nutrients, but only in vary small amounts. For example, the amount of copper removed by 9500 kg of grain will only be 68 g. This is easily replaced by the normal weathering of minerals in the soil.

It would be misleading to assert that all farmers use chemical fertilizers. In reality, most of the nutrients used by crops come from sources other than chemical fertilizers. Chemical fertilizers supplement natural sources. As stated previously, natural soil-forming processes release enough of most nutrients (copper, zinc, etc.) that we do not normally have to worry about adding these to soils. For some soil types, natural soil-forming processes release enough potassium to supply plant needs, but most other soils require additions. Nitrogen and phosphorus are almost always limiting, and farmers must constantly work to maintain these nutrients at adequate levels.

There are several nonchemical sources of nitrogen and phosphorus commonly used in agriculture. Most farmers follow a 2- to 4-year rotation, repeating the same crop after 3 to 4 years. This means that one year a field will be planted to maize, the next year to

Figure 10.16. Dry chemical fertilizer spreader.

wheat, the next to clover, and so on Nearly all rotations include a legume crop. A legume is a type of plant that forms a symbiotic relationship with bacteria (*Rhizobium* spp.) that fix nitrogen from the air into a form usable by plants. A leguminous crop such as red clover or soybean will produce forage or grain and leave 100 to 150 kg N/ha in the soil after harvest.

Compost is another source of nitrogen as a fertilizer. In tropical climates composting is a rapid process, completing the cycle from raw plant material to finished compost in 6 to 8 weeks. It is enhanced by the use of earthworms. In cooler climates, like that of the Andes in Ecuador (Fig. 10.17), composting may last from 3 to 5 months. An advantage of compost is that the nutrients are less concentrated and there is less danger of damaging the crop. It also supplies organic matter that helps build up the soil and keep it from eroding. Most importantly for the low income farmer, no cash is needed to buy it.

A disadvantage of compost is that it has a relatively low nutrient content. If we take a typical municipal compost (made from yard clippings, leaves, sewage, etc.), we can expect to find about 1.5 percent nitrogen, 1.5 percent phosphate, and 1 percent potassium. Let's make the same calculations we made for the compound chemical fertilizer. To replace the 125 kg of nitrogen per hectare that our maize crop will remove, we will need to apply 8330 kg of compost on each hectare. Unfortunately, this will give twice as much phosphorus as the maize crop needs, and we risk contaminating groundwater with the excess phosphorus. The other disadvantage will be to transport 83 tons of compost and spread it on our 10 ha field. This assumes we can afford to purchase 83 tons of good-quality compost. This is one of the reasons most commercial farmers use chemical fertilizers. The cost of transport and application can become very expensive.

It is possible to reduce the amount of applied fertilizer and maintain, or improve, the nutrient status of soil. As previously mentioned, leguminous crops fix nitrogen, and

Figure 10.17. Compost pile for a greenhouse in Ecuador.

this is released when the crop dies. Some forage crops produce a large amount of green matter that serves as a green manure. These nutrients do not require transport or application. They are already in the soil where the crop is to be planted. The disadvantage is that an entire growing season is needed to form this extra reserve of nutrients.

10.13   WEED, INSECT, AND DISEASE CONTROL

Weeds, insects, and diseases are three dangers that can rob a farmer of all the work and investment needed to plant a crop. Typically, soil preparation and planting require about 60 percent of the total investment in a crop. All of this can be completely lost by poor weed management or by unkind winds bringing insects or a disease to the field. There are three types of response a farmer may have to these threats: do nothing, preventive treatments, or curative treatments.

10.13.1   Do Nothing

This may seem foolish at first glance but for most insect and disease problems there is an "economic threshold." There are always insects and diseases present in any given field. It is impractical, and sometimes even unwise, to attempt to eradicate the problem insects or disease. For example, any attempt to control the European corn borer will involve some cost and/or labor. If chemical insecticides are used, the chemicals must be purchased and applied to the field. If traditional methods are used, they still involve the time to prepare the treatment and to apply it. With very light infestations (below the economic threshold), the loss of yield will be less than the cost of chemicals to prevent the loss. It is more economical not to control a few borers than to treat the entire field.

Chemical treatments applied to leaves also have the disadvantage of being indiscriminate: They kill beneficial and damaging insects. Many insects in a field are predators on damaging insects. When the predators are killed, other insects they had controlled will multiply without checks and may cause other problems. So a farmer may be economically better off not to control a light infestation of a damaging insect with chemical insecticides.

Treatment of diseases is handled in a manner similar to that for insects. Diseases are very dependent on a source of spores, wind to spread the spores, and humidity for the spores to infect susceptible plants. If these conditions are not met, very little disease will develop. There are normally spores, or sources of infection, of many diseases present in a field. A disease becomes economically important when the weather conditions are right for its spread and multiplication. As with insects, farmers should only worry about most diseases when the economic threshold of infection is reached.

Doing nothing for a weed problem is normally not a solution. Once weeds have become established they will continue to grow and reduce crop yields. Weeds become less of a problem after the crop has grown large enough so that the leaves completely shade the soil. Once the leaf canopy has closed over weeds, vital sunlight is reduced and most weeds slowly die. In the critical stage between germination and canopy closure, weeds often get ahead of the crop. For situations like the one shown in Figure 10.18, the only solution is to kill the weeds.

10.13.2 Preventive Treatments

From past experience a farmer will know if there is a good chance of insects or diseases attacking the crop. Some insects are endemic to an entire region as are soil-borne diseases. For these it is not a question of waiting to see if a problem will develop: it is certain. In this case it is better to prevent the attack than to wait until the insect or disease has built up in the field. There are several preventive tactics a farmer can use to diminish yield losses from insects and diseases.

Reduce Source of Spores or Insects. Prevention is always cheaper than trying to cure a problem. Knowing the life cycle of a disease or insect can help to reduce its effect on a crop. For example, the wild barberry bush is often infected by wheat rust. The disease does not kill the barberry but remains dormant over the winter in the northern United States. In spring spores develop and are carried by winds to wheat plants in fields. Soon many wheat plants are producing more spores to infect other plants in the field and an epidemic can destroy the crop. Removing wild barberry plants from areas near fields is one way to reduce the severity of rust attacks in wheat fields. This has been practiced since the 1600s in Europe. Starting in 1918 in the United States a program was put into place to eradicate wild barberry as a means of reducing wheat rust. This was effective in preventing early infections of wheat fields in the northern United States but did not completely eliminate wheat rust. Enough spores were blown north

Figure 10.18. Weeds competing with maize.

from infected fields in the southern United States to start the next cycle of rust infections in the north.

Another method commonly used to reduce disease and insect problems is to destroy the stems and leaves from the previous crops. The preferred way is to cover these residues with soil (plowing) so that decomposition will destroy the eggs and spores. In some areas of the world burning is still practiced. This is effective but causes loss of soil organic matter and nutrients as well as increasing erosion. In spite of these disadvantages, it is still practiced in parts of the Sahel, Asia, and the United States.

A common way to reduce insect and disease problems is by crop rotation. If corn borers are a problem, planting soybeans for 2 years will deprive them of food and reduce the population to manageable levels. Nurseries with nematode problems (a microscopic worm in the soil) will often plant marigolds to cause the nematode cysts to hatch. Since nematodes cannot feed on marigolds they will die by starvation and the total population in a field is dramatically reduced.

Use Varieties of Crop Resistant to Insect or Disease. Using a variety of the crop that is resistant to the insect or disease is an extremely cost-effective way to manage a problem. Normally, the seed costs the same or slightly more than varieties that are not resistant. Just planting a different seed solves the problem. This has been the most successful approach to control diseases. Improved wheat varieties sold around the world have resistance to wheat rust. Browsing through any sales brochure for varieties of a crop reveals a long list of diseases for which each variety is more or less resistant. Because diseases mutate, new resistant varieties must be constantly developed to maintain production.

Development of insect resistance has been less successful using conventional breeding techniques. With the release of genetically modified organism (GMO) crops, insect resistance has become something farmers can expect. The release of cotton varieties in China that make their own insecticide (Bt trait) has dramatically reduced use of insecticides and costs of production. The farmers only need to buy a different seed and their boll weevil problems are greatly diminished but not eliminated. A problem with insect-resistant crops is that 20 percent of the field must be planted to non-GMO crop plants as a "refuge." This prevents a buildup of Bt-resistant insects by giving them something to eat and survive. So there is no magic bullet to prevent insect damage to crops. The best solution to date is to reduce damage to minimal levels.

Apply Preventive Treatments. Certain insecticides are applied to seeds before planting. These work only against insects in the soil that eat the roots of the seedling. The corn rootworm is an example. These insects will eat away much of the root system of a young maize plant. If the seed is treated with insecticide, the worm will only eat enough of the insecticide to kill it. If an insect does not eat the roots of the seedling, there is no effect. Seed treatment is an efficient way to prevent damage to young plants but does not protect mature plants.

Applying chemicals to prevent a disease is possible but may be expensive. One chemical that has been used for many years to control disease is Bordeaux mixture. This was developed in France in the 1860s to control leaf diseases in vineyards. It is still a valuable disease control measure today. Bordeaux mixture is a simple mixture of copper sulfate, hydrated lime, and water.[7] The mixture is affordable and effective for many types of diseases infecting leaves through spores. It will not cure a disease within a leaf or stem.

10.13.3 Curative Treatments

Once an insect or disease has established itself in a field and the threat for economic damage is clear, the farmer must attempt to limit the damage. At this stage the farmer almost always resorts to chemicals to kill the insects or disease.

Insecticides can be divided into three groups based on the feeding habits of the insects to be controlled. If an insect eats the leaf of the plant (most caterpillars), then a poison can be applied to the surface of the leaf. When the insect eats the leaf, it will also eat the poison and be killed. If the insect damages the plant by sucking the sap from the leaves or stems (aphids), a systemic insecticide must be applied. These penetrate the plant and circulate in the sap within the stems and leaves. A few insecticides depend on direct contact with the insect.

In all cases, use of chemical insecticides is a last resort. Continuous use of chemicals to control an insect problem often results in a buildup of insecticide-resistant insect populations. A small valley in the Andes of Ecuador had a problem with whitefly on tomatoes. Farmers had used various chemicals to kill this sucking insect, but after a few years it seemed the insects came back even sooner after the spraying. Then the farmers tried mixing two, three, even four insecticides together in a formula they called *la bomba atómica* (shown in Fig. 10.19) in an attempt to finally eradicate the whitefly. Of course, the insects finally evolved resistance to this mixture and the problem was worse than before.

There are few chemicals available to cure bacterial or fungal diseases after they have infected the plant. With diseases caused by a virus, the only remedy is destruction of the infected plants. This usually consists of pulling out the diseased plants, gathering them together, and burning the remains.

10.14 ELECTRICITY

While crops and animals do not directly use electricity, their production often depends on a dependable supply of electric current. The warmth provided by an electric lightbulb shown in Figure 10.20 is enough to keep young chicks warm while their feathers develop. Without electricity the farmer would need to use a kerosene heat source that is a fire danger and could possibly asphyxiate the birds.

Electricity is used in the farmer's home to provide light at night and in the farm buildings to feed animals after dark. In Ecuador the sun sets between 6:15 and 6:30 p.m. (it is on the equator) and nearly 12 hours of each day are dark. Electric lights help lengthen the day to finish chores. In the United States we are much more dependent on electricity. Farmers use electric pumps for watering livestock, to charge batteries, to power milking machines, repair machinery, and to make their

Figure 10.19. Applying curative insecticide in Ecuador.

Figure 10.20. Electricity used to keep young chicks warm.

homes more comfortable. In irrigated areas electric pumps are used to pump water from deep wells or reservoirs. In India approximately 20 percent of the nation's electricity is used for farm irrigation.[8] The cost of electricity is significant, even when the government subsidizes agricultural users.

In the United States we take electricity for granted. It is rare to lose electric power for more than a few minutes. In developing countries electricity is highly valued because it is often difficult to get a reliable connection. During the hot season in India (April and May) the grid is overloaded because of added fans and air conditioners. The power stations respond by rationing electricity, cutting service to areas for 5 to 6 hours each day. Many other national systems use hydroelectric generators that must go on reduced schedules in dry seasons when irrigation is most needed. Keeping medicines and vaccines for livestock is a problem without reliable refrigeration. Provision of a reliable electrical supply is essential for modern farming.

10.15 WATER: IRRIGATION

Irrigated lands are often the most productive. It has been estimated that the practice of irrigation results in a 100 to 400 percent increase in crop yields.[9] An estimated 15 percent of the cultivated land in the world is irrigated. This relatively small area produces 35 to 40 percent[10] of the world's food! The proper use of irrigation greatly decreases risks of crop failure from drought, increasing food security in a country. Many national plans for increasing food production include expansion of irrigated areas.

Irrigation is not a new technology. It was practiced by Egyptians as early as 5000 BC. Most societies have used irrigation to improve crop yields or to prevent losses from drought.

The quality of irrigation water depends largely on its source. Water taken from lakes or rivers will vary depending on the conditions in the watershed. A watershed, or catchment area, is all of the land area contributing water to a river or lake. If the land is forested with little human activity, the water is likely to be of good quality for irrigation. If the land is cultivated farmland, there is a chance of irrigation water carrying eroded soil with nutrients, chemicals, and diseases from the watershed. If the watershed is an urban area, there will be runoff from yards, streets, and industries carrying pollutants detrimental to growing plants.

Water taken from wells may seem to be pure and ideal for irrigation. In most cases, well water is biologically pure. A serious problem for irrigation is the accumulation of salts such as sodium, calcium, magnesium, and potassium. These come from rocks and soils as part of the natural breakdown process of weathering and soil formation. Other toxic elements such as boron or arsenic may also be present. When the concentration of salts in irrigation water is high, use of this water may result in their accumulation in the soil eventually causing a dramatic reduction in yields. The field shown in Figure 10.21 has a serious salt accumulation on the tops of the ridges between furrows. It is estimated that half of the irrigated fields in the world suffer from some degree of salt accumulation.

Figure 10.21. Field showing salt accumulation.

There are three types of irrigation systems found in nearly all parts of the world: surface, sprinkler, and drip. The most common is Surface irrigation. Examples are rice paddies. Sprinkler is normally an expensive type of system and most commonly found on high value crops in industrialized countries. Drip irrigation is the most efficient type of system used for high value crops. Each type of irrigation system can use water from any source.

10.15.1 Surface Irrigation

Surface irrigation includes irrigation by furrows, one of the most common irrigation methods. The basic system is one of using smaller and smaller canals until the water ends up in a small furrow between plants in the field. An advantage of furrow irrigation is that it is the cheapest to install, but it is also one of the most difficult to install properly. The most important requirement is that each section of canal and furrow must have a uniform slope. As with rice paddies, land leveling is essential. Even in rice paddies the field must have some slope to permit drainage in preparation for harvest. Fields irrigated by furrows must have slightly more slope to keep the water flowing. Even though the furrows in Figure 10.22 appear to be flat, they actually have a slight slope toward the bottom of the figure. This permits the farmer to fill the furrow with water quickly and move on to the next. The result is a uniform application of water and all plants in the field receive the right amount. If the furrows are not uniformly sloped, there will be wet and dry spots with overall losses in yields.

A disadvantage of surface irrigation is that the farmer must be present to manage the water. If too much water is let into the furrows, it will build up and spill over into other fields, causing erosion and loss of valuable water. If not enough is put on the field, another irrigation will be needed too soon.

10.15.2 Sprinkler Irrigation

Sprinkler irrigation consists of spraying water into the air to simulate natural rainfall. When properly done, it will distribute a uniform amount of water over a crop with

Figure 10.22. Irrigation furrows.

little over- or underapplication. There is no need to level a field. Some sprinkler systems are used on orchards in hilly fields where furrow irrigation would be extremely difficult. In vegetable crops, fertilizers may be mixed with the irrigation water to supply nutrients with the water. In general, sprinkler irrigation requires a bit less attention in the field than furrow irrigation. A farmer will set a section of pipe with sprinklers and let it operate for a few hours before moving it to the next location in the field. In fields with a sandy soil, furrow irrigation results in too much loss of water and sprinkler irrigation must be used.

A disadvantage of sprinkler irrigation is the equipment needed to install a system. It requires a high-volume, high-pressure pump, long sections of easily disassembled aluminum pipe, and special nozzles to uniformly apply water to the crop. Some very large commercial farmers use a type of sprinkler system called center pivot. In these systems the field is circular and the irrigation pipe is mounted on large wheels. The entire assembly turns around a well and pump located in the center of the field. These systems need even less attention than the sprinkler system shown in Figure 10.23. In very dry areas, the loss of water by evaporation may be significant. Sprinkler irrigation may not be recommended for some crops in certain areas since it wets the leaves, creating an ideal environment for growth of fungal diseases.

10.15.3 Drip Irrigation

The system of drip irrigation is a relatively recent invention, being invented in 1965 in Israel. The basic system consists of small tubes carrying water to evenly spaced "emitters," or drippers. These emitters release water drop by drop to the base of each plant. The result is a very steady application of just enough water to each plant to permit maximum growth with no excess of water.

Drip irrigation systems are normally used on high value vegetable and fruit crops such as tomatoes or peaches. An advantage of drip irrigation is that crops yield more using less water than other irrigation methods. This is especially important in dry

Figure 10.23. Sprinkler irrigation in Mexico.

areas where the cost of irrigation water is very high. Even in regions where water cost is not high, use of drip irrigation gives higher, and more consistent, yields.

Basic drip irrigation systems are relatively simple and inexpensive. Some farmers make their own systems from materials found in local stores. An advantage of drip irrigation is the low volume of water used, better yields, less disease caused by high humidity, and the application of liquid fertilizers with the water. Once the system is installed, it needs minimal attention. In the greenhouse tomatoes shown in Figure 10.24 daily harvest does not need to be stopped because of wet soil. The system may be left on for several days or turned on only as needed by the crop.

Disadvantages of drip irrigation are the expense of the pump, water lines, and individual plant emitters. Installation may be tedious and time consuming. Individual emitters may clog, stopping water flow to a plant. When an emitter becomes clogged, the only solution is usually to replace the entire assembly. This can become expensive on large systems. As a preventive measure, most drip irrigation systems have elaborate filters to prevent particles from entering the irrigation lines.

10.16   VACCINES AND MEDICINES

A fact of life is that all animals get sick, just like people. The most damaging illnesses are due to bacterial infections, parasites, and insects. Sometimes the illness is so brief that the farmer doesn't notice it. Occasionally a sudden illness will kill an animal over-night. Some illnesses or infections are mild and long term, reducing weight gain or milk production. The majority of animals that pass through a farm are there for less than 1 year before they are sold. From the farmer's point of view the animal is an investment that needs to be protected until sold.

Treatment of animals can be classed as preventive or curative. Although preven-tion is less costly than curing a problem, it is a fact that most small farmers in the

Figure 10.24. Drip irrigation of tomatoes in Ecuador.

world wait until they see an obvious problem before treating it. This almost always results in a more costly treatment than prevention.

Preventive treatments are designed to reduce disease losses in animals. Some preventive treatments are very inexpensive and easily adopted by small farmers. An example is treatment for foot rot in sheep and goats. Some animals seem to be more susceptible than others. When they are infected, they spread the infection to others in the same flock. Effective prevention consists of trimming their hooves to remove infected material, running them through a medicated dip, as in Figure 10.25, and finally keeping their pens clean and dry.

Preventive vaccination has been extremely useful in the control of some diseases. A vaccine for foot-and-mouth disease is available to prevent this viral disease. Other animal diseases caused by bacteria that can be easily prevented by vaccines include pneumonic pasteurellosis and infectious bovine rhinotracheitis in cattle.

Some preventive treatments have little direct benefit to the farmer but may have significant benefits for the consumer. One example of vaccination benefiting consumers is the reduction of salmonella-contaminated eggs in England. In 1996 government testing showed that one box in every 100 boxes of eggs had salmonella contamination. After 8 years of voluntary producer vaccination of poultry against salmonella, another survey in 2004 found only one box in every 290 boxes of eggs

Figure 10.25. Sheep in foot dip to prevent foot rot.

was contaminated, a 60 percent reduction.[11] Nationwide, salmonella poisoning cases had also declined by 50 percent in the same period, possibly due to lower contamination in eggs.

10.17   CREDIT

Few people think of credit as a raw material of agriculture. For Octavio in Ecuador a loan would be a risk he is unwilling to take. He would have to promise his land as collateral. If anything happened to the crop or animals, he could lose everything. For Donio in the Philippines, credit from a bank is impossible since he has no collateral. He has a form of credit from the landowners whose fields he farms. They give him use of their land until harvest when he is expected to pay with part of the harvest. Steve in the United States must use credit. All of his raw materials such as fertilizer and seed are purchased. Over 77 percent of the cost of a crop is incurred before planting and shortly thereafter.[12] There is no return on this investment for 3 to 4 months.

Without credit, modern commercial agriculture would be very difficult. Few farmers keep enough cash on hand to pay for their investment in a crop or animal enterprise. In fact, it is often more profitable to borrow money and pay interest in order to gain the high rate of return on a profitable farm operation.

10.18   LAW OF DIMINISHING RETURNS

One of the problems with inputs faced by every farmer is how much to use. If 10 kg of fertilizer gives an extra 200 kg of maize, then won't 20 kg of fertilizer give an extra 400 kg of grain?

Unfortunately, farmers are faced with the law of diminishing returns. This law simply states that with each additional unit of an input, the resulting additional output (or yield) decreases. This is controlled by nature. The biology of the crop or animal dictates this response. In the case shown in Figure 10.26, the first 18 kg of fertilizer applied to 1 ha gives a yield increase of 246 kg of soybeans per hectare. The next 18 kg of fertilizer applied to the same 1 ha field gives an additional 345 kg of soybeans. The next 18 kg of fertilizer applied to the same hectare gives an additional 394 kg of soybeans. At this point each additional amount of fertilizer gives ever-increasing amounts of soybeans. In other words the rate of increase is positive. However, the next (fourth) 18-kg addition of fertilizer only gives an additional 246 kg of soybeans. The fifth addition of 18 kg of fertilizer only increases total yield by 148 kg of soybeans. The sixth and seventh 18-kg additions of fertilizer actually decrease rather than increase yield. The highest yield occurs with the fifth addition of fertilizer, or 88 kg/ha.

Does this mean the farmer should apply 88 kg of fertilizer per hectare? Probably not. The fifth addition of fertilizer only increased yield by 148 kg per hectare. If the fertilizer cost US$5.50 each kilogram and the value of soybeans was only US$6.00 per kilogram, we would advise the farmer not to apply this last increment. The extra income of US$0.50 per hectare would not be enough to pay for the cost of application. A more conservative strategy would be to apply only 70 kg fertilizer per hectare. This still gives a profit with enough "cushion" to protect the farmer from bad weather, disease, or last minute changes in market prices.

This does not mean that we are forever doomed to stay below the maximum yield point for our crops. This example only describes the response to fertilizer. By changing the variety, using a different planting date, treating seed with fungicide, and other factors, the farmer will be able to reach the optimum production level for other inputs. There are easily 20 to 25 inputs that a farmer can vary to change

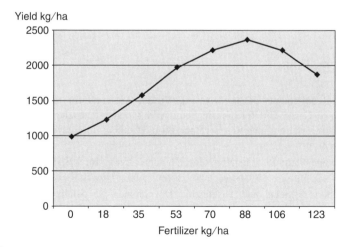

Figure 10.26. Diminishing rate of return on fertilizer applied to soybeans.

yields, each one having a unique curve of diminishing returns. To make matters more complicated, these responses interact. A small change in one input may cause a much larger response to another input. This is commonly seen where the application of phosphate fertilizers cause a much greater response to application of nitrogen fertilizers.

10.19   CONCLUSIONS

Agriculture requires a large number of raw materials to produce a profitable crop. All must be present in adequate amounts, not too much nor too little. Some are under the farmers control while others (like rain and sunlight) cannot be controlled. Other factors such as roads are very costly. National or regional governments frequently provide funds to construct roads that farmers can use to transport produce. The provisions of these raw materials at the proper times results in a productive agriculture.

QUESTIONS

1. Why are roads so important to farmers?
2. What is the difference between hybrid and open pollinated maize varieties?
3. What is an "organic" fertilizer?
4. What is the source of phosphate fertilizer?
5. What is the source of most nitrogen fertilizers?
6. Name three types of irrigation. What are their advantages and disadvantages?
7. Why would a farmer do nothing to control a damaging insect?
8. What is the law of diminishing returns?
9. How does land tenure affect agricultural production?
10. How is electricity used on a farm?

REFERENCES

1. Climatic Requirements of Some Crops, Virtual Academy for the Semi-Arid Tropics, International Center for Research in Semi Arid Tropics, Patancheru, India. Available at: http://www.vasat.org/learning/weather/crop_weather/climatic_requirement/climatic_requirement.htm.
2. U.S. Average Farm Real Estate Value, National Agricultural Statistics Service, USDA. Available at: http://www.nass.usda.gov/Charts_and_Maps/index.asp.
3. Land Tenure Systems and Sustainable Development in Southern Africa, ECA/SA/ EGM.Land/2003/2, Economic Commission for Africa, Southern Africa Office, Lusaka, Zambia, FAO, December 2003.
4. Special Alert No. 307, FAO Global Information and Early Warning System on Food and Agriculture, FAO, Rome, April, 2000.

5. J. M. F. J. Silveira and I. C. Borges, *An Overview of the Current State of Agricultural Biotechnology in Brazil*, Institute of Economics, University of Campinas, Brazil, April, 2005.

6. W. M. Stewart, D. W. Dibb, A. E. Johnston, and T. J. Smyth, The Contribution of Commercial Fertilizer Nutrients to Food Production, *Agron. J.*, **97**, 1–6, 2005.

7. B. Ohlendorf, *Bordeaux Mixture*, Pest Notes Publication 7481, University of California, Berkeley, November, 2000.

8. M. Gregory, India Struggles with Power Theft, BBC News, March, 2006.

9. Water at a Glance, FAO. Available at: http://www.fao.org/AG/agl/aglw/WaterTour/index_en.htm.

10. Frequently Asked Questions about Salinity, Agricultural Research Service, USDA. Available at: http://www.ars.usda.gov/Aboutus/docs.htm?docid = 10201&pf = 1&cg_id = 0.

11. Salmonella in Eggs Down, Survey Shows, Food Standards Agency, UK Government, March 2004. Available at: http://www.food.gov.uk/news/newsarchive/2004/mar/salmonellaeggnews.

12. M. Duffy and D. Smith, Iowa Crop Production Cost Budgets, 2006. Available at: http://www.extension.iastate.edu/AgDM/crops/html/al-20.html.

BIBLIOGRAPHY

J. Davison, Evaluating the Potential of an Alternative Crop, Cooperative Extension Fact Sheet 02-54, University of Nevada, Reno. Available at: http://www.unce.unr.edu/publications/FS02/FS0254.pdf.

Gene Savers—India. Available at: http://tve.org/ho/doc.cfm?aid = 891.

Global Precipitation Time Series, Climate Prediction Center, National Weather Service. Available at: http://www.cpc.ncep.noaa.gov/products/global_monitoring/precipitation/global_ precip_accum.shtml.

11

INCREASING FOOD SUPPLIES

To obtain the latest information about food and food security, the authors suggest consulting the following organization websites:

CGIAR (Consultative Group on International Agricultural Research) http://www.cgiar.org/

IFPRI (International Food Policy Research Institute) http://www.ifpri.org/

When considering the state of world crop and animal production, we sometimes forget why we should be concerned with global agriculture. The obvious answer is that all humans must eat. If someone does not have enough to eat, why cannot we just send some of our surplus food? But what if we do not have surplus food? Human history is full of stories of widespread death by starvation. Highly variable food supplies have been the norm for traditional agriculture throughout most of human history. Can we change our own history? We begin with a quick look at that history.

World Food: Production and Use. By Alfred R. Conklin, Jr. and Thomas Stilwell
Copyright © 2007 John Wiley & Sons, Inc.

11.1   MALTHUS: SCAREMONGER OR PROPHET?

One of the best known scholarly publications about food supply and population was written by Thomas Malthus in 1798.[1] At the time there was a debate about the general nature of humans and how we would survive in the future. He responded to speculations by his colleagues with convincing arguments using statistics and mathematics to show that while food supply could optimistically be assumed to increase arithmetically, population increases geometrically: "Population, when unchecked, increases in a geometrical ratio. Subsistence increases only in an arithmetical ratio." (Malthus, Chapter 1, p. 4). He presented a numerical example summarized in Table 11.1.

Under conditions where food was not limiting, the population doubled every 25 years, as happened in the United States during the last part of the 1700s. Even with the most optimistic views of agricultural production, it was highly improbable that food supplies could be increased by more than a fixed amount during the same 25 years. The result of this simple calculation was that millions of people would be without food within a hundred years. Even after taking into account diseases, wars, and other causes of mortality, his predictions still forecast a time when the number of humans would be greater than the food supply could support.

11.2   FAMINES

Malthus predicted that disastrous famines would take place if population were not controlled. The human race has a long history of famines, and we should not lightly ignore this history. Famine is generally defined as "a prolonged food shortage that causes widespread hunger and death."[2] A regional famine may affect only certain population groups in a region, normally, the poorest. In a general famine all classes of a society in an affected region suffer. Famines are also classified by their cause: natural or human. Natural causes include drought, flooding, cold, plant diseases, and crop damage by insects. Human causes include hoarding, military blockades, governmental interference, and price differentials.

TABLE 11.1.   Arithmetic and Geometric Increases Predicted by Malthus

Year	Subsistence	Population in Thousands	Food Available for Thousands	Food Not Available for Thousands
1800	100	7,000	7,000	0
1825	200	14,000	14,000	0
1850	300	28,000	21,000	7,000
1875	400	56,000	28,000	28,000
1900	500	112,000	35,000	77,000
1925	600	224,000	42,000	182,000
1950	700	448,000	49,000	399,000
1975	800	896,000	56,000	840,000
2000	900	1,792,000	63,000	1,729,000

11.2.1 Great Famine of China

The worst famine on record was the great famine of China, which occurred from 1959 to 1961.[3] It is estimated that up to 40 million people died as a result of bad agricultural practices forced on farmers. In 1955 Mao Tse-tung ordered the formation of collectives of 100 to 300 families each. The following year grain production fell by 40 percent.

Attempting to recover the loss, the Chinese government ordered these collectives to adopt several unproven crop production practices. They ordered planting four to five times more seeds per hectare than normal in the belief that there would be four to five times greater yield. Instead, the close planting caused overpopulation and very few seedlings survived, resulting in even less grain yield.

Believing that birds ate a large amount of grain, they ordered the destruction of birds such as the English sparrow. So many sparrows were killed that insect populations exploded, seriously damaging the remaining crops in the field.

Some officials prohibited the use of chemical fertilizers in the belief that they poisoned the soil. This also reduced yields. Commune members were ordered to leave up to 30 percent of their fields in fallow (without crops) with the aim of recovering the natural soil fertility. The result was that around 20 percent of the cropland was not planted, further reducing national grain production.

Officials who reported drops in grain production were fired. As a result most officials reported large increases in grain production, and the central government continued to export grain until 1958. Villages unable to send grain for sale were accused of hoarding, and the army slaughtered many in retaliation. Finally in 1961 provincial governors started to abandon the practices ordered by the central government, and farmers were able to produce enough grain to end the famine.

Estimates of the number of deaths range from 30 to 40 million people. Demographers using modern data estimate that as many as 25 percent of the dead were young girls due to their lower social status in Chinese society.

11.2.2 Great Famine of 1315–1317

Some famines are caused solely by nature. An example of such a famine was the great famine of 1315–1317 in Europe.[4] Small, local famines were relatively common in Medieval Europe, but the great famine affected large areas and populations. The Medieval warm period saw increases in population, but declining yields per hectare of grains, possibly due to declining soil fertility. Starting in 1315, there were cold wet summers followed by severe winters. Grain could not ripen in the wet weather, and livestock had no dry hay to eat in the cold winters. Even the king of England was affected. This pattern persisted until 1317 when normal weather returned. Several million deaths were caused.

11.2.3 Irish Potato Famine

More often, famines are caused by a combination of human and natural events. An example is The Irish potato famine.[5] Britain prohibited Irish Catholics from owning land so that most were landless laborers working for absentee British landlords. In the summer of 1845, potatoes started to rot after being dug. The cause was a fungus disease that affected all potatoes being grown in Ireland. Potatoes were the staple of Irish diets, being eaten three times a day. Soon people had no food. The result was widespread starvation and mass emigration to North America. In 1846 Britain permitted free sale of grain to the Irish, but few people had enough cash to buy it. In 1847, Britain set up feeding programs and emergency relief programs, but when a banking crisis hit London, these were suspended. By 1855 an estimated 750,000 people died by direct and indirect effects of the famine. Another 2 million emigrated to North America.

11.2.4 Modern-Day Famines

Modern-day famines still occur but are not widespread. This is due largely to the efforts of wealthy nations to provide food and to the existence of transportation systems able to move food from one region to another. There are still limits to the degree we can ameliorate natural disasters. The food shipments to north India in the 1960s were insufficient largely due to the limitations of the port facilities for unloading grain.

11.3 EFFECTS OF FAMINE RELIEF

When people are starving, it is natural to wish to relieve the suffering by providing food. It is hard to argue that this is a bad thing, and yet it can lead to dislocation of local food-producing systems and sometimes can worsen the overall food situation in a region. A secondary problem is lack of knowledge of use of food being supplied.

Supplying grain to a local area will keep people from starving but will also depress local markets. Why buy food when it is being given away free? In such a situation there is little incentive for farmers to plant, let alone harvest. Thus famine relief should continue for the shortest period of time possible and no longer than when a new crop can be produced. The local population must be made aware that the supply of free food will end during the harvest of crops. However, it is hard to make people act on the basis of a threat that free food will end.

Two examples of the sorts of problems associated with free food are cheese production in Costa Rica and grain in the Sahel. Costa Rica has a vibrant cheese-producing sector that can provide all the cheese needed by the country. However, the United States shipped excess cheese to Costa Rica in the 1980s depressing the market and stemming the development of cheese production in the country.

During a famine in Tanzania, bulgur wheat was shipped to starving people. The problem was that the local population did not know how to cook bulgur wheat, and so its usefulness in stemming starvation was limited.

11.4 GREEN REVOLUTION

Traditionally, Nobel prizes are granted to outstanding individuals working in physics, chemistry, medicine, literature, economics, and peace but not agriculture. However, there has been one Nobel Prize recipient, Dr. Norman Borlaug, who is an agriculturalist. Dr. Borlaug received the Nobel Prize for Peace in 1970 for his contribution to the so-called Green Revolution. We will examine that Green Revolution briefly.

Prior to the 1800s increases in food production were brought about by expanding the cultivated area. Agricultural production was managed by farmers with only minor intervention by other trades. Starting in the mid-1800s Justus von Leibig laid down a basic theoretical foundation for plant nutrition, and Jean-Baptiste Boussingault developed the basis for scientific crop production. Superphosphate fertilizer had been produced in England, and Chilean nitrates had started to arrive in Europe and North America. This led to the development of other chemical fertilizers. The rediscovery of Gregor Mendel's work on genetics started the field of crop improvement. The combination of these developments in the early 1900s started European and North American crop yields on a steady increase that has not yet stopped.

In spite of these breakthroughs benefiting temperate region farmers, tropical regions were nearly untouched. Crop and animal production remained close to subsistence levels in spite of rapidly increasing populations. Part of the reason for this difference was that specific varieties and fertilizer practices were not suitable for tropical areas.

Starting in the 1960s scientific production techniques were developed for wheat and rice in tropical areas. The results of these new techniques can be seen in Figure 11.1. Since 1961 there have been steady and significant increases in the production of wheat and rice, two staple foods of India.[6]

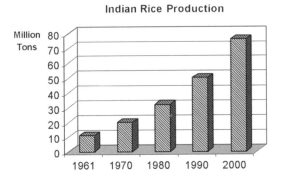

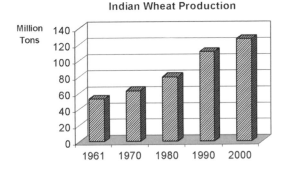

Figure 11.1. Indian rice and wheat production since 1961.

What happened in India to allow these increases to take place? Although many factors affected this increase, three are notable: irrigation, fertilizer, and tractors. The trends for each of these factors can be easily seen in Figure 11.2. There were significant increases in all three factors during the years 1961–2000 corresponding to the increases in wheat and rice production.

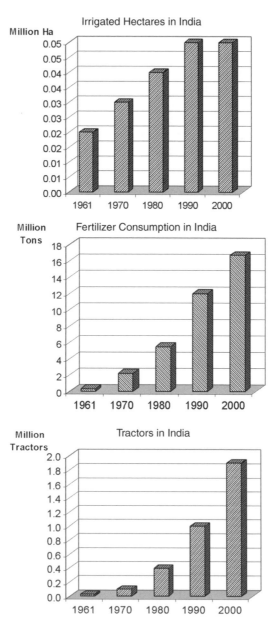

Figure 11.2. Irrigated hectares, fertilizer use, and tractor use in India since 1961.

There are many other characteristics of the Green Revolution that could be pointed out as its salient character: new high-yielding, stiff-straw rice varieties, varieties responsive to fertilization, a suite of improved cultural practices, use of pesticides and insecticides, and improved harvest and postharvest handling of rice grain. In reality, it is better viewed as either a mindset, in terms of developing crops for increased productivity, or as a package of practices that when applied to rice production leads to dramatically increased yields. The former idea eventually led to the development of varieties and packages of practices that resulted in increased yields in grains other than rice, particularly maize and wheat, but also to a lesser extent other grains and even some nongrain foods.

Intensive research is continuing in the development of improved rice varieties. Two directions in this development are hybrid rice and transgenic rice. In terms of current rice production most effort today seems to be toward the development and dispersal of hybrid rice.

The concepts developed during the Green Revolution can be applied to any crop and will result in increased yields. Many of the ideas are inexpensive to research and put into practice. For instance, one of the big pushes of the Green Revolution was determining the proper spacing of rice plants to maximize grain production. To this end extensive spacing experiments were done throughout Asia. Eventually, the best spacing for different varieties of rice was determined and recommended to farmers.

Many of the activities characteristic of the Green Revolution can be carried out in any location by any group with the result of increased food production. Spacing experiments can be done with any crop, with little expense and with little loss of yield. They can be repeated in every locality by any interested person including farmers themselves. Similarly amendments of compost, manure, compost–manure mixtures, or commercial fertilizers can be applied at different rates to determine the best application method and rate. These basic concepts can be applied to all areas of food production.

Extensive on-farm experiment programs have been developed in Latin America and parts of Asia. Each field may only have 4 to 5 plots demonstrating contrasting practices of levels of fertilizer. When the yield data from 40 to 50 locations is combined for several years, concrete predictions can easily be made for profitability and risk of losses for an entire region. New practices can then be recommended to farmers knowing the probability of success.

Getting farmers started on this type of continuing improvement requires education and positive results. However, once started, it can be self-sustaining, done at little or no additional cost with significant beneficial results for the local farmers and all the local residents.

11.5 FACTORS AFFECTING FOOD PRODUCTION

There are many factors that affect national or regional food production. Some, such as weather, cannot be easily controlled or modified by humans. However, a surprising number are subject to change by governments or even individuals. A few are even very inexpensive and have significant impact. We will examine some of the factors that can be modified and have been shown to have a positive impact on agricultural production.

11.5.1 Transport

The ability to move raw or processed foods from the farm to the consumer is an essential link in the food supply chain. From the farmer's point of view, a good, all-weather road is needed to transport fertilizers and seed from the stores to the farm. A good road is also needed to carry the harvest or mature animals to market. In some areas of the Andes, similar to the area of Octavio Tipán, farmers regularly band together to repair roads leading into their valleys so that commercial truck drivers can drive to their fields and purchase the harvest. If the roads are too rough or difficult, the drivers will pass them by and go to other valleys to buy produce. Then they are faced with the unpleasant task of carrying their produce out of the valley on horseback or on their own backs.

An obvious limitation of poor road infrastructure is seen in Figure 11.3. The type of locally constructed bridge shown is barely adequate for a light vehicle. There is no possibility for a fully loaded truck to cross this bridge. This effectively cuts off farmers from supplies and markets.

The importance of a transportation network was verified by a study of the relationship of roads and electricity on agricultural production in 83 countries.[7] This study verified that two of the most important types of infrastructure needed for high agricultural production are transportation and electricity. The high density of roads in an area gives residents access to information both in print and presented in stores and field demonstrations. Better transportation obviously gives farmers low-cost access to inputs such as fertilizer and chemicals needed for crop production and to distant markets for sale of their produce.

11.5.2 Markets

Efficient markets are directly related to agricultural productivity. The existence of well-functioning markets help farmers acquire and use improved production inputs and to sell their products.

Figure 11.3. Poor transportation networks limit markets.

It seems obvious that commercial farmers must have access to a market to sell their products. In many parts of the world this critical factor is lacking. The small fruit stand by the road shown in Figure 11.4 is a market, but not an important one for most farmers. Fruit farmers need a guaranteed buyer for their produce at the time of harvest. The fruit shown in Figure 11.4 cannot be stored until a buyer is found. It must be sold when ripe or left to rot in the field. Even dry grains, such as wheat, cannot be stored for long periods due to damage by mold and insects. If the farmer has taken out a loan to produce the crop, the production must be sold to repay the loan. Any delays incur mounting interest costs. Farmers must be able to sell their production at harvest at a fair and profitable price.

To be an important outlet for farm products, a market must provide price information to farmers. This price information should be freely available by newspaper or radio for several purchasing points in a region. When price information is available, farmers are able to move their products toward the better paying market and help maintain the stability of prices. In some markets controlled by a few large buyers, the spread of price information is actively discouraged since this will enable other buyers to pay more and draw farmer production to them. These large monopolies depend on intermediaries or farmers bringing products to the warehouse and accepting the price offered. If the sellers have knowledge beforehand of prices, they can simply change their route to the best paying buyer.

Even in the open economy of United States, markets restrict or encourage what farmers can produce and sell. When a farmer grows a bumper crop of sorghum, a buyer is needed to convert the grain into money. Grain markets in the United States are limited by the size of railroad grain cars and storage facilities. A single grain car

Figure 11.4. Typical small fruit market in Ecuador.

is filled with 80 to 120 tons of only one grade of one type of grain. While waiting for an empty grain car, an elevator owner must store grain from several farmers in large concrete silos holding as much as 50,000 tons of grain. It is not economical to use these large storage bins to hold a mere 5000 tons, the yield from 1000 ha of specialty maize. The result is that all farmers in a region grow only a few crops that the local grain buyers can accept. Other grains must enter local, specialty markets.

When a farm supply business enters a region, it must provide fertilizer, chemicals, and equipment that give an obvious increase in yield or income to the local farmers. Frequently, these products are accompanied by technicians trained in their use. These technicians set up demonstration fields and visitation days for farmers to see the differences in productivity offered by their products. Radio and newspaper advertisements encourage farmers to buy their products. When farmers decide to buy the fertilizer or hybrid seed, it must be available at that moment. If the business does not have the seed and fertilizer ready during the 2- to 3-week-long planting season, farmers will plant other seed and their potential market is lost until next year. Having the right product available at the right time is critical. An efficient market system is able to provide the raw materials of agriculture when and where they are needed. This means there must be an efficient communication system as well as a safe and speedy transport system in the country.

11.5.3 Technology

Technology can greatly increase food production efficiency, but there are three major problems with the wholesale application of technology or rapid changes in technology use. First, up to 90 percent of the population of some countries are employed on the farm or part of a farming operation. Application of technology can thus lead to large-scale unemployment. Unemployed people cannot afford food, and thus the general food and nutritional situation in the country is not improved and may be gravely worsened. While the planting of a new, high-yielding variety of maize or rice may increase a farmer's income, it may also lead to oversupply and thus a decrease in the price the farmer gets for his produce. This is particularly the situation where it is hard to ship the produce to distant markets.

The second major problem is that in many cases technological innovations are not available to the farmers in a location. This could be seed but might also be any other agricultural inputs including education and capital. It may be a relatively simple problem such as the lack of roads or transportation or more complex infrastructure problems. As part of this problem the technology must be able to be supported locally. If tractors are brought in to an area that does not have fuel, oil, or spare parts, then their usefulness will be short lived. Also if the tractors are too big to fit in the fields, they will be of limited value and may be used in very inefficient ways such as transportation.

Technological innovations with certain characteristics can be very useful in improving agricultural productivity. A piece of equipment that increases farmer efficiency and is produced from locally available materials by local craftsmen, such as the field liners used by Donio to determine plant spacing in his rice field, would be an excellent example. They decrease the need for hand labor during planting, they are produced locally, and persons displaced from planting are employed in producing the liners. These jobs can be expected to be higher paying and thus benefit both the farmer and the local economy positively.

The third major problem is education. It seems to be widely accepted that new technology can be used by farmers without the necessity of their learning how to use it.* For example, when the no-till method was first introduced, it was said that it could only be used on sandy soils and only to grow maize. As farmers learned how to no-till, it became used on other soil types and for other crops. Today some farmers no-till all their crops and do so on soils that are high in clay.

11.5.4 Land Tenure and Land Use Patterns

Land tenure and use patterns vary greatly around the world from land ownership with title, that is, Steve's situation, land assigned by a tribal chief, land assigned by the government, and a number of land rental systems from payment in cash, cash rent, to payment being in the form of produce from the land, as in the three bags of rice Donio "pays" as rent for his rice field. In some cases agricultural or food production may be as small as a single person owning a single fruit tree and another person "renting" the tree and harvesting and selling the produce.

Renting can be of several types. It is common to have a yearly rental agreement where the rent may change from year to year or the agreement must be renewed each year. Rent can be a longer term contract where the owner allows the renter to rent the land for 5 to 10 years at a time. The rental payment will be on a yearly basis, but the rental amount may be negotiated each year. Long-term agreements allow the farmer to invest in the productivity of the land such as fertility plans and the like.

The exact nature of the relationship of the farmer to the land and land ownership is not as important in terms of agricultural productivity as is the secure or assured use of the land. However, there are some severe drawbacks to lack of ownership of land. One is the intrusiveness of the owner who may interfere with farming operations or dictate crops to be planted or production methods. One example of this is the owner who refused to continue to rent his land to a farmer because the farmer was no-tilling the land. In no-till the land is left with plant cover all the time, and the owner saw this as not taking care of the land even though soil conservationists agree that this method of farming is preferred because it conserves the soil by stopping erosion.

A long-term problem with lack of ownership is that the farmer cannot use the land as collateral for loans. The inability to obtain capital for investing in improved farming methods, that is, fertilizer, seed, and equipment, severely limits the ability of the farmer to increase productivity.

All systems put some constraints on the farmer and what and how he or she produces food. In the case of Steve, although he owns half the land he farms, he still

---

*The unwillingness to put money and resources into a new technology has been the major impediment to its successful use in all areas of human endeavor.

must pay a yearly tax on this land. Donio is constrained by the fact that his rent is rice and so he must grow rice. In many African countries where the chief assigns land to farmers, they are constrained by the necessity of always planting the land to a crop or lose the land. While this may not seem like an imposition, it does mean that the farmer is not free to experiment with crops that are very different from the main crop grown. For instance, if the normal crop is millet, growing faba beans may be seen as not growing a crop! Aída and Octavio own their land, and this gives them the freedom to grow a wide variety of crops and animals, thus producing a balanced diet for the family as well as a continuing income.

11.5.5 Weather

Climate is not controllable—rainfall and temperature are variable from year to year, however, they fall within some common ranges, and these can be exploited by farmers in two ways. It is easy to find, from local research done by local governmental or educational organizations, the average dates for the start and end of climatic conditions, rainfall, and temperature that mark the growing season. Optimum dates or conditions for crop planting and harvesting can then be used to plan farming operations. Planting at the optimum time is one of the most important farming operations in terms of obtaining the best yields.

In areas with excessive rainfall farmers can either use drainage to remove excess water or plant crops, such as rice and taro, that can grow in flooded conditions. There is even a rice variety that grows taller as water becomes deeper such that it can produce grain even under flooding conditions.

In areas of insufficient rainfall two things can be done: conserve rainwater or carry out irrigation. A number of different approaches can be used to conserve water. One is to prevent water leaving the field and thus infiltrating into the soil. In this way the soil holds the most water for crop production. To go along with this approach, maintaining or increasing soil organic matter will maintain or increase the soil's ability to take water in and hold it.

Another aspect of this is that conservation of water is accomplished using the same methods used to conserve soil. It is not possible to separate soil and water conservation. Soil erosion occurs in two steps: destruction of soil structure by raindrops hitting bare soil structures, also called splash erosion. This is the initial step in soil erosion. The second step is movement of soil particles off or out of the field by water flowing over the soil surface. Note that preventing water from moving over the soil surface prevents erosion, but if water cannot move out of the field by moving across the surface, it must either infiltrate into the soil, thus conserving it, or evaporate.

The second way is to prevent water escaping from soil either from the surface or down into the lower portions of the soil where roots cannot penetrate. Mulch can be used to limit evaporation from soil surface, thus conserving soil water. Any type of material may be used as mulch such as straw, leaves, paper, or plastic. Mulch will also affect soil temperature, so this effect must also be kept in mind.

A layer of water-impermeable material, such as plastic, can be placed in the soil at some depth, such as 30 cm deep. The depth used must be determined by the expected maximum rainfall because, if a barrier is too shallow, saturation of the soil that will be detrimental to plant growth could occur. When rain falls it can penetrate to this barrier, but then it cannot percolate deeper into the soil. The water is thus captured in the upper soil for plant use. This approach has been investigated and can be effective. In most cases mulch should be used to prevent water loss from the soil surface. However, installing such a layer is cost and labor intensive and thus is not be useful in most cases.

Temperature is a critical factor in crop production. In most cases it is low temperature that is of the most concern. Temperatures of small areas can be controlled in a number of different ways. Cold frames have been used in many places to get plants started earlier in the spring. In this case a frame, made of 15-cm-wide boards is placed on an area and a piece of plastic or glass placed over it. This works as a shallow greenhouse keeping young plants from being adversely affected by low temperatures in the spring and allowing plants to be started earlier.

The more commonly seen are greenhouses, made of glass or plastic, which can be heated or cooled such that plants can be grown all year long. Although greenhouses are common throughout the world, they are generally only economical for high-value crops such as flowers. They are also commonly used to start plants early in the growing season. These plants are then transplanted outside when the weather permits, thus allowing for earlier and longer production. The cost of building a greenhouse is not great, but the expense of heating and cooling it can be excessive in cold climates.

Conservation or control of water available to plants will result in dramatic increases in crop production. For this reason water and soil conservation, drainage, irrigation, and mulching are all activities that farmers can undertake to increase productivity in all environmental conditions. Many of these activities, such as keeping the soil covered at all times, are not expensive and can be done anywhere.

Note that leaving organic matter in the field on the soil surface means that there will be more organic matter in soil, and this will also increase the soil's ability to hold water for plants. Thus another thing all farmers can do to increase agricultural production is increase soil organic matter by leaving as much organic matter in the field as possible, applying animal manure and compost, and by growing green manure crops. Also the use of organic mulches will result in increased organic matter in soil and thus increase the soil's water holding and supplying capacity. Mulches can also protect the soil surface between crops if it is not totally decomposed during the growing season, which is generally the case.

Farm animals are adversely affected by both high and low temperatures and must have readily available water and food. Animals must be protected from low temperatures and will not produce well in very hot climates. Thus barns or similar structures are used to house animals in the winter, and shade is provided for animals during hot weather. Animals need drinking water, and there must also be enough rainfall to allow for the growth of plants they can eat.

In many countries animals are constantly kept in buildings, for example, Steve's swine operation, where water and food are provided either at certain times or on a continuous basis. In these cases the building must be either heated or cooled, depending on the outside weather. In some cases all three, that is, water, food, and climate are controlled automatically, and the farmer need only make sure that all systems are functional.

Except in those cases where a new enterprise is being developed and thus buildings needed, improvements or development of new structures will have very little effect on increasing existing animal productivity.

11.5.6 Education

Education is an essential and critical part of any activity designed to bring about change. How the education is accomplished, formal, informal, and practical, hands on, schooling, workshops, field days, or any other are all appropriate and a mix of approaches is usually the best. For agriculture this must include demonstration plots or other demonstrations because farmers will not accept and apply a new practice unless they have seen it work in the field.

One of the major failings in education has been the general unwillingness to educate the educators or teachers. Teachers can only be effective when they have a thorough background, understanding, and experience in using that which they teach. This must include hands-on experience.

A second major mistake people tend to make is to call education an expense and to say that education is too expensive. This is never the case. First and foremost, education is an investment not an expense. As an investment it results in the best return to invested money of any productive activity. Second, the monetary investment is relatively small compared to most activities that people undertake.

One other aspect of education that is often overlooked is research, which can be viewed as the researcher learning how to do something. When the researcher learns how to do something, that information is shared with his or her colleagues and eventually with the community and the society as a whole. Thus, research into the best planting times, best fertilizer use, best varieties of crops to plant, and the like must be carried out and the results shared with the agricultural community. These types of research need to be long term and ongoing. Climate and society need time to change, and so research needs to be continual so that recommendations pertaining to the present situation can be made.

Often technology must be adapted to an area where it has not been used before. In these cases it needs to be investigated, studied, and changed to make sure it is appropriate to the local situation. Even in small areas such as a state or on an island, there may be different climatic situations in different places. For example, one can look at the variability in the climate in Ohio and on the island of Leyte in the Philippines described in Chapter 8. Once technologies are well studied and found to be beneficial to a local situation, then the farmers of the area need to be educated in the use of these technologies.

11.5.7 Government Interventions and Subsidies

Government interventions and subsidies can have both positive and negative effects on food and food production. In general, governments wish to have a positive effect on agricultural production. This stems from several sources. First, agriculture is always a major part of an economic system. Second, self-sufficiency leads to stability and increases independence. Third, food in large quantities is needed to support the army that is used to protect the country. Fourth, food production is a source of pride for many countries.

Positive interventions can be of several types: support of farming inputs to keep their prices low, support of food prices to keep the money paid to the farmer higher than their market value, or governments may support programs such as soil or water conservation programs or projects that ultimately lead to improved or increased agricultural production or increased food security. For instance, water conservation and reservoir construction projects can result in water availability for crop irrigation allowing food production during drought conditions.

Programs that also control importation of food will have a general positive effect on local food production except in the case where the food imported cannot be grown in country and is not in competition with a locally grown food source.

A negative impact will occur when a government does not do any of the above. It can also occur when the government allows unrestricted importation of food from any outside source.

11.5.8 Irrigation

For many countries irrigation is a possible route to increased food production. When water is available, constantly replenished, and the technology available, it can result in a more balanced availability of food throughout the year. Even those with minimal resources can practice some irrigation technology, involving simple canals and movement of water by gravity. More complex systems requiring more sophisticated equipment are restricted to those with both the resources to purchase the equipment and maintain it. This also requires that the return on invested monies be sufficient to justify the expense.

Irrigation has two problems that must always be addressed. The first is salt. All irrigation, except in areas with very high rainfall such as the humid tropics, results in the buildup of salt in the soil. This salt must be removed by installing a suitable drainage system and by practicing irrigation methods that limit salt buildup. If not controlled, salt buildup will eventually render a soil nonproductive. Because of the salt problem, irrigation schemes need to be supported by technically competent individuals who can advise farmers on how to control salt in soil.

The second major problem is the long-term availability of water for irrigation. For a sustainable irrigation project the water used must be replaced on a yearly basis by precipitation that can be in any form, that is, rain, snow, or ice. For this reason both the climate and the hydrology of an area to be irrigated must be well studied and understood so that only sustainable amounts of water are used.

11.6   AGRICULTURE FOR THE LONG TERM

We must look forward to ensure that our present comfortable situation does not evolve into a replay of historical food shortages and famines. How can we ensure a food supply for the next 1000 years? Is it even possible? Let's look at some possibilities.

11.6.1 Renewable Inputs

The whole concept of renewable inputs is quite complex. In many real aspects an example of a renewable input would be carbon dioxide in the atmosphere. It is taken up by plants and converted to compounds and food that is eaten by animals, including humans, and eventually returned to the atmosphere as carbon dioxide. A nonrenewable input would be phosphate mined from deposits that will eventually be totally consumed.

This seems simple, however, it is not. Nitrogen fixed by nitrogen fixing plants would be seen as a renewable source of nitrogen for crop production because it results in cycling of nitrogen in the environment much as carbon dioxide is cycled. However, nitrogen fixed by a chemical process can also be seen as renewable if the source of energy for the chemical process is renewable. That is, the use of electricity generated by wind, water, geothermal sources, or from sunlight should be considered renewable.

This brings up the knotty problem of energy for agriculture. Fossil fuel is a nonrenewable fuel source. Electricity may be either renewable or nonrenewable depending on how it is generated. It is often assumed that human or animal power does not require energy or that the energy is free. Animal or human power requires that the animal or human consume additional calories to be able to work the land. This means that additional food must be produced to supply these calories. In this sense animal and human power, while it may be renewable, is not free. In some sense machines are more energy efficient because they can be shut down when their power is not needed, while an animal must be kept "fueled" whether it is working or not.

For farmers recycling of organic matter whether through use of manure, compost, no-till, or green manure crops will decrease the need for outside inputs and make the farm more sustainable.

One area of particular importance to sustainability, which cannot be overemphasized, is the use of water. Water is either a renewable or nonrenewable resource depending on how it is used. When water use does not exceed water production, in terms of rain, for most areas of the world, then it is a renewable resource. Exceptions would be water from snow or ice. On the other hand, water drawn from an underground aquifer may be renewed only very slowly. The result is a lowering of the water table in areas subject to excessive pumping. In all cases considerable margin of error must always be included in water use if it is to be sustainable.

11.6.2 Organic Agriculture

Organic agriculture, or organic farming, has been defined in many ways. Basically, it started with the idea of eliminating the use of chemically produced inputs, particularly insecticides and herbicides. It also excludes the use of chemically produced fertilizers. This would be nitrogen and phosphate fertilizers. For most organic farmers potassium fertilizers are acceptable because they can be used as they come from the mine, without chemical modification. Note that potassium fertilizers are considered suitable for organic agriculture but are not renewable.

Organic farming of animals means that no chemically produced inputs are used. Thus no insecticides, antibiotics, or growth hormones are used.

In its purest form organic farming would involve use of neither fossil fuels nor chemical inputs. The crops would be grown using animal power and green manure, nitrogen-fixing crops, and composting. Such a farming operation is usually significantly more expensive than traditional agriculture and thus a problem for commercial farmers to institute. Organic farming of animals involves allowing animals to pasture or have free space to roam and eat feed that is produced organically.

Most farmers in the world will benefit from incorporation of organic farming concepts into their farming system even if they are not or do not intend to become organic farmers. Returning all organic matter, including compost, manure, and plant residues to the soil will result in improved crop production. Likewise the incorporation of crop rotation, particularly when it includes nitrogen-fixing, leguminous, crops, will increase yields and productivity.

11.6.3 Erosion Control

Erosion is the loss of soil by water or wind from a field. Both types of erosion are controlled by keeping the soil surface covered at all times. Both growing plants and plant residue on the soil surface are effective in controlling erosion. Controlling erosion does not have to involve expensive application of amendments to the soil but simply making sure that there is always some plant cover on it.

In addition to keeping soil covered, it is important to stop or control movement of water or wind over the soil surface. In the case of water this is done by changing the slope of the field, by terracing, or by impeding the movement of water, which can be done by many different methods. Planting on the contour will impede water movement across the field and can be effective on fields with only small slopes. Alternating cropped and sod strips across a field will inhibit movement of water down the field with steeper slopes. In some case it may be necessary to remove water from a slope using drainage ditches or underground drains. No-till, which always leaves plant residues on soil, is also effective in reducing erosion. In severe cases fields may need to be terraced to prevent erosion.

In a similar fashion wind erosion is prevented by slowing movement of wind across the field. Plant residues are effective in accomplishing this, particularly if crop stubble, stalks left after grain has been harvested, is left standing in the field. In some cases it may be necessary to have windbreaks at strategic places, perpendicular to the wind direction, across the field. Usually these are plantings of trees that break or slow the velocity of wind.

In all cases erosion control conserves water, and this is important in improving crop production. It also preserves the soil fertility by preventing it from being removed, with the soil, from the field. Erosion control will increase food production and is not capital intensive and does not have to be equipment intensive. It does involve changing some cropping habits and procedures.

11.7   DEALING WITH POPULATION CHANGES

It is obvious that Earth's population is increasing. The present structure of world population has a large number of young men and women who will enter their child-bearing years soon. They will probably follow their parent's social mores and produce more than the two children needed to replace them. The most realistic projections of world population predicts about 9 billion humans will be alive on the planet by 2050.[8] How will this affect our food supply?

11.7.1   Increasing Numbers of People

We are faced with constantly increasing numbers of people on Earth such that we may feel like one of the people in the crowd of Figure 11.5. There must be some point when the Earth's population increase comes into equilibrium with deaths so that the overall number of humans is relatively stable. The birth rate of 2.1 children per woman is calculated as the number of children needed to replace both parents and non-child-bearing adults in a society. Because of the present population distribution in many countries, even a birth rate of 2.1 children per couple will guarantee significant population increases in the near future.

A case in point is India. Figure 11.6 shows that a large part of the total population is under 20 years of age. As this segment of the population reaches child-bearing age, they will contribute to a large increase in the total population even if they only bear 2.1 children per couple by the year 2020. A dramatic population increase is already guaranteed for India. Although China is now the most populous country, by 2050 India will be the most populous country.[8] India will probably not be able to "put on the brakes" to its birth rate and is already committed to a huge population increase.

This increase in total population is expected to occur in many other countries, even if they achieve the replacement birth rate of 2.1 children. An example is China. In 1950 the fertility rate was 6.2 children. After drastic measures limiting

Figure 11.5. Threat of overpopulation.

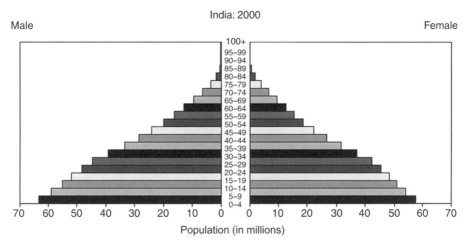

Figure 11.6. Population distribution in India in 2005. (*Source:* U.S. Census Bureau, International database.)

married couples to only one child, the fertility rate dropped to 1.9 children in 1995. During this same period the total population of China increased from 555 million to 1.2 billion in 1995.[8] The population increase was already built into the age distribution of the population in 1950. It could not be avoided.

This same increase is already starting to occur in many developing countries, most notably in Africa and Asia. The five countries that will contribute most to world population growth before 2050 are expected to be India, China, Pakistan, Nigeria, and Ethiopia. This will occur even with significant reductions in birth rates.[9]

The most obvious result of such large population increases is that citizens of the so-called developed countries will be in a distinct minority. Japan is already experiencing a decline in total population numbers. Other European countries are heading in the same direction. In 2006 Europe and Africa each had about 12 percent of the world's population. By 2050 Europe will only have about 7 percent of the world's population, while Africa will have over 21 percent.[8] How will this affect global politics?

How will HIV affect the population trends of countries? Experts think there will be little long-term effect. Even the plagues of the Middle Ages had little long-term effect on populations. Survivors simply had more children to replace those who died. It is expected that the same will happen in AIDS-affected countries.

The short-term effects for some countries will, however, be dramatic with some countries in southern Africa actually experiencing a decreasing population and many others experiencing a dramatic reduction in population growth. One thing that makes AIDS different from other diseases is that it can dramatically decrease the portion of the population that is most productive of both food and people. People 20 to 30 years of age will be lost in large numbers, thus dramatically affecting the ability of the society to sustain itself. Included in this is the ability to maintain a cadre of highly educated individuals needed to carry on research and education in agriculture.

11.7.2   Changes in Consumption Patterns

As populations increase and national incomes rise, there are changes in what and how much people eat. Obviously, humans will increase caloric intake to reach a comfortable and sustainable weight and energy level. The total quantity of food humans will voluntarily eat is normally limited and increases little with changes in income. However, people do not readily change the types of foods they eat. That is, rice eating populations will not readily change to a diet consisting primarily of potatoes.

In a similar fashion introduction of new foods into the diet, to improve nutrition or to improve the soil by crop rotation, is difficult particularly if the taste of the introduced food is significantly different from those tastes to which the population is used. In some cases this problem can be overcome by judicious use of herbs and spices. In other cases the new foods can be incorporated into existing dishes or types of dishes and thus into the diet with less problem.

Often the problem with introduced new foods or food sources is that people do not know how to prepare the food for a meal. How much cooking and for how long is needed? Thus any program to introduce new sources of nutrition into a population has two required essential phases. One is to have the farmers learn how to grow the food, harvest, store, and transport it. The second phase is to have the local food preparers learn how to prepare the food in ways that the local population will eat. Trying to introduce a new food source without vigorous education of the local population in both phases will lead to failure.

11.7.3   Predicting Problems

Predicting problems with agricultural production or food availability is problematical. While the effects of government programs are predictable and some programs, such as support of research and conservation, can be predicted to increase the stability of food supply, many things affecting food supply, are not predictable. The effect of weather is not very predictable for time periods needed to produce crops. Likewise storms that disrupt harvesting, storage, and transportation of food are not predictable. On the other hand pest pressures such as grasshoppers may be somewhat more predictable than weather by following population changes and egg laying, for example.

Sometimes the occurrence of crop or animal diseases can be overcome by careful monitoring of crops. This allows development of strategies for overcoming these diseases through disease-resistant animals or crops or raising other types of animals or crops that are not affected by the disease until the problem can be fixed in a more suitable way.

Preparations for global warming are already underway in IRRI (International Rice Research Institute) for higher rainfall in some rice-growing areas. Rice varieties have been developed that can yield well under deeper than normal water. When weather changes produce higher rainfall, farmers will be able to maintain their yields.

11.8 FOOD SECURITY

All countries must plan for food security, including plans for overcoming or dealing with natural phenomena that interfere with food production and natural disasters. Emergency preparedness is essential and must include plans and facilities for storage of food for emergencies.

Ideally, the government of each country should have enough food in reserve to feed the population for 12 months. This would be a worst-case scenario where no crops are harvested during a single cropping year. A 12-month supply of essential food would avoid starvation until the next (hopefully normal) harvest is collected. Only a few countries actually have such reserves. Most depend on private traders to have grains in storage to provide part of the national emergency food supply. As population increases, demand for food increases, and traders will sell some of their stocks to gain an immediate profit. The result is that many small countries have less than 2 to 3 months of grain in storage at any one time. Any interruption of the food supply longer than 3 months will be a disaster.

11.8.1 Germplasm Banks

Germplasm banks are typically storage facilities for seed from every type of cultivated plant, including variation within a single plant species. The seeds are stored under controlled conditions, including humidity and temperature designed to prevent deterioration of the genetic material. There are approximately 1400 germplasm banks in various parts of the world. Any time a variation, natural or induced, of a plant is found a sample of the plant and its seed is placed in the bank so that it will be available when needed.

Germplasm banks are the source of genetic material needed to produce new varieties capable of overcoming environmental constraints to crop production. Today the emphasis is on transgenic engineering, which makes use of germplasm banks. However, traditional breeding programs that produce hybrids or new pure strains of crops also use material from germplasm banks.

The U.S. government maintains a network of germplasm banks of all types of crops, even those not grown commercially in the United States. The main storage facility is at Fort Collins, Colorado, with smaller cold-storage facilities in each agricultural region of the country. Seeds must be regularly tested for viability and replanted to maintain the collection.

Unfortunately, some germplasm banks are in countries with unstable political situations. Amaranth is an essential grain for many poor farmers in the Horn of Africa. During a drought, a local government was unable to protect an agricultural experiment station. Looters broke into the germplasm bank and stole several hundred cans of grain for food, destroying years of collection efforts.

To protect against this type of loss and even larger calamities the Norwegian government is building a "doomsday vault" inside a mountain on the island of Spitsbergen. It will hold samples of all seed held in all germplasm banks around the world. Permafrost will keep the seeds cool even if power is lost during a catastrophe. The isolation of the island in the far north will further protect the contents from raiders.[10]

11.8.2   Transportation Networks

Transportation is essential in creating food security. If there is no way to get food to people, then having a store of it is useless. Because different methods of travel are vulnerable to different types of interruption, it is best to have a transportation network that includes water, land, and air links. Railroads are the most cost effective for transporting large quantities of materials while air transport is the fastest and truck transport is potentially the most versatile because trucks can go to the locality where the cargo is most needed and are not restricted to a particular route. However, access can be prevented or limited by water or lack of roads.

An example of the limitations imposed by transportation was seen in India in the drought of the 1960s. There was widespread hunger in north and central India. In order to get imported wheat to these regions it was necessary to unload freighters in Mumbai and Kolkota. As soon as the grain was unloaded, it was put into railroad cars to be sent to the areas of greatest need. The railroad network was good and could deliver grain to almost any area of the country. The locomotives and rail cars needed to carry the wheat were in short supply. The most serious bottlenecks were the ports. Ships would wait several days to unload their grain even when the ports were operating 24 hours a day. Even though the distribution network existed, it was impossible to get enough grain through the ports in the time needed.

11.9   CONCLUSIONS

There is no country in the world where farmers are producing food at the maximum level possible. This is due to several factors. In most cases it is not economical to produce at the highest possible level (yield per hectare). In other cases resources, seed, fertilizer, and other inputs may not be available. Lastly, we have not exhausted the potential for further increases in food production.

In many countries grains are produced at very low yields that can be increased dramatically (2- to 10-fold) by relatively simple and inexpensive means. In some areas of the Philippines farmers grow barely 300 kg of maize per hectare while it is possible to grow more than 10,000 kg/ha! However, to go from 300 to 600 kg/ha could be done by simple methods such as crop rotation, application of manure or compost, or the use of improved varieties.

In Niger a farmer may only produce 300 kg/ha of millet when it is possible to produce 3000 kg/ha. In this case it may only be necessary to keep wandering animals out of the young millet to double production to 600 kg/ha. Both of these cases point out the fact that in many countries it is possible to dramatically increase agricultural output by application of simple and inexpensive improved production practices.

For these reasons the outlook for feeding the world's population, present and future, is not as bleak as many would believe. However, there is much work to be done to make areas of the world famine-proof.

QUESTIONS

1. Explain the basis for the expectation that world population will out grow food production in the near future. What might lead you to think this will not happen?

2. Famines can be local or widespread. Explain how this can happen.

3. Describe the basic concepts behind the Green Revolution. Are these concepts still useful today? Explain.

4. Giving examples from this and other chapters, explain how land tenure can dramatically affect a farmers well being and food production.

5. Explain the difference between renewable and nonrenewable agricultural inputs.

6. Explain the roles of education in providing increases in food and food security.

7. Explain how governmental interventions can either increase or decrease food availability and security.

8. Discuss renewable inputs and organic agriculture describing how the two are similar and the ways in which they are different.

9. In most places in the world population is increasing. However, in some places it is or will be decreasing. Describe these places and explain the effects of this decrease on food production.

10. What are germplasm banks and how are they important in terms of food security?

REFERENCES

1. T. Malthus, An Essay on the Principle of Population, 1798. Available at: http://www.ac. wwu.edu/~stephan/malthus/malthus.0.html.

2. Science Net, Singapore Science Centre. Available at: http://www.science.edu.sg/ssc/ detailed.jsp?artid=3455&type=6&root=4&parent=4&cat=49.

3. Guiness World Records, Worst Famine Death Toll. Available at: http://www.guinessworld records.com/.

4. Great Famine of 1315–1317, Wikipedia, the free encyclopedia. Available at: http://en. wikipedia.org/wiki/Great_Famine_of_1315–1317.

5. Irish Potato Famine, Digital History. Available at: http://www.digitalhistory.uh.edu/ historyonline/irish_potato_famine.cfm.

6. FAOSTAT database. Available at: http://faostat.fao.org/site/408/DesktopDefault.aspx? PageID=408.

7. F. Felloni, T. Wahl, P. Wandschneider, and J. Gilbert, *Infrastructure and Agricultural Production: Cross-Country Evidence and Implications for China*, TWP-01-103, Impact Center, Washington State University, Pullman, Jan. 2002.

8. G. K. Heilig, *World Population Prospects: Analyzing the 1996 UN Population Projections*, WP-96-146, International Institute for Applied Systems Analysis, Laxenburg, Austria, 1996.

9. U.S. Census Bureau, IDB Population Pyramids. Available at: http://www.census.gov/ipc/www/idbpyr.html, 2006.

10. The Global Crop Diversity Trust, Scientists to Employ Arctic Ice and Polar Bears to Protect Diversity of World's Crops, FAO, Rome, Press release, June, 2006.

BIBLIOGRAPHY

T. Barnett and A. Whiteside, *AIDS in the Twenty-First Century: Disease and Globalization*, 2nd rev ed., Palgrave Macmillan, New York, 2003.

J. Bongaarts and R. A. Bulatao, eds., *Beyond Six Billion: Projecting the World's Population*, National Academies Press, Washington, D.C., 2000.

Chinese Famine of 1958–1961. Available at: http://www.overpopulation.com/faq/health/hunger/famine/chinese_famine.html.

S. Gillespie, HIV/AIDS and Food and Nutrition Security: From Evidence to Action, International Conference, International Food Policy Research Institute, Durban, South Africa, 2005. Available at: http://www.ifpri.org/events/conferences/2005/20050414/HIVAAIDS.htm.

J. Potter, *Anthropology of Food: The Social Dynamics of Food Security*, Poly Press, New York, 1999.

C. F. Runge, B. Senauer, P. G. Pardey, and M. W. Rosegrant, *Ending Hunger in Our Lifetime: Food Security and Globalization*, International Food Policy Research Institute, Washington, D.C., 2003.

Worldwatch Institute, State of the World 2001, Washington, D.C., 2001.

World Health Organization, Global and Regional Food Consumption Patterns and Trends, 2006. Available at: http://www.who.int/nutrition/topics/3_foodconsumption/en/index.html.

GENETICALLY MODIFIED CROPS AND ANIMALS

12.1 THREE FARMERS AND THEIR CROPS

Subsistence Farmer. Donio does not plant transgenic crops. First, he does not know what transgenic crops are. Second, even if the seeds were available, he would be unable to afford their cost. He uses maize seed saved from his previous crop. The idea of paying money for maize seed is strange and extravagant. Rice seed is different. There are varieties of rice that he knows will yield more than others. Since rice is a money-making crop, he is willing to pay a reasonable price for good seed to have the crop shown in Figure 12.1.

Family Farmer. Octavio and Aída have not heard of transgenic potatoes or maize. They use their crops primarily for feeding the family and farm animals. Their most important commercial crop is potato. For potato they will buy seed potatoes only if they do not have enough saved from the previous crop. Sometimes Octavio will pay cash to buy a small quantity of seed potatoes if it is a new variety with promise of disease or insect resistance. He usually depends on nontransgenic maize seed saved from the previous crop shown in Figure 12.2.

World Food: Production and Use. By Alfred R. Conklin, Jr. and Thomas Stilwell
Copyright © 2007 John Wiley & Sons, Inc.

Figure 12.1. Nontransgenic rice in the Philippines.

Figure 12.2. Nontransgenic maize in Ecuador.

Commercial Farmer. Steve Murphy and his brother raise around 1000 ha of maize and soybeans each year. About 500 ha are planted to nontransgenic maize, 240 ha to herbicide-tolerant transgenic soybeans for sale, 80 ha of herbicide-tolerant transgenic soybeans for sale as seed, and 120 ha of nontransgenic tofu soybeans. They are experimenting with herbicide-tolerant maize (Fig. 12.3) to see if the lowered cost of herbicide application will pay for the extra cost of the transgenic seed. Their decision will depend strictly on economics.

Figure 12.3. Transgenic maize in the United States.

12.2 WHAT DOES GENETICALLY MODIFIED MEAN?

In strict scientific terms, genetically modified has been defined as: "All human-designed changes in a plant or animal, whether done through traditional breeding or genetic engineering. Genetically modified and genetically engineered are sometimes used interchangeably."[1] In current, popular language, genetically modified refers to the intentional insertion of a gene sequence from one species into the DNA (deoxyribonucleic acid) of another species. Transgenic "refers to plants and other organisms that have been changed by adding genetic material from another species."[1] This is a more accurate description of the popular concept of genetically modified. In actual fact, most of our foods have undergone extensive genetic modification since their domestication thousands of years ago.

Genetic modification is not the same as cloning. A clone is an individual or group of cells derived from a single ancestor. In popular language the term often refers to an individual derived from a single ancestor through human intervention instead of sexual propagation. In agriculture, many crops are cloned. Banana and pineapple are only propagated vegetatively, that is, cloned, because they do not produce viable seeds. Many forage grasses naturally reproduce by new plants growing from rhizomes to form clones.

Conventional breeding "is used to describe traditional methods of breeding, or crossing, plants, animals, or microbes with certain desired characteristics for the purpose of generating offspring that express those characteristics."[2] Most of our present crop and animal species have been modified by conventional breeding.

Transgenic crops and animals have generated much controversy with some groups strongly opposed and others in favor of this technology. This chapter will attempt to describe which transgenic crops and animals may enter into the food chain.

12.3   BRIEF HISTORY OF GENETIC MODIFICATION

It is useful to describe three stages of plant and animal modification:

1. Domestication
2. Conventional breeding
3. Transgenic modification

12.3.1   Domestication

Domestication involves the adaptation of a wild plant or animal to serve human needs. For example, domestication of dogs probably involved humans raising the progeny of wild wolves or wolflike species. Domestication of wheat meant that humans gathered, stored, and intentionally replanted seeds of plants that eventually became what we call wheat today.

We normally think of domestication as a benign process. In reality, it is an event signaling a major loss of functions in the plant or animal concerned.[3] Most wild plants have characteristics that make them undesirable for human food. The seed may be too small, too difficult to collect, too bitter or even poisonous, to hard to preserve, or just not taste good. For example, domestication of the potato involved selection of plants with levels of bitter glycoalkaloids below 20 mg/100 g.[4] If the potatoes had higher levels, they were rejected. This resulted in a selection of only a few types of wild potatoes for human use and cultivation. All others were considered unfit for human use. Early farmers selected a potato with less resistance to insects in the wild but which could be used for food by humans. Further selection was made for color, size of tuber, and time of harvest.

Similar events certainly occurred for wheat and rice.[5] For plants to survive in nature, they must have a means of dispersing their seed. Wild relatives of rice accomplished this by readily dropping their seeds when mature. This also helped avoid being eaten and destroyed by birds. However, for humans wanting to use this seed it was not a desirable trait. Stooping over to pick small seeds on the ground was too hard. It was easier to pick the heads of rice that still held seeds. This meant only plants that did not drop their seeds at maturity were collected and preserved for planting. Plants that had dropped their seeds were not saved. Naturally, the early farmers selected the largest, most visible seeds to collect. This also meant that plants with small or dark-colored seeds were not selected. The effect this had on our present-day crops is easily seen in Figure 12.4. The seeds inside the circle are domesticated crops and those outside the circle are their wild relatives. The seeds are, clockwise from top, peanuts, maize, rice, coffee, soybean, hops, pistachio, and sorghum.

In some cases this selection occurred at the same time as a mutation in the species. Maize is an example of a crop with no clear wild progenitor. We can find relatives of maize in the wild, but nothing that appears like maize. Scientists theorize that a single mutation in teosinte or pod corn resulted in the now familiar ear of maize. This made harvesting and utilization of the grains much easier than that of other species. If early

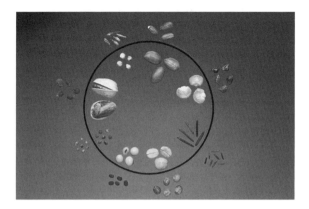

<u>Figure</u> 12.4. Crop seeds and their wild relatives. (Used by permission of ARE, USDA. Photo courtesy of Stephen Ausmus.)

humans collected and replanted this mutated species, it could be used and spread to other communities. Maize is a unique species that needs human intervention to reproduce. The grains are tightly clustered together on the cob and covered by sheaths of leaves. If the ear falls to the ground, over a hundred plants will germinate, but they are so weak that few survive. Humans must remove the individual grains from the husk and ear before planting them in places where each one can grow and reproduce. This change is actually detrimental to the species in the wild but is useful to humans.

The domestication of animals has favored important changes in the species. Sheep were selected that had fewer bristly outer hairs (the kemp) in favor of sheep that retained more of the soft inner hairs (the wool).[6] Dogs were selected for specific abilities such as racing (greyhounds), digging out badgers (dachsunds), or just as ornaments (Pekingese). Many of these modifications make them desirable to humans but are detrimental to the survival of the animal in the wild.

Though we consider domestication to be useful and desirable, it actually involves the loss of characteristics and functions that are useful in the wild. Sometimes these losses are intentional, but other times they occur as the accidental result of other selections. These loss-of-function changes have been so great that many domesticated plants cannot easily survive without cultivation by humans. Because of these losses, plant breeders often search for wild relatives of a crop to recover characteristics (disease resistance, drought tolerance) lost during domestication.

12.3.2 Conventional Breeding

Conventional breeding of plants and animals is generally considered to have begun in 1900 with the rediscovery of Mendel's works on trait inheritance.[7] This enabled plant and animal breeders to make crosses of parents selected for special traits, such as disease resistance, yield, color, taste, speed of maturation, and the like.

Conventional plant breeding programs have made extensive use of varieties that have hybrid vigor. *Hybrid vigor*, or heterosis, is the increased vigor or yield of progeny over that of their parents. Hybrid vigor is greatest when the two parents are genetically dissimilar. The effect is most easily seen when the two parents each come from a genetically uniform line.

The first step in making a hybrid is to form at least two lines of *inbreds*. These are plants that are self-pollinated by the plant breeder to eventually have a genetically uniform makeup. The step of self-pollination is labor intensive, involving placing paper bags on the tassel, as shown in Figure 12.5. When the inbred lines are crossed, the result is a first-generation (F1) hybrid that typically grows and yields much better than either parent. A commercial maize breeding program is able to produce a new hybrid in a period of 10 to 12 planting cycles. This includes time for development of the inbred lines, testing the lines, making crosses, and testing the resulting hybrid in field trials. This can be shortened to 5 to 6 years if winter plantings are made in the Southern Hemisphere, usually Argentina or Chile.

The result is a variety that yields better, resists disease, or has other desirable characteristics. However, in the rush to accomplish a specific goal, some unknown traits may be included or selected out. This occurred in the United States in 1970 when race T of the Southern corn leaf blight appeared and infected a large number of fields in the Corn Belt. It spread so quickly because most hybrids in use at the time incorporated Texas male cytoplasm in their inbred lines. This line was very susceptible to the new race of disease. At some point during the hybrid development

Figure 12.5. Inbred line of maize.

process, resistance to this disease had been lost. Within 2 years the maize breeding industry was able to remove this susceptibility from their hybrid lines.

Conventional breeding of animals usually involves mating animals that have desirable visible traits or established production records. Livestock judging is an established art in animal science programs. It enables trained persons to select animals for breeding on the basis of their appearance. In the dairy industry, extensive production records are kept of each cow during their lifetime. It is easy to select the most productive cows to mate with a bull from a productive bloodline. Many farms have no bulls, relying entirely on artificial insemination to upgrade their livestock bloodline.

An unconventional method of reproducing animals has been pioneered by Texas A & M University. Normal practice consists of freezing semen of productive bulls to continue their bloodline long after the death of the donor. An unusual animal was identified in Texas as being naturally resistant to brucellosis, tuberculosis, and salmonellosis. All of these diseases can be passed to humans by unpasteurized milk, uncooked beef, or simple contamination. Since this animal was naturally resistant to these diseases, it could be raised with little or no antibiotics and still produce good beef. The problem was that this animal was a steer. Steer 86 had been castrated before its unique genetic makeup was discovered. There was no semen to preserve! It finally died at the ripe old age of 21 years without producing any offspring.

In 2000 researchers succeeded in cloning steer 86 to produce a young bull named 86 Squared shown in Figure 12.6.[8] It has all the disease-resistant characteristics of steer 86 but will be able to donate semen for insemination of large numbers of cows. The result will be the preservation and multiplication of this unique genetic makeup. The long-term goal of the cloning steer 86 project is to reduce or even eliminate dependence on antibiotics for beef production.

In some cases, naturally occurring traits in plants have been discovered that mimic the traits of transgenic plants. Clearfield* rice is an example. Clearfield rice varieties are

Figure 12.6. Angus bull 86 Squared. (Used with permission of Larry Wadsworth, Texas A&M University.)

*Clearfield is a registered trademark of BASF.

herbicide tolerant but are not considered to be transgenic. The difference is the way the herbicide-tolerant rice was developed. By spraying over 1 billion rice seedlings with a herbicide, researchers finally found one plant that did not die. That plant was moved to a greenhouse and raised to maturity. The grain from that one plant was then replanted to give many more plants. The genes of this one tolerant plant have been spread to many rice varieties by conventional breeding to yield a new type of rice that is herbicide tolerant but not transgenic.

Conventional breeding depends heavily on natural variability existing in plants and animals to provide raw genetic material to combine needed traits with highly productive animals or plants to form new varieties or bloodlines. Many plant breeding programs incorporate genetic traits from wild relatives that were lost during domestication. This is the reason why so much effort has been put into collection and preservation of seeds of crops used by humans.

12.3.3 Transgenic Modification

For agricultural purposes, the history of transgenic crops began in 1994 with the FDA approval of the Flavr-Savr[†] tomato for sale as food. This modification reduced the level of the ACC synthase (aminocyclopropanecarboxylate) enzyme that causes ethylene to form and trigger ripening in tomatoes. A second modification reduced the formation of the polygalactourease enzyme that breaks down cell walls. The result was that tomatoes could be picked ripe when they reached peak flavor, instead of green, and kept on grocery store shelves for as much as 3 weeks without spoiling. Flavr-Savr tomato paste was marketed in the United Kingdom at reduced prices since the delayed ripening made manufacturing less costly. It was removed from the market in 1999 due to declining sales.

In 1996 Monsanto received approval to sell seed of herbicide-tolerant soybeans, borer-resistant maize, and weevil-resistant cotton in the United States. Herbicide-tolerant cotton was released in 1997. Herbicide-tolerant maize was first sold in the United States in 1998.[9]

When looking at the history of transgenic crops, it is useful to divide their development into three stages, or generations:

First Generation   Useful production traits
Second Generation   Useful output traits
Third Generation   Specific products for industry

First-generation transgenic crops possess traits to enhance the production process. Examples are herbicide tolerance and insect resistance. These traits provide significant economic advantages for the farmer but have no obvious advantages for the consumer. To date, almost all transgenic crops in production are of this type.

Second-generation transgenic crops possess traits of direct benefit to the consumer, whether human or animal. The development of vitamin-A-containing "golden rice" has

[†]Flavr-Saver is a registered trademark of Calgene, Inc.

few benefits to the farmer, but it has great potential to reduce blindness in children in Africa. Conventional breeding has often been able to select for color or size preferred by consumers, but transgenic technology greatly expands the modifications that can be made.

Third-generation transgenic crops have been designed to produce pharmaceutical compounds or products with specific qualities for industry. Sometimes called "pharming," the production of these crops offers economic advantages for the manufacture of products that are difficult to obtain through normal production methods. An example is human growth hormone normally obtained from pituitary glands of large numbers of cadavers. This hormone can be produced in significant quantities by transgenic maize at low cost without risk of transmissible diseases or contamination.

Today's crop plants are not the same as their wild progenitors. Humans have made changes that have resulted in loss of functions for the plant concerned. In some cases our crops are so different they depend on humans for their survival. Modifications now being made in crop plants are significantly different from those made over the past several thousand years. Current transgenic changes involve addition of traits to domesticated plants. How this will affect future crops remains to be seen.

12.4 TRANSGENIC CROPS

There are currently many transgenic crops being grown by farmers. Many more are in the development stage or ready for release to producers. Transgenic crops produced in 2004 or approved for marketing in various countries of the world are shown in Table 12.1. Most of these crops have been modified with traits to make their cultivation easier (herbicide tolerance) or to resist diseases or insects. Canola, rice, soybean, and sugar beet have been modified to make modified, or very different, end products.

TABLE 12.1. Transgenic Crops Produced in 2004

Crop	Modified Traits
Canola	Herbicide tolerance, high laurate content, and high oleic acid content
Carnation	Herbicide resistance, delayed senescence, modified color
Cotton	Herbicide tolerance, insect resistance
Maize	Herbicide resistance, insect resistance
Papaya	Resistance to papaya ringspot virus
Potato	Colorado potato beetle resistance, potato leafroll virus, potato virus Y and potato virus X resistance
Rice	Vitamin A content
Soybean	Herbicide tolerance, high oleic acid content
Squash	Resistance to watermelon mosaic virus and zucchini yellow mosaic virus
Sugar beet	Resistance to beet necrotic yellow vein virus, herbicide tolerance, fructose production
Tobacco	Herbicide tolerance
Tomato	Herbicide tolerance, delayed ripening

Source: Adapted from Ref. 10.

Delayed ripening tomato is a quality enhancement and carnation color changes enhance marketability. Other crops and traits are being released for cultivation each year.[10]

In addition to the crops and traits listed in Table 12.1, there are others in various stages of development. The crops and traits shown in Table 12.2 are only those for which preliminary approvals for field trials have been granted by one or more countries. Table 12.2 is not complete since many companies withhold news of new crop traits to prevent copy or imitation by competitors. A review of the current list

TABLE 12.2. Partial List of Transgenic Crops in Development from 1991 to 2006

Crop	Traits
Alfalfa	Herbicide tolerance
Apple	Delayed ripening, resistance to apple scab, fire blight, and coddling moth, self-fertilization
Banana	Delayed ripening, resistance to black sigatoka, fusarium wilt, banana bunchy top virus, and nematodes, vaccines against human diseases
Barley	Resistance to barley head blight, barley yellow dwarf virus, and production of glucanases and xylases to enhance grain digestion
Cabbage	Insect resistance
Cassava[a]	Increased starch storage
Chicory	Herbicide tolerance
Coconut	Increased lauric acid content
Cotton	Insect resistance
Creeping bentgrass	Herbicide tolerance
Eggplant	Insect resistance
Flax	Herbicide tolerance
Grape	Resistance to virus
Kentucky bluegrass	Herbicide tolerance
Lentil	Herbicide tolerance
Maize	Increased lysine level
Mango	Delayed ripening, enhanced flavor, disease resistance
Olive	Disease resistance
Pineapple	Delayed ripening, nematode resistance
Rice	Herbicide tolerance
Soybean	Low linolenic acid content
Strawberry	Fruit formation without pollination
Sunflower	Herbicide tolerance
Sweet potato	Resistance to sweet potato virus disease
Tobacco	Reduced nicotine
Tomato	Insect resistance
Watermelon	Increased yield
Wheat	Herbicide tolerance

[a]From Ref. 24.
Source: Adapted from Ref. 11.

of species under modification in Europe shows over 70 crops in field trials.[11] A quick glance at these tables shows that the most frequently added trait is herbicide tolerance. This relates to the expense and time farmers dedicate to management of weeds in field crops. Next in importance are damages due to insects and disease. A few unusual traits under development are noted for banana. Many human diseases are endemic to tropical areas simply because vaccines cannot be distributed under refrigerated conditions. By incorporating a vaccine into food that every child eats, it is theoretically possible to inoculate large populations against a common disease at very low cost.[10,11]

It is clear that we are seeing more transgenic crops in production and on our table. Rather than commenting on the desirability of these new traits, this chapter will attempt to describe only what is happening and place it in historical context. An idea of the interest in transgenic crops can be gained by the number of applications for field trials of transgenic crops. The trends from 1991 through 2006 are shown in Figure 12.7. There was an obvious surge in field work during the 1997 to 1999 period. There are still significant numbers of field trials of transgenic crops in Europe, even in countries thought to be antitransgenic.

Similar trends can be seen in the United States, as shown in Figure 12.8. These figures show a rapid increase in field trials starting in 1993, with a peak in 1998 to 2002.[12] Most likely the declines in field trials reflect a more careful selection of crops to test in the field.

The large investment in developing transgenic plants is a response to the overwhelming adoption of certain varieties of transgenic crop plants. Within one year after the introduction of Roundup[§]-tolerant soybeans (1996) in the United States, 17 percent of the total area planted to soybeans was transgenic. This jumped to

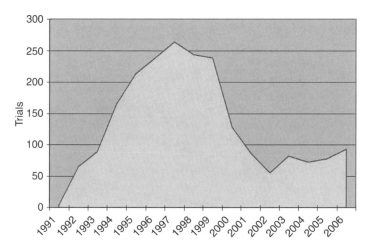

Figure 12.7. Permits for transgenic crop trials in EEC.

§Roundup is a registered trademark of Monsanto Corp.

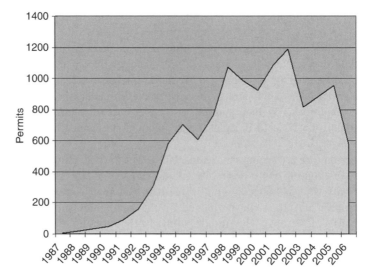

Figure 12.8. Transgenic field trial permits in the United States.

50 percent in the year 2000. The area planted to European-corn-borer-resistant maize in 1996 rose from 1 percent in that year to 26 percent in 1999 but then fell to 19 percent in 2000. The worldwide trend of adoption of transgenic crops is shown in Figure 12.9. Although only 21 countries have significant areas under transgenic crops, the trend is significant. Even developing countries are adopting transgenic crops at the same rate as the industrialized countries.

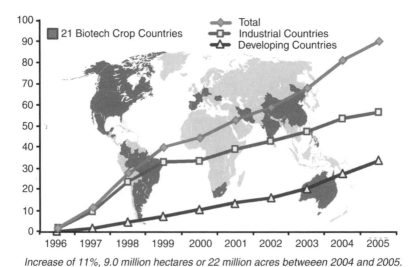

Increase of 11%, 9.0 million hectares or 22 million acres betweeen 2004 and 2005.

Figure 12.9. Global area of biotech crops, million hectares (1996–2005). (From Ref. 23. Used by permission of International Service for the Acquisition of Agri-Biotech Applications.)

> The PLU (price lookup) code found on most fresh produce can be used to identify organic or transgenic produce. If the first number in the code is 9, this indicates the product is organically produced. If the fist number in the code is 8, this indicates the product is transgenic. If the first number is a zero (0), then the product is neither organic nor transgenic.

The important question is: "Why are farmers changing their practices so rapidly?" This rate of adoption is even more surprising when we realize that it often requires purchase of more expensive seed. Herbicide-tolerant crops are very specific as to which herbicide they will tolerate. Purchase of the seed requires a contract promise by the farmer not to save and use seed for the next planting. This also increases costs. Despite higher seed costs, overall production costs remain at similar levels or can be lower with the use of transgenic crops. Examples of realized reductions on inputs may include a reduced need for tillage, crop scouting, and a potential for an overall reduction in the amount of pesticides required for the control of diseases, insects, and weeds. An increase in net economic returns can be realized with transgenic crops due to better protection of the plants favoring higher yields. A second view of the total hectares planted in Figure 12.10 gives us an idea of the types of transgenic traits being adopted. Herbicide resistance is the most popular trait of the new crops, followed by insect resistance, and then by the combination of the two traits, often referred to as *stacked traits*.

One practice many U.S. farmers have been adopting for crops such as maize is called no-till. Conventional tillage, practiced for thousands of years, requires that the farmer first plow the field, then disk the field to break up clods or clumps of soil, and to destroy weeds. Sometimes the farmer must pass over the same field three to four times before the soil is sufficiently pulverized to permit planting. After planting,

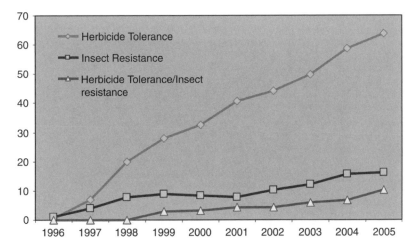

Figure 12.10. Global area of biotech crops by trait (1996–2005). (From Ref. 23. Used by permission of International Service for the Acquisition of Agri-Biotech Applications.)

the farmer must again go through the field two to three times to destroy weeds. If weeds are not controlled, as much as 75 percent of the grain yield may be lost. Insect attacks are not always spectacular. More frequently the European corn borer will attack 10 to 15 percent of the field, leaving stalks without ears or with ears partially eaten by worms. Rootworms will attack the crop below the surface causing slow, but measurable, yield losses. Controlling weeds before World War II was done by tractor-mounted cultivators driving slowly through maize fields to remove weeds (see Figure 12.11). Hand weeding was needed since the tractor could not remove weeds in the row between the maize plants. This was expensive because of the cost of gasoline or diesel fuel used each time the tractor traveled over a field. It was also time consuming since each tractor needed a driver and each hoe needed a person handling it.

The practice of no-till was developed in the United States in the 1960s when herbicide use for maize became widespread. Since weeds could be controlled with one application of a chemical, time and costs were drastically reduced. This also meant that one farmer could plant several hundred hectares of maize without help, further reducing production costs. Research done in the 1960s showed that when weeds could be controlled by herbicides, tillage (plowing and disking) could be reduced or even eliminated with no loss of yield. This meant that the farmer could plant a field without previously plowing and disking. Weeds could be removed in one single operation without disturbing the soil and erosion was greatly reduced. With a slightly modified planter, the farmer was able to raise a crop of maize with only three operations: planting, spraying, and harvesting. Costs were further reduced. The only problem was that increasingly potent and complex combinations of herbicides were needed to kill the thousands of weed species in maize fields. One of the most common herbicides was atrazine. At low concentrations it had little effect on maize but controlled many weeds. Disadvantages were that it only controlled certain broadleaf and grassy weeds. Atrazine also had limited residual activity at typical rates, controlling weeds

Figure 12.11. Mechanical weed control in maize.

for only 8 to 10 weeks. This often made additional weed control necessary with either row cultivation or more herbicides.

The discovery of glyphosate, the active ingredient in Roundup, furthered the adoption of no-till practices around the world. Glyphosate is a very effective nonselective herbicide. It kills most plants that it touches but is readily deactivated in the soil. The development of transgenic crops that are tolerant to glyphosate promoted the adoption of no-till agriculture as improved weed control became easier to achieve. Now farmers could plant and produce a crop without tillage. Preplant tillage was replaced with an application of glyphosate combined with a residual herbicide, such as atrazine, to kill existing weeds and prevent early-season weed competition. After emergence the crop can then be sprayed with glyphosate to control weeds emerging later in the season. The development of these two technologies has worked to greatly increase adoption of herbicide-tolerant maize in Argentina,[13] shown in Figure 12.12. The use of atrazine continues to be an important tool for weed control and is applied at reduced rates. Glyphosate has replaced atrazine in most no-till fields in Argentina. Clearly, the use of herbicide-tolerant maize has facilitated the use of no-tillage practices that have lowered costs for the individual farmer.

Even greater advantages have helped push the adoption of insect-tolerant cotton in China.[14] A government program recommended adoption of insect-resistant (Bt) cotton starting in 1999. Some large farms adopted this new variety, but most of the seed was purchased by very small farmers. In 2001 nearly 30 percent of the total area planted to cotton was transgenic Bt cotton. Average yields were increased using the Bt cotton

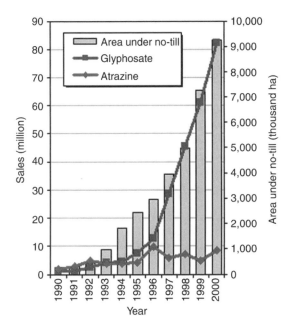

Figure 12.12. Adoption of no-till and Roundup in Argentina. (Used by permission of AgBioForum, cited in Ref. 23.)

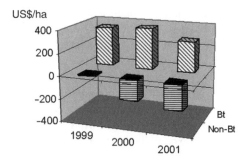

Figure 12.13. Net revenue of Bt and non-Bt cotton in China. (Adapted from work cited in Ref. 14.)

variety. Analysis of the economic benefits in Figure 12.13 shows the real reason for this rapid adoption. Farmers using Bt cotton had net profits of US$277 to US$367 per hectare during the period studied. Increased yields were only part of the reason for increased profitability. Insecticide costs were reduced more than half, as shown in Figure 12.14. In addition to the increased profitability, the number of reported insecticide poisonings by farmers using Bt cotton were significantly reduced.

The adoption of transgenic crops has been slowed by the refusal of some markets to accept transgenic food grains. In most cases these countries have initiated a study of the safety of transgenic foods before they are permitted to enter the human food chain. There are signs that official restrictions are slowly being reduced and markets will open for more transgenic crops. Various international accords have been reached to help supervise or control the introduction of modified crops and animals.[15] The Convention on Biological Diversity is concerned with the conservation and the sustainable utilization of ecosystems. This convention adopted the Cartagena Protocol on Biosafety in 2003 to give advance notice before the introduction of transgenic plants that may have adverse environmental effects. This notice is registered with the international Biosafety Clearing House. Another international agreement is the International Plant Protection Convention. This agreement attempts to prevent the spread of pests affecting cultivated plants. Specifically, it seeks to prevent the introduction of traits that may cause some plants to become invasive (unwanted spreading to cultivated areas).

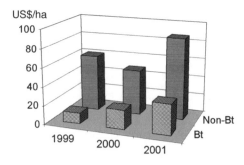

Figure 12.14. Pesticide costs on cotton in China. (Adapted from work cited in Ref. 14.)

It also seeks to prevent gene flow from transgenics to wild species and to prevent unwanted effects on wildlife.

Reviewing the rapid adoption of transgenic herbicide- and insect-resistant crops, it appears that farmers are adopting these new crops for reasons of economics and safety. In most cases there are long-term benefits from reduced soil erosion.

12.5 FUTURE OF TRANSGENIC CROPS

The trend in transgenic crops is easily seen in Table 12.2. Additional crops are being developed to increase profitability and streamline production. A few crops are being developed with enhanced grain composition for more efficient animal gains or reduced environmental pollution.

An example of meeting both goals is the development of low-phytate maize. Up to 80 percent of the phosphorus in maize grain is bound as phytic acid. Animals that eat this maize are not able to digest this phosphorus due to a lack of the phytase enzyme in their stomach. As a result, most of the phosphorus contained in normal maize passes through the animal and is excreted. This in turn produces high levels of phosphorus in the manure collected in lagoons and spread on crop fields. Phytic acid also complexes other nutrient elements, such as zinc, causing potential deficiencies in the diet of swine. Varieties of maize with low levels of phytic acid are more readily digested by swine and poultry, resulting in more complete utilization of the feed and less contamination by high phosphorus manures.

One problem being addressed by regulatory agencies is the potential for escape of transgenic crops into the wild and competing with native plants. In most cases domesticated crops have lost many traits needed to compete in a wild, or feral, state. An example being debated at the time of this writing is that of herbicide-tolerant creeping bentgrass. This grass is widely used on golf courses. Pure, weed-free stands of bentgrass could be easily maintained by simple applications of herbicide once a year. However, Creeping Bentgrass can become feral, or grow in the wild. If this variety escapes, it will not be controllable by at least one herbicide. There are fears that it could hybridize with near relatives in the wild and pass on herbicide resistance to other types of grasses that are more invasive. Other crops that can survive in the wild are also unlikely to be released in transgenic forms.

12.6 TRANSGENIC ANIMALS

In addition to the selection of domesticated animals for specific traits, some animals have been bred to develop traits useful in medicine. The most notable case is the laboratory mouse. Over many years, researchers have identified mutants with debilities or qualities that make them especially useful for studies of human diseases. The first transgenic mice were born in 1980. By selecting specific gene sequences to insert, it is possible to produce a mouse with natural hypertension, diabetes, premature aging, and many other defects. It then becomes a problem of developing treatments for the condition without complications of associated problems.

Transgenic mice are widely used in all types of medical research. A glance at the Induced Mutant database[16] shows over 320 strains of transgenic mice. With the wide use of transgenic animals in the medical field, it may seem surprising to find that, in contrast to plants, relatively few transgenic animals have been developed to benefit producers. We will review a few transgenic animals that may someday contribute to increased, or better, production for farmers.

The most numerous farm animal, the chicken, has been the subject of transgenic efforts, again for medical purposes. Since chicken eggs are used to produce many human vaccines, efforts have concentrated on developing transgenic chickens that produce proteins in eggs. To date, none have been marketed. There appear to be few, if any, transgenic chickens under development for production purposes.

Cattle have received more attention for transgenic modification of production factors. A major problem with many milk-producing mammals, especially cows, is mastitis. Mastitis is an infection of the udders, usually by *Staphylococcus aureus*. If untreated, it causes decreased milk production, damage to the mammary glands, and contamination of the milk produced. Normal treatment consists of injecting antibiotics into the udder and partial immunization by vaccines. These treatments inevitably pass some proteins into the milk. A transgenic Jersey cow (GEM) has been developed that secretes a specific protein, lysostaphin, to kill *S. aureus*.[17] This is not an antibiotic produced by fungi but a naturally occurring protein. The cow, shown in Figure 12.15, did not develop mastitis compared to 71 percent of test animals infected with *S. aureus*.[17]

Figure 12.15. GEM, a transgenic cow. (Photo courtesy of U.S. Department of Agriculture.)

This holds potential for reducing antibiotic use in milk-producing cows. Current research on cows is directed toward reducing methane emission from their digestive tract and the production of naturally low-fat milk.

Sheep have been produced to make pharmaceutical proteins. A transgenic ewe, Tracy, produced a human protein called α-antitrypsin. This is a potential treatment for cystic fibrosis. Unfortunately, she died in 1997. The first cloned animal was Dolly, made from cells of Tracy. Another transgenic sheep was produced in 1996, resulting in a small but significant increase in wool weight.[18] This was not enough of an increase to warrant adoption by farmers but shows the potential.

Swine have received much attention because of their potential to grow organs for transplant to humans. Some programs are concentrating on environmental effects of large swine fattening farms. A major problem with large swine farms is that they produce a lot of manure. This manure tends to be high in phosphorus. When this manure is applied to crop fields, the phosphorus builds up and eventually leaches into streams, an environmental hazard. A transgenic swine, dubbed Enviropig, has been developed to reduce contamination from manure. Genes were inserted to enable the swine to digest phytate, thereby eliminating the need to add phytase to their rations and reducing phosphorus excretion by 60 percent.[19] By digesting phytase, the animals have improved digestion and increased weight gain.

Another transgenic swine has been produced that makes increased amounts of insulin-like growth factor I (IGF-I).[20] Production of more IGF-I stimulates muscle production and results in a less fat, more lean animal. This extra lean animal was worth an extra $6 at market prices in 2004.

Goats have also been the focus of transgenic modification. Most attention has been focused on the production of pharmaceuticals such as malaria antigens and monoclonal antibodies useful against cancer and AIDS (acquired immunodeficiency syndrome). One unique application is the insertion of genes in a goat to produce spider fibers via the milk.[21] Spider silk is extremely strong and flexible but cannot be manufactured in large quantities. Unlike silk worms, spiders are not social creatures and tend to eat each other when raised in groups. The transgenic goats produce spider fibers in their milk, serving as a source of large quantities of spider fibers.

Transgenic fish have been developed in several countries. Research for development of common carp is underway in Canada, rainbow trout in China, and striped bass and various shellfish in Israel.[22] Many of these fish have an increased level of growth hormone resulting in larger sized fish with increased feed conversion efficiency. A major problem with moving these transgenic fish into production is the danger of escapes. Some fish production facilities are basically cages in open water. Others are raised in ponds adjacent to rivers and lakes. The risk of escape, and competition with wild species, is great and little is known about their potential for multiplication in the wild.

12.7 FUTURE OF TRANSGENIC ANIMALS

At the current stage of development, the widespread use of transgenic animals in farm production appears unlikely for many years. Their use has been adopted most readily in applications where they are confined and where the products do not enter the food chain.

12.8 CONCLUSIONS

Humans have a long history of genetically modifying crops and animals to suit our needs. In most cases these modifications have made the crop or animal less suitable for survival without human intervention. Recent efforts in genetic modification have involved the introduction of genes from other species to give the plant or animal traits not easily obtained by traditional breeding methods. These methods have caused much controversy in many parts of the world and remain to be resolved.

QUESTIONS

1. What is the difference between genetically modified and transgenic?
2. What is domestication?
3. What is conventional breeding?
4. Describe a crop trait that is typical of first-generation transgenic crops.
5. How does net revenue of cotton change when Bt cotton varieties are used in China?
6. What is so unique about 86 Squared?
7. What is unique about the transgenic cow, GEM?
8. How does cloning differ from genetic modification?
9. What is hybrid maize?
10. What is a third-generation transgenic crop?

REFERENCES

1. EcoHealth: Environmental Change and Our Health, Johns Hopkins Bloomberg School of Public Health. Available at: http://www.ecohealth101.org.
2. *Safety of Genetically Engineered Foods*, Food and Nutrition Board, Institute of Medicine, Board on Agriculture and Natural Resources, Board on Life Sciences, National Academies Press, Washington, D.C., 2004.
3. P. Gepts, A Comparison between Crop Domestication, Classical Plant Breeding, and Genetic Engineering, *Crop Sci.*, **42**, 1780–1790, 2002.
4. T. Johns and S. L. Keen, Taste Evaluation of Potato Glycoalkaloids by the Aymara: A Case Study in Human Chemical Ecology, *Human Ecology*, **14**, 437–452, 1986.
5. S. Konishi, T. Izawa, S. Y. Lin, K. Ebana, Y. Fukuta, T. Sasaki, and M. Yano, An SNP Caused Loss of Seed Shattering During Rice Domestication, *Science*, **312**, 1392–1396, 2006.
6. The History and Prevalence of Genetically Modified Organisms, PowerPoint presentation. Available at: Northwest Association for Biomedical Research web page, http://www.nwabr.org/education/articles/2005Lessons/GMO_05/256,1, The History and Prevalence of Genetically Modified Organisms.
7. J. Sapp, The Nine Lives of Gregor Mendel. Available at: http://www.mendelweb.org/MWsapp.html.

8. Third Animal Species Cloned At Texas A & M University, *Science Daily*, Sept. 10, 2001. Available at: http://www.sciencedaily.com/releases/2001/09/010906071658.htm.

9. Brief Biotech Timeline, Monsanto Company. Available at: http://www.biotechknowledge. com/biotech/bbasics.nsf/timeline.html.

10. Transgenic Crops, International Service for the Acquisition of Agro-Biotech Applications. Available at: http://www.isaaa.org/kc/Global%20Status/crops.htm.

11. List of SNIFs circulated under Article 9 of Directive 90/220/EEC, 11/05/2006. Available at: http://biotech.jrc.it/deliberate/doc/snifs.pdf.

12. Field Test Releases in the U.S., Information Systems for Biotechnology, Agricultural Experiment Station at Virginia Tech. Available at: http://www.nbiap.vt.edu/cfdocs/ fieldtests1.cfm.

13. E. Trigo and E. Cap, The Impact of the Introduction of Transgenic Crops in Argentinean Agriculture, *AgBioForum*, **5**(3), 87–94, 2003.

14. C. Pray, J. Huang, R. Hu, and S. Rozelle, Five Years of Bt Cotton in China—The Benefits Continue, *Plant J.*, **31**(4), 423–430, 2002.

15. *The State of Food and Agriculture, Health and Environmental Impacts of Transgenic Crops, 2003–2004*, FAO, Rome, 2004.

16. Induced Mutant Resource database. Available at: http://www.jax.org/imr/notes.html.

17. Transgenic Cows Resist Mastitis-Causing Bacteria, Press Release April 4, 2005, ARS, USDA, 2005. Available at: http://www.ars.usda.gov/is/pr/2005/050404.htm.

18. S. Damak, Hung-yi Su, N. P. Jay, and D. W. Bullock, Improved Wool Production in Transgenic Sheep Expressing Insulin-like Growth Factor 1, *Bio/Tech.*, **14**, 185–188, 1996.

19. Enviropig™ an Environmentally Friendly Breed of Pigs That Utilizes Plant Phosphorus Efficiently, University of Guelph, Canada. Available at: http://www.uoguelph.ca/enviropig/.

20. New Transgenic Pigs with Lean Pork Potential, Press Release, ARS, USDA, 1998. Available at: http://www.ars.usda.gov/is/pr/1998/980218.htm.

21. GM Goat Spins Web Based Future, BBC News, 21 August, 2000. Available at: http:// newsbbc.co.uk/1/hi/sci/tech/889951.stm.

22. E. Hallerman, Status of Development of Transgenic Aquatic Animals, ISB News Report, April, 2003. Available at: http://www.mindfully.org/Water/2003/Transgenic-Aquatic-AnimalsApr03.htm.

23. C. James, Global Status of Commercialized Biotech/GM Crops: 2005, Brief 34, International Service for the Acquisition of Agri-Biotech Applications, Ithaca, NY, 2005.

24. U. Ihemere, D. Arias-Garzon, S. Lawrence, and R. Sayre, Genetic Modification of Cassava for Enhanced Starch Production, *Plant Biotech. J.*, **4**, 453, July 2006.

BIBLIOGRAPHY

H. Baldassarre, B. Wang, C. L. Keefer, A. Lazaris, and C. N. Karatzas, State of the Art in the Production of Transgenic Goats, *Reproduction, Fertility and Development*, **16**, 465–470, 2004.

Consensus Document on Compositional Considerations for New Varieties of Soybean: Key Food and Feed Nutrients and Anti-Nutrients, Organisation for Economic Co-operation and Development, JT00117705, November, 2001.

J. Fernandez-Cornejo, GM Crop Adoption and Changing Farm Practices, Economic Research Service, USDA. Available at: http://www.riskassess.org.

C. Fu, W. Hu, Y. Wang, and Z. Zhu, Developments in Transgenic Fish in the People's Republic of China, *Rev. Sci. Tech., Office Int. Epizootie*, **24**(1), 299–307, 2005.

GM Database, Agbios. Available at: http://www.agbios.com.

J. R. Harlan, *Crops & Man*, American Society of Agronomy, 1975.

Pharming the Field: A Look at the Benefits of Bioengineering Plants to Produce Pharmaceuticals, Pew Initiative on Food and Biotechnology. Available at: http://pewagbiotech.org/events/0717/ConferenceReport.pdf.

C. A. Pinkert and J. D. Murray, *Transgenic Animals in Agriculture*, CABI Publishing, Oxfordshire, England, 1999.

Scientists' Transgenic Chicken Aids Embryo Research, *Science Daily*, March, 11, 2003. Available at: http://www.sciencedaily.com/releases/2003/03/030311074337.htm.

GLOSSARY

A horizon   Upper layer of soil, most usually darker and higher in organic matter.

Allelopathy   Release of specific organic compounds by a plant to discourage germination or establishment of other plant species nearby.

Aquic   Soil has high water contents that limit oxygen content of soil.

Arable land   Land that can be used to grow crops.

Aridic   Dry for more than half the year.

B horizon   Subsurface layer(s), usually higher in clay and redder in color, of soil.

Barley, two-rowed   Grains of *Hordeum vulgare* are arranged in only two rows on the spike. Generally, this type of barley is used for brewing due to its better diastatic quality.

Barley, six-rowed   Grains of *Hordeum vulgare* are arranged in six rows on the spike. This type of barley is used for animal and human food.

Barleycorn   Grain of barley (*Hordeum vulgare*).

Barrow   Male swine (*Sus scrofa*) that has been castrated.

Beef   Type of cattle (*Bos taurus*) specifically destined for meat.

Billy   Mature male goat (*Capra hircus*).

Boar   Mature male swine (*Sus scrofa*).

Bread wheat   Wheat used primarily for baking loaf bread. Usually includes hard red spring wheat, hard red winter wheat, and hard white wheat.

Buck   Mature male goat (*Capra hircus*).

Bull   Mature male cattle (*Bos taurus*).

Bullock   Type of cattle (*Bos taurus*) used for pulling farm implements such as plows, carts, etc. Usually a male that has been castrated after the hump develops at the base of the neck.

Calf   Young cattle (*Bos taurus*).

Camote   Name used in central Philippines for bonito or sweet potato.

Carambola (W), **star fruit** (NA)   A fruit (*Averrhoa carambola*).

Cassava (W), **yuca** (NA, SA)   A root crop (*Manihot esculenta*) (may also be spelled yucca).

World Food: Production and Use. By Alfred R. Conklin, Jr. and Thomas Stilwell
Copyright © 2007 John Wiley & Sons, Inc.

Cerrado　Area of central Brazil characterized by soils with low pH.

CGIAR　Consultative Group for International Agricultural Research.

Chick　Young, baby chicken (*Gallus gallus*).

Chicken corn　Another name used for sorghum (*Sorghum bicolor*).

CIMMYT　Centro Internacional para el Mejoramiento del Maíz y Trigo (International Maize and Wheat Improvement Center).

Clone　Individual plant or animal derived from a single ancestor.

Cock　Male chicken (*Gallus gallus*). *See* Rooster.

Cockerel　Young (less than one year old) male chicken (*Gallus gallus*).

Coco yam (A), **cocoyam** (A), **taro** (PI)　A root crop (*Colocasia esculenta*).

Common rice　Term used in trading to denote any type of *Oryza sativa*.

Common wheat　Term used in trading to denote any type of *Triticum aestivum*.

Corn　In the United States, this refers specifically to *Zea mays*. In Europe corn is more generic referring to grains of many cereals such as wheat, barley, or rye.

Corn (NA), **maize** (W)　A grain crop (*Zea mays*).

Cow　Mature female cattle (*Bos taurus*).

Cranberry (NA)　A fruit (*Vaccinium* spp.).

Cultivar　Variations, size, shape, color, etc. of the same crop.

Cultivated rice　Any domesticated type of rice contrasted to wild rice.

Denitrification　Loss of soil nitrogen through a biological process that releases N_2 and ammonia forms of nitrogen to the atmosphere.

Doe　Female goat (*Capra hircus*).

Drake　Male duck (*Anas platyrhynchos*).

Duckling　Young duck (*Anas platyrhynchos*).

Durum wheat　Type of wheat (*Triticum durum*) with hard grain used for pasta.

English wheat　General term for *Triticum aestivum* to distinguish it from German wheat (*Triticum spelta*).

Evaporation　Loss of water from any surface. Going from the liquid state to the gaseous state.

Evapotranspiration　Combined loss of water through leaves and from the soil surface.

Ewe　Female sheep (*Ovis aries*).

Fallow　Leaving field without a crop for a period of time, i.e., growing season.

FAO　Food and Agriculture Organization of the United Nations.

Filbert nut (E), **hazel nut** (A)　A type of nut (*Corylus maxima*).

Flint corn　This refers specifically to a type of *Zea mays* with flinty or vitreous endosperm.

Gabe　Name for taro commonly used in central Philippines.

Gander　Male goose (*Anser anser*).

German wheat   Also known as spelt (*Triticum spelta*). This is considered a possible ancestor of modern wheat.

Gilt   Immature female swine (*Sus scrofa*).

Gosling   Young goose (*Anser anser* or *Anser cygnoides*).

Ground nuts (A), **peanuts** (NA)   A nut as defined herein (*Arachis hypoqaea*).

Guinea corn   Another name used for sorghum (*Sorghum bicolor*).

Hardy   Applied to either plants or animals and refers to an ability to survive adverse conditions.

Hectare   Metric unit of measurement used throughout the world (except in the United States) for land area. One hectare (ha) equals 10,000 square meters (m^2).

Heifer   Young female cattle (*Bos taurus*) that has not yet given birth to a calf.

Hen   Female chicken (*Gallus gallus*) or a female duck (*Anas platyrhynchos*).

Hog   Any type of swine (*Sus scrofa*).

ICRISAT   International Center for Research in the Semi-Arid Tropics.

IITA   International Institute of Tropical Agriculture.

Indian maize   In the United States this refers to types of *Zea mays* with multicolored kernels. In fact most Native American tribes preferred types of maize with uniform coloration.

Interfrost   Soil has mix of permanent and nonpermanent frozen soils.

Japan pea   Another name for soybean (*Glycine max*).

Jicama (A), **yam bean** (A)   A root crop (*Pachyrhizus erosus*).

Kid   Young goat (*Capra hircus*) of either sex.

Lac   Resinous secretion by certain insects used in making shellac.

Lamb   Young sheep (*Ovis aries*).

Lipids   Fats and oils.

Litchi (A), **lychee** (A)   A fruit (*Litchi chinensis*).

Lodging   For grain crops this refers to plants falling over to lie on the ground. Usually results in loss of grain.

Lowland rice   Term used to indicate a type of rice grown under partially flooded conditions. It contrasts with upland rice.

Lychee or **litchi** (A)   A fruit (*Litchi chinensis*).

Maize (W), **corn** (NA)   A grain crop (*Zea mays*).

Mealies   In southern Africa this term refers to ears of maize for human consumption.

Milo   Name used for sorghum (*Sorghum bicolor*) in the Great Plains area of the United States.

Minerals   Inorganic atoms, ions, and molecules needed in the diet.

Naked barley   Grain of a hull-less variety of barley.

Nanny   Mature female goat (*Capra hircus*).

Nectarines   Peaches without fuzz.

New cocoyam tanier (A, C), **tannia** (A, C), **taina** (A, C), **yautia** (A, C)   A root crop (*Xanthosoma sagittifolium*).

Nitrogen   Essential plant element. In agricultural usage, application rates are generally expressed as kilogram of nitrogen per hectare even though crops use nitrate (NO_3^-) or ammonium (NH_4^+) forms.

OIE   World Organization for Animal Health (Office International des Epizooties).

Ox   Type of cattle (*Bos taurus*) used for pulling farm implements such as plows, carts, etc. Usually a male that has been castrated after the hump develops at the base of the neck.

Paddy   In Asia a paddy refers to the field where rice is grown under flooded conditions. It also refers to the threshed rice grain before any milling is done to remove the hull.

Paddy rice   Term for rice grown under partially flooded conditions.

Papaya (W), **fruta bomba** (C)   A fruit (*Carica papaya*).

Passion fruit, (W) **granadilla** (A)   A fruit (*Passiflora edulis*).

Peanuts (NA), **ground nuts** (A)   A nut as defined herein (*Arachis hypogaea*).

Permafrost   Soil that has permanently frozen layer.

Phosphate   Essential plant element. In agricultural usage, application rates are generally expressed as kilogram of P_2O_5 per hectare, not the element P.

Phosphorus   Pure form of the element.

Pig   Any type of swine (*Sus scrofa*).

Pollenizer   Plant producing or providing the pollen for another plant.

Pollinator   Pollen-transferring agent.

Potash   Essential plant element. In agricultural usage, application rates are generally expressed as kilogram of K_2O per hectare, not the element K.

Potassium   Pure form of the element.

Pullet   Young (less than one year old) female chicken (*Gallus gallus*).

Pummeo (A), **pomelo** (A), **pommelo** (A), **pamplemousse** (A), **bali lemon** (PI), **Limau besar** (A), **shaddock** (C)   A fruit (*Citrus maxima*).

Ram   Male sheep (*Ovis aries*).

Red wheat   Any type of wheat with a reddish coloration such as hard red winter wheat or soft red winter wheat.

Refined   Refers to foods that have been processed to take away some components, commonly fiber but may also include vitamins and minerals.

Rhizobium   Bacteria genus that forms a symbiotic relationship with certain plant species. The bacteria obtain N_2 from the atmosphere and convert it to forms usable by the plant.

Rooster   Male chicken (*Gallus gallus*).

Rough rice   In the United States this is the name applied to threshed rice grain before any milling is done to remove the hull.

Scotch barley   Grains of barley that have been husked and coarsely ground.

Soft wheat   Any wheat with a relatively soft endosperm such as soft red winter or soft white wheat.

Soil horizon   Distinct layers below a soil's surface.

Soja bean   Another name for soybean (*Glycine max*).

Sorgo   Name used for sorghum (*Sorghum bicolor*) originating from Spanish.

Sow   Mature female swine (*Sus scrofa*).

Soy pea   Another name for soybean (*Glycine max*).

Soya bean   Another name for soybean (*Glycine max*).

Spring wheat   Any type of wheat planted in the spring.

Star fruit (NA), **carambola** (W) a fruit (*Averrhoa carambola*)

Steer   Male cattle (*Bos taurus*) that has been castrated while young to improve weight gain.

Taina (A, C), **tannia** (A, C), **yautia** (A, C), **new cocoyam tanier** (A, C)   A root crop (*Xanthosoma sagittifolium*).

Taro, **coco yam**, **cocoyam**, **tannia** (PI)   A root crop (*Colocasia esculenta*).

Temperature   °C, degrees centigrade, a measure of temperature, 0°C being the freezing point of water (32°F) and 100°C the boiling point of water (212°F).

Tiller   (a) Mechanical tillage device used to stir up soil with rotating blades. (b) Secondary stems arising from the crown of a grain crop of the grass family. A single plant can have many tillers, each bearing a seed head.

Tissue culture   Producing new plants from a few cells of an existing, usually disease-free, plant.

Transgenic   Plants and other organisms that have been changed by adding genetic material from another species.

Transpiration   Loss of water through the leaves of plants.

Triglycerides   One component of lipids made up of three fatty acids bonded to a glycerin molecule.

Udic   Soil is not dry for more than 90 days.

Upland rice   Rice grown without standing water in the field.

Ustic   Intermediate between udic and aridic.

Vitamin   Essential organic compounds needed in the diet.

Wether   Male sheep (*Ovis aries*) that has been castrated. Also, a male goat (*Capra hircus*) that has been castrated.

Wheat berry   Grain of wheat (*Triticum aestivum*).

White wheat   Type of wheat (*Triticum aestivum*) that gives a very light colored flour used for biscuits, cakes, crackers, and cookies.

Winter wheat   Any type of wheat planted in the fall and that remains dormant during the winter to be harvested the following spring/summer.

Xeric Winters moist and cool, summers warm and dry.

Yam bean (A), **jicama** (A, W) A root crop (*Pachyrhizus erosus*).

Yautia (A, C), **tannia** (A, C), **taina** (A, C), **new cocoyam tanier** (A, C) A root crop (*Xanthosoma sagittifolium*).

Yuca (NA, SA), **cassava** A root crop (*Manihot esculenta*).

Common names in various parts of the world as indicated: W = World, NA = North America, SA = South America, A = Asia and Africa, PI = Pacific Islands, E = Europe, C = Caribbean.

INDEX